RECENT INTERFEROMETRY APPLICATIONS IN TOPOGRAPHY AND ASTRONOMY

Edited by **Ivan Padron**

Recent Interferometry Applications in Topography and Astronomy
Edited by Ivan Padron

Published by InTech

Janeza Trdine 9, 51000 Rijeka, Croatia

Publishing Process Manager Petra Nenadic

Technical Editor Teodora Smiljanic

Cover Designer InTech Design Team

Additional hard copies can be obtained from orders@intechopen.com

Recent Interferometry Applications in Topography and Astronomy, Edited by Ivan Padron
p. cm.
ISBN 978-953-51-0404-9

For EU product safety concerns:
IN TECH d.o.o., Prolaz Marije Krucifikse Kozulić 3, 51000 Rijeka, Croatia,
info@intechopen.com or visit our website at intechopen.com.

Meet the editor

Ivan Padron was born in Camajuani, Cuba in 1970. He received his bachelor degree in Physics from the University of Havana in 1993, and his Ph. D in Applied Physics from New Jersey Institute of Technology in 2010. His research has been directed towards diaphragm based Fabry-Perot interferometric sensors. His critical thinking and enterprising personality where acknowledged in two U.S. patents and one patent application, and offered him the opportunity of publish his work in varies peer-reviewed scientific journals, participates in numerous international conferences and to serve as a peer reviewer of the IEEE Sensor Journal. He is a member of the American Physical Society and IEEE Professional Association.
His research interests include optical sensors and the application of physics to life sciences and medicine.

Contents

Preface

This book is an actualized reference of some interferometry applications. It provides a current overview of the theoretical and experimental aspects of interferometry techniques applied to Topography and Astronomy.

The first two chapters comprise interferometry techniques used for precise measurement of surface topography in engineering applications. Chapter 1 shows the advantages of the coherence correlation interferometry in comparison with phase shifting methods for precise measurement of surface topography. Chapter 2 includes a summary of recent developments of interferometry techniques for surface micro topography measurement and introduces a new approach to reduce coherent noise in interferometry phase-contrast image. Also, different numerical reconstruction algorithms are reviewed and compared. The principle of multiple-beam interferometry and its ability to feature extremely small height objects are described and experimental verified.

Chapters three through eight are dedicated to interferometry applications related to Earth's topography. Chapter 3 presents a short review of some advanced multi-temporal phase unwrapping approaches for the generation of surface deformation time series through the application of the small baseline subset Differential Synthetic Aperture Radar interferometry (DInSAR) technique and includes experimental results, which clearly demonstrate, the validity of these approaches and their valuable applicability in real cases. Chapter 4 demonstrates the 3-D spit reconstruction from DInSAR using three C-band Synthetic Aperture Radar (SAR) images acquired by RADARSAT-1 SAR F 1 mode data. The chapter presents the integration between the conventional DInSAR method and the fuzzy B-spline as an excellent tool for 3-D coastal geomorphology reconstruction from SAR data under the circumstance of temporal decorrelation. Chapter 5 proposes a new method of co-registration, namely the SPEC method, which uses the number of singular points in the temporary interferogram as the evaluation criterion to co-register the master slave map locally and nonlinearly. The SPEC method successfully improves the quality of the digital elevation model, and it makes ambiguous fringes clearer in its appearance. Chapter 6 presents the interferometric phase estimation method based on subspace projection and its modified method. A fast algorithm is developed to implement the modified method, which can significantly reduce the computational burden including theoretical analysis and simulations. Chapter 7 discusses the functional relation between Doppler shift and phase interferometry and presents some methods which can synthesize these two methods directly from mathematical formulations. Based on this fusion in a physical relationship, Doppler shift is obtained by measuring its phase demonstrating that Doppler-phase interference help to develop airborne passive location techniques, as well as correlation techniques. Chapter 8 describes methods and results giving insight into the applicability of C-band InSAR coherence data for retrieval of stem volume in boreal forests. It includes an analysis of a large dataset of European Remote Sensing Satellites (ERS) tandem interferometric synthetic aperture radar (InSAR) data with a focus on the accuracy by which stem volume can be estimated in boreal forest. The introduction of a method without the need for in situ data for estimating model parameters demonstrates that the estimation of stem volume is characterized by the same accuracy as when the model is trained by means of in situ measurements. The technique is based on a method by means of consistency plots. Such plots could also illustrate for what forest stands the coherence observations are less stable and the expected accuracy worse than for those closer to the ridge in the plots.

Chapter 9 is an application of interferometry in Astronomy, directed specifically to detection of planets outside our solar system. This chapter demonstrates theoretically that the existence of the planet is optically directly confirmed by detecting faint straight fringes, using a one aperture interferometer, and develops the theoretical basis for the argument that the radiation from the star is parallel to the optical axis, while that coming from the planet has a tilt. Also, a new signal-to-noise ratio of 10-3 is established.

The book has been made possible as the outcome of the exceptional work done by all authors and the professional assistance of Publishing Process Manager, Ms. Petra Nenadic, during all phases of editing.

Ivan Padron
New Jersey institute of Technology
USA

1

Coherence Correlation Interferometry in Surface Topography Measurements

Wojciech Kaplonek and Czeslaw Lukianowicz
Koszalin University of Technology
Poland

1. Introduction

Assessment of surface topography can be done by various methods, especially stylus and optical methods as well as utilising scanning tunneling microscopes and atomic force microscopes (Whitehouse, 1994, 2003; Thomas, 1999; Wieczorowski, 2009). The most accurate techniques for assessing surface topography include optical methods (Leach, 2011), especially methods of interference (Pluta, 1993; Hariharan, 2007). In the last two decades, there have been rapid developments in interferometry, as a result of the new possibilities for digital recording and analysis of interference images. The fastest growing interference methods include Phase Stepping Interferometry (PSI) (Creath, 1988; Stahl, 1990; Kujawinska, 1993; Creath & Schmit, 2004) and methods based on coherence analysis of light reflected from the test and reference surfaces (Harasaki et al., 2000; Blunt & Jiang, 2003; Schmit, 2005; Blunt, 2006; Petzing et al., 2010). This second group of methods is defined in different terms in English (Petzing et al., 2010; Leach, 2011), for example:

- Coherence Correlation Interferometry (CCI),
- Coherence Probe Microscopy (CPM),
- Coherence Scanning Microscopy (CSM),
- Coherence Radar (CR),
- Coherence Scanning Interferometry (CSI),
- Coherence Scanning Microscopy (CSM),
- Scanning White Light Interferometry (SWLI),
- Vertical Scanning Interferometry (VSI),
- White Light Scanning Interferometry (WLSI).

Modern methods of interferential microscopy used for the assessment of surface topography are based on automated interference image analysis and coherence analysis (Deck & de Grot, 1994; Patorski et al., 2005; de Grot & Colonna de Lega, 2004; Niehues et al., 2007). Assessment of surface topography requires that all of the information contained in one or more interference images, be porcessed in a sufficiently brief period of time. This process was made possible by:

- progress in the development of photodetectors,
- several hundred-fold increase in computing power,
- technological improvements in effective data storage,

- dynamic development of photonic engineering,
- development of new algorithms for digital interference image processing,
- development of advanced specialized computer software.

PSI methods (Hariharan & Roy, 1995) consist in the analysis of the light phase distribution on the measured surface. They enable the highly accurate measuring of the height of surface irregularities. The measuring range of these methods, however, is relatively small, while methods based on tracking and analysing the degree of temporal coherence of the interfering waves offer the possibility to measure the height irregularities of the surface over a greater range. Some of the measuring instruments used have the capacity to execute both methods. The principles and use of Coherence Correlation Interferometry (CCI) in surface topography measurements are presented in this chapter.

2. General characteristics of CCI

A brief introduction will present the characteristics of interference-based methods used in surface topography measurement, forming a background that highlights the advantages of CCI. The second section will include:

- presentation of the general characteristics of CCI,
- an overview of the general idea of the method and its key features.

2.1 General principle of CCI

The CCI methods are based on the cross-coherence analysis of two low-coherence light beams, the object beam being reflected from the object t, whilst the reference beams is reflected from a reference mirror. The general idea of this method is shown in Figure 1. The high-contrast interference pattern arises if the optical path length in the object arm is equal to the optical path length in the reference arm. For each object point a correlogram is recorded during the movement of the object. The position of the corresponding object point along the x-axis can be measured by another measuring system for the maximum of the correlogram (step B in Figure 1). An interference signal is helpful for accurate determining of this position.

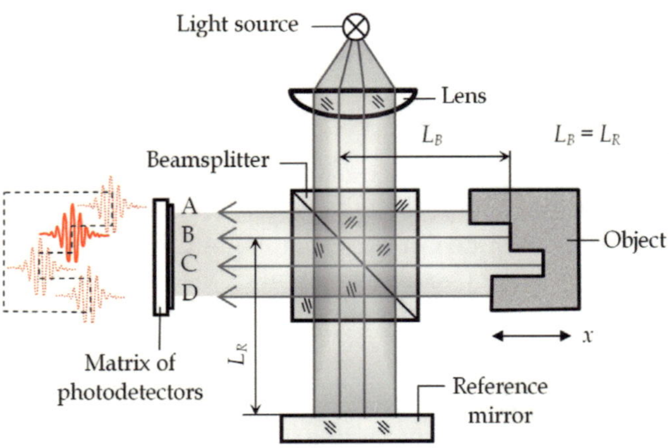

Fig. 1. Principle of the coherence correlation interferometry.

In the methods based on an assessment of the degree of temporal coherence of two interfering beams white light or low coherence light is used. These methods are incorporated in many types of measuring instrument. One of the most commonly used is the white-light interference microscope. This instrument combines the three most important features: white- light illumination, interference objective lenses and a precision scanner. The principle of operation applied in many white-light interference microscopes is based on the CCI technique.

2.2 Features of CCI

Features of CCI include:

- wide measuring range and high resolution,
- a large measurement area on the object,
- short time of measurement,
- exceptional versatility for for measuring objects made from different materials.

Similarly, the modern white-light interference microscopes have a number of advantages, e.g. wide measuring range, high accuracy and short time of measurement. The range of applications of white-light interference microscopes is very wide and includes:

- material sciences and engineering (analysis of structures of the polymer materials),
- mechanics (measurement of surface topography of the precision machined parts and elements, measurements of the smooth and super-smooth surfaces),
- electronics (measurements of silicon wafers),
- analysis of the surface structures of micro electro-mechanical systems (MEMS) and micro-opto-electro-mechanical systems (MOEMS),
- optics (measurements of optical elements e.g. microlenses, diffractive optics).

3. Theory

The third section will focus on theoretical basis of coherence correlation interferometry. This section will also cover a discussion of partially coherent wave interference and an analysis of a dependence upon the complex degree of coherence in interfering waves from a surface topography.

3.1 Coherence of light

The temporal coherence of light is connected with the phase correlation of the light waves at a given point in space at two different instants of time. It is characterized by coherence length, i.e. the propagation distance over which coherence of the beam is kept. Temporal coherence is a measure of the correlation between the phases of a light wave at different points along the direction of propagation. When we deal with two light waves then mutual temporal coherence can be analysed. This is a measure of the cross-correlation between the phases of these waves.

In coherence theory a complex degree of coherence is a measure of the coherence between two waves. This measure is equal to the cross-correlation function between the normalized amplitude of one wave and the complex conjugate of the normalized amplitude of the other:

$$\gamma_{12}(\tau) = \frac{\Gamma_{12}(\tau)}{\left[\Gamma_{11}(0)\Gamma_{22}(0)\right]^{1/2}}, \tag{1}$$

where: $\gamma_{12}(\tau)$ - complex degree of mutual coherence, τ - time delay, $\Gamma_{12}(\tau)$ - function of mutual coherence of light, $\Gamma_{11}(0)$, $\Gamma_{22}(0)$ - self-coherence functions of light.

3.2 Use of CCI for surface topography measurements

Typically a two -beam interferometer (e.g. Michelson's, Linnik's or Mirau's interferometer) is used in coherence correlation interferometry. The two -beam interferometer is illuminated by a light source with a low degree of temporal coherence light. The optical path length in one arm of the interferometer changes during measuring and the interference signal is then analysed.Advanced analysis of the signal is carried out on every single point of the evaluated surface. In most cases this consists in determining the interference signal's envelope maximum. The shape of the interference signal depends on the complex degree of cross-coherence of the interfering waves.

Let the two partially coherent light waves with an average frequency ν_0 and equal intensities I_0 interfere in the system interference microscope. One of the waves reflected from the surface of the test with $z(x, y)$, and the other from the reference mirror. The light intensity I in the selected point interference field is determined by the equation:

$$I = 2I_0\left[1 + \left|\gamma_{12}(\tau)\right|\cos\,\alpha_{12}(\tau)\right], \tag{2}$$

where: I_0 – intensity of interfering waves, α_{12} – function, which defines phase difference of interfering waves.

As shown in equation (2) an interference signal depends on the complex degree of mutual coherence γ_{12} in interfering waves. The complex degree of mutual coherence γ_{12} can be interpreted as a normalized cross-correlation function of the light wave reflected from the analysed surface and the wave reflected from the reference mirror.

$$\gamma_{12}(\tau) = \left|\gamma_{12}(\tau)\right|\exp\left\{-i\left[2\pi\nu_0\tau - \alpha_{12}(\tau)\right]\right\}, \tag{3}$$

where: i – imaginary unit, ν_0 – average frequency of interfering waves.

The complex degree of mutual coherence γ_{12} is analyzed separately at each point of the surface. If a test surface $z = f(x, y)$ is rough, it can be assumed that the complex degree of mutual coherence is a function depending not only on the parameter τ, but also on the coordinates x and y. Time delay τ is zero at the considered point of the surface for such position z_0 of scanning system at which the lengths of optical paths in both arms of the interferometer are equal. The parameter value of τ for the position of z is determined from the relation:

$$\tau = \frac{2(z - z_0)}{c}, \tag{4}$$

where: z - position of the scanning system of interferometer, z_0 - position of the scanning system at which the lengths of optical paths in both arms of the interferometer are equal, c -speed of light.

Measurement of surface topography using the CCI is based on analyzing the time delay values $\tau(x, y)$ between the interfering waves at different points of the surface. Time delay is function x and y which is changing during the scanning process. It is possible for each point of the surface to determine the position of $z = z_0$, for which $\tau = 0$. As a result, such a measurement is obtained through discrete function $z = z_0(x, y)$ that describes a set of points. Each point is defined by three coordinates x, y, z. This imaging of the shape of the surface possible.

Fig. 2 shows a graph of relative light intensity at some point of the interference image depending on the position of the scanning system. The physical quantity λ_0 shown in Figure 1 is the average wavelength of light and can be determined from the relationship:

$$\lambda_0 = \frac{c}{\nu_0}. \tag{5}$$

The graph shown in Figure 2, in fact, describes the cross-correlation function of the interfering waves and the envelope of this function.

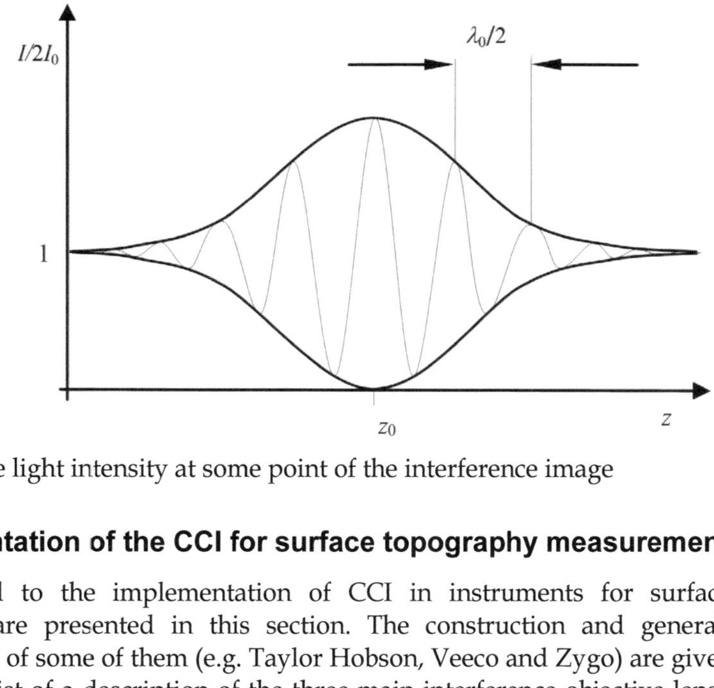

Fig. 2. Relative light intensity at some point of the interference image

4. Implementation of the CCI for surface topography measurements

Issues related to the implementation of CCI in instruments for surface topography assessments are presented in this section. The construction and general metrological characteristics of some of them (e.g. Taylor Hobson, Veeco and Zygo) are given. This section will also consist of a description of the three main interference objective lenses (Michelson, Mirau and Linnik) used in CCI instruments, as well as relevant diagrams and technical specifications characterizing these objective lenses. In the final part of the section, selected algorithms used in the practical implementation of surface topography measurements, using CCI, are discussed.

4.1 Construction of typical CCI instruments

Typical CCI instruments are equipped with the measurement base supported on an antivibration table. The antivibration table is used for separating the instrument from external vibration sources. Usually the granite slab can fulfill the role of the anti-vibration table. In other instances more advanced solutions are used, such as active vibration isolation systems. The translating table and a vertical column are mounted upon the measurement base. The translating table enables tilt adjustment (by several degrees) around x and y axes (in some instruments also with z axis rotation). Tilt adjustment is one of the pre-measurement procedures. This procedure is especially useful for samples, which are characterized by high angles of slope and high values of surface irregularity. Movement and adjustment of the stage can be realized manually or motorized by joystick control. The main elements of a typical CCI instrument (Talysurf CCI 6000 produced by Taylor Hobson Ltd., (UK)) are shown in Fig. 3.

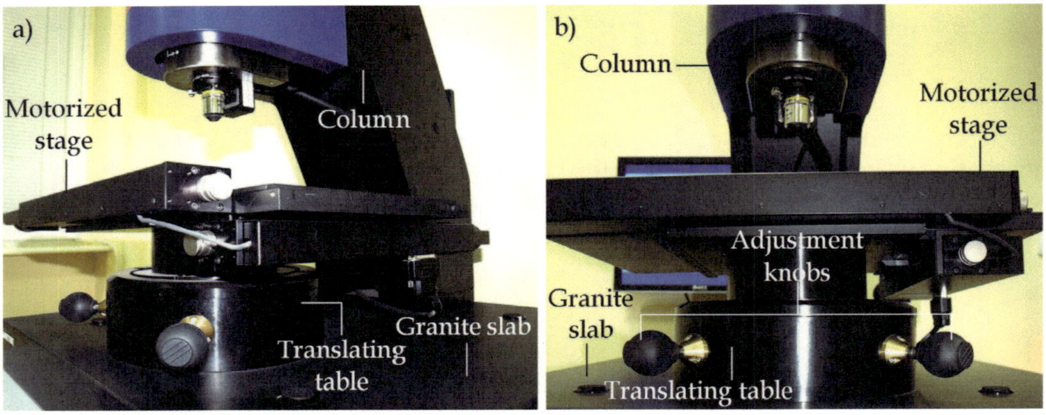

Fig. 3. White-light interference microscope Talysurf CCI 6000 produced by Taylor Hobson Ltd., (UK): a), b) front view and side view of the instrument with the main elements.

The optical interferometry and vertical scanning systems are located in a column. A source of light can be installed in compact form in the optical interferometry system or can be located outside the instrument. In this case a specially designed illuminatore was usually utilised. Light from the illuminator to the interferometer's optical system is delivered by an optical fibre.

An integral part of the CCI instrument is a stationary computer and a one or more computer monitors. Computer must be characterized by a high speed computing and adequate HDD free storage space for the archiving of recorded measurement data. In addition to controlling the CCI instrument and recording measurement data, the computer is used for processing and analysis of data. This process is realized by the use of specialist software, which is usually supplied by the producer of the CCI instrument.

4.2 General characteristics of the CCI instruments

The CCI instruments are characterized by certain metrological parameters (Leach et al., 2008; Petzing et al., 2010), which are specified in the technical specifications. Metrological parameters are a necessary source of information for determining the requirements for the purchase and later operation of instruments. A description of the selected metrological parameters is given below.

- Vertical resolution (*z* axis) is the ability of the CCI instrument to resolve a surface profile vertically (in z axis). Vertical resolution for commercial instruments is in the range from 0.01 nm to 0.5 nm.
- Lateral resolution (*x* and *y* axis) is defined as one-half the spatial period for which the instrument response (measured height map compared to the actual surface topography map) falls to 50 %. Lateral resolution for commercial instruments depends on the magnification and can be between 0.5 µm to 7 µm.
- Repeatability is the ability of an instrument to provide similar indications for repeated evaluation of the same measurement under the same conditions of measurement. Repeatability for commercial instruments is between 0.002 nm to 50 nm. For example: for the white-ligt interference microscope TMS-1200 TopMap µ.Lab produced by Polytec (Germany) repeatability (at 10 nm sampling increment) is 0.25 nm (smooth surface) as well as 2.5 nm (rough surface) and (at 87 nm sampling increment) is 0.5 nm (smooth surface) and 20 nm (rough surface).
- Instrument noise is typically a small random error term superimposed on the measurement data. The source of this noise can be electronic components and devices mounted on the instrument (e.g. amplifiers, stabilizers, detectors). Some manufacturers provide software procedures, which neutralise this effect.

Selected metrological parameters, which characterize some of the commercially interferential microscopes available on the market are presented in Tab. 1. Similar metrological parameters for selected four instruments of the Talysurf CCI line manufactured by Taylor Hobson Ltd., are presented in Tab. 2.

Instrument	Producer	Method	Vertical range (mm)	Vertical resolution (nm)	Measurement area (max.) (mm)
MarSurf WS1	Mahr	SWLI	0.1	0.1	1.6 × 1.2
TMS-300-214 TopMap In.Line	Polytec	SWLI	0.5	<0.55	18.6 × 13.9
Wyko NT1100	Vecco	PSI/VSI	1	0.01	8.24 × 8.24
WLI	MRC	VSI	0.15	1	0.25 × 0.25

Table 1. Selected metrological parameters of commercially available interferential microscopes.

The CCI instruments are a large group of advanced optical measurement systems currently used in many fields of modern science and technology. There are a number of commercial CCI instruments available on the market. They are produced by Fogale, Mahr, MRC, Phase Shift, Polytec, Taylor Hobson Ltd., Vecco, and Zygo. Some examples of commercially available white-light interference microscopes are shown in Fig. 4.

Instrument	Method	Vertical range* (mm)	Vertical resolution (nm)	Repeatability of surface (nm)	Measurement area** (mm^2)
Talysurf CCI Lite	CCI	2.2	0.01	0.002	6.6
Talysurf CCI 9150	CCI	0.1	0.01	0.003	0.25 - 7.0
Talysurf CCI 6000	CCI	0.1	0.01	0.003	0.36 – 7.0
Talysurf CCI 3000	CCI	0.1	0.01	0.003	0.36 – 7.2

*Standard range (optional, can be extended), **Square area

Table 2. Selected metrological parameters for instruments of the CCI line manufactured by Taylor Hobson Ltd., (UK)

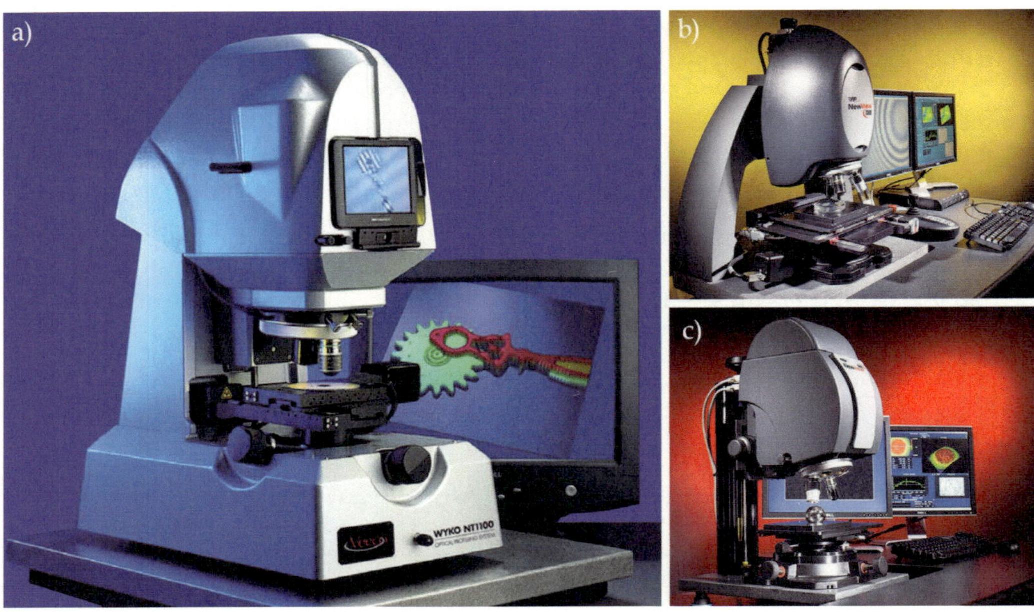

Fig. 4. The commercially available white-light interference microscopes: a) Wyko NT1100 produced by Veeco (Vecco Instruments, 2002), b) NewView™ 7300 produced by Zygo (Zygo Corporation, 2010), c) NewView™ 600 produced by Zygo (Zygo Corporation, 2007)

4.3 Interference objective lenses

Modern interference microscopes typically use standard microscope objective lenses, containing one of three types of classic two-beam interferometer, the configuration of which has been appropriately modified to enable the measurement of such instruments:

- Linnik interferometer (Linnik, 1933),
- Mirau interferometer (Delaunay, 1953; Kino & Chim, 1990),
- Michelson interferometer (Michelson, 1882).

The last two of these interferometers are most commonly used in CCI instruments. In a Mirau interferometer the basic elements are a microscope objective lens, a reference surface and a beamsplitter. All the elements are on the same optical axis of the objective lens. In this configuration, there are two paths from the light source to the detector. In the first path a light beam reflects off the beamsplitter, passes to the reference mirror and then reflects back, before passing through the beamsplitter to the microscope objective lens and then on to the detector. In the second path, a light beam passes through the beamsplitter, to the reference mirror, reflects back to the beamsplitter and then reflects from the half mirror into the microscope objective lens before proceeding to the detector. The light reflected from these surfaces recombines and a fringe interference pattern is formed. The objective lenses with Mirau interferometer are usually used in applications requiring higher magnification (>10×) or numerical apertures. In such instances objective lenses can be used following typical 10×, 20×, 50× and 100× magnification.

A Michelson interferometer is configured in a similar way to a Mirau with the exception that the reference mirror is in a different position (outside the optical axis). This configuration provides a longer working distance, wider fields of view and larger depth of focus at small magnifications (<5×). The following magnification can be typically used for objective lenses with a Michelson interferometer: 1×, 1.5×, 2×, 2.5× and 5×.

Commercial microscope objective lenses with interferometers are manufactured by Leica, Nikon, Olympus, Seiwa Optical, Veeco, Zeiss and Zygo. Availability of wide range of interference objective lenses give a great opportunities concerned with selection of the suitable lens for carrying out scientific works. Selected issues related with suitable selection of interference objective lenses were described in the work (Petzing et al., 2010). Schematic diagrams of the typical configurations of interference objective lenses are shown in Fig. 5, whereas the general characteristics of interference objective lenses produced by Nikon (Japan) are presented in Tab. 3.

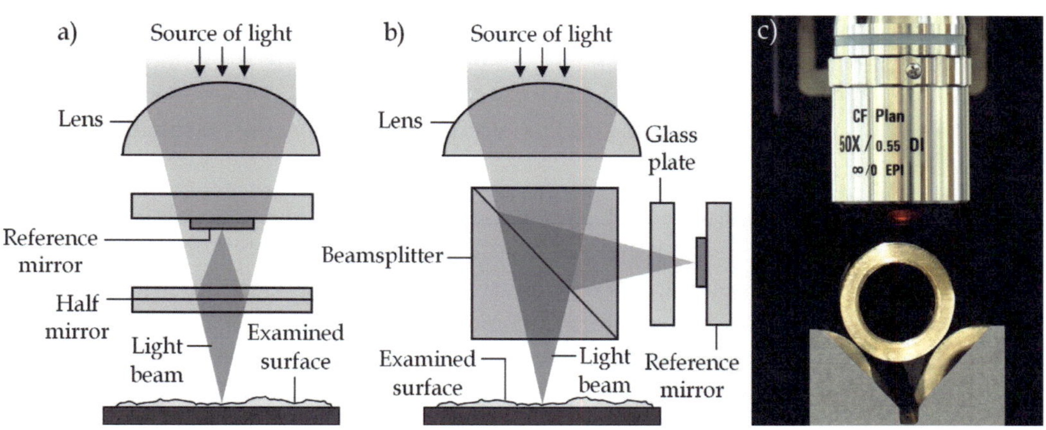

Fig. 5. Schematic diagrams of the typical configurations of interference objective lenses: a) Mirau, b) Michelson, c) commercially produced microscope objective lens with Mirau interferometer (Nikon 50× CF Plan IC EPI).

Designation of objective lens	Magnification	Numerical aperture	Working distance (mm)	Focal lenght (mm)	Weight (g)
2.5× CF Plan IC EPI	2.5×	0.075	10.3	80	440
5× CF Plan IC EPI	5×	0.13	9.3	40	280
10× CF Plan IC EPI	10×	0.3	7.4	20	125
20× CF Plan IC EPI	20×	0.4	4.7	10	130
50× CF Plan IC EPI	50×	0.55	3.4	4	150
100× CF Plan IC EPI	100×	0.7	2	2	200

CF - Chrome-Free, Plan - Flat Field Correction, IC - Infinity Corrected, EPI - Epi Illumination

Table 3. General characteristics of interference objective lenses produced by Nikon (Japan)

4.4 Algorithms

The accuracy of surface topography measurements using the CCI is dependent on many factors. One of the most important factors is the algorithms used during the scan and the interference signal analysis algorithms. Analysis of signal interference generally requires the use of algorithms determining the maximum envelope of the signal (CCI mode) an the phase image-based interference (PSI mode).

An algorithm for digital envelope detection in white-light interferograms was described in the work (Larkin, 1996). It was compared with other algorithms. It was developed based on algorithms used in the PSI methods (Creath, 1988; Stahl, 1990; Kujawinska, 1993). The new algorithm was assessed in terms of accuracy and computational efficiency. It can be interpreted as a second-order nonlinear filter. Use of quadrature modulation and Hilbert transformation as a new measurement strategy, based on a direct reconstruction of the envelope of the correlogram, was presented in the paper (Hybel et al., 2008).

At work (Park & Kim, 2000) proposed the use of a new algorithm for detecting the true peak of the interference fringe. This two-step algorithm is an efficient means of computation to obtain a good measuring accuracy with greater insulation against disturbances and noise.

In order to obtain high accuracy and performance out of CCI and PSI techniques, many algorithms have been proposed (Ailing et al., 2008; Buchta et al., 2011; Kim et al., 2008). Among those algorithms analyzed were: phase-shifting, phase-crossing (Pawłowski et al., 2006), zero-crossing and Fourier-transform techniques. Algorithms that give an estimate of the envelope peak position from only a fraction of the interferogram (Sato & Ando, 2009) were also proposed.

Within the CCI techniques there were developments in scanning algorithms, merge and stitch images and error correction (Viotti et al., 2008; Kim et al., 2008), many of which have been patented and used in commercial systems (Harasaki & Schmit, 2002; Bankhead & McDonnell, 2009).

5. Measurements of engineering surfaces

The section is dedicated to some selected issues related to the applictaion of CCI for precise surface topography measurements (Conroy, 2008). A brief description of the white-light interference microscope Talysurf CCI 6000 by Taylor-Hobson Ltd., is presented. Subsequently issues related to the pre-measurements and measurements procedure, as well as processing and analysis of measurement data, are described in detail. The selected results of measurements of surface topography obtained by CCI for objects machined by various techniques are presented and discussed in the final part of this section.

5.1 Interference microscope with Mirau objective lenses

One of the commercially available CCI instruments is a white light interference microscope produced by Taylor Hobson Ltd. The firm produces this line of instruments under the designation Talysurf CCI. One of these instruments is Talysurf CCI6000 (Conroy & Armstrong, 2005; Cincio et al., 2008; Lukianowicz, 2010). The principle of operation of this instrument, shown in Fig. 6, is based on coherence correlation interferometry.

White light generated by an external source (150 W quartz lamp installed in Fiber-Lite® DC-950 illuminator manufactured by Dolan-Jenner's Industries) is supplied to the instrument by optical fibre. In the measuring head the light beam is directed to a beamsplitter, where it is separated into two parallel beams. The first beam is directed towards the sample and the second beam is directed towards an internal reference mirror. The two beams recombine and give a local interference image, which is sent to the CCD detector.

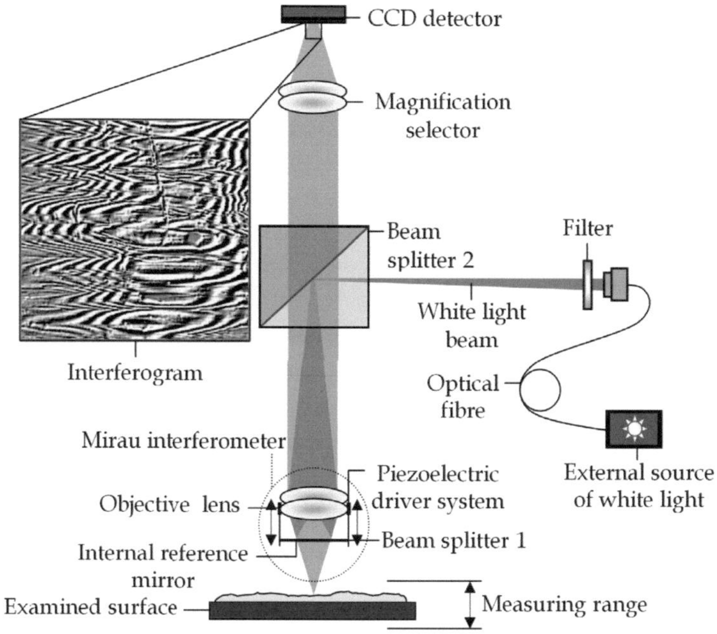

Fig. 6. Schematic diagram showing the principle of operation of the white-light interference microscope Talysurf CCI 6000 produced by Taylor Hobson Ltd.

The optical measuring head of the microscope is coupled with precision piezo-electric actuators (PZTs). The PZTs move the measuring head vertically above the measured sample.

The CCD detector measures the intensity of the light as the interferometric lens (Mirau type, 10×, 20× and 50× magnification) is actuated in the vertical direction (*z* axis) and finds the interference maximum. The location of individual points on the test surface is determined based on the analysis of the mutual temporal coherence of the interfering waves, conducted separately for each surface point.

In this way an image is generated and acquired, before being processed on a computer manufactured by Dell, with an Intel Xenon class processor. Using this recorded data a high-resolution surface topography is then generated. The instrument enables the obtainment of a vertical resolution to 10 pm (0.01 nm) at the measuring range (in the z axis) to 10 mm. Regardless of the magnification the surface topography contains more than one million data points (1024 × 1024 points).

The manufacturer of the instrument has supplied dedicated computer software. These are two applications: Talysurf CCI v. 2.0.7.3 (control elements of the system and measurements) and TalyMap Platinum v. 4.0.5.3985 (visualization and advanced data analysis). The TalyMap software uses Mountain Technology™ provided by DigitalSurf (France).

The general view of the white-light interference microscope Talysurf CCI 6000 produced by Taylor Hobson Ltd., is shown in Fig. 7.

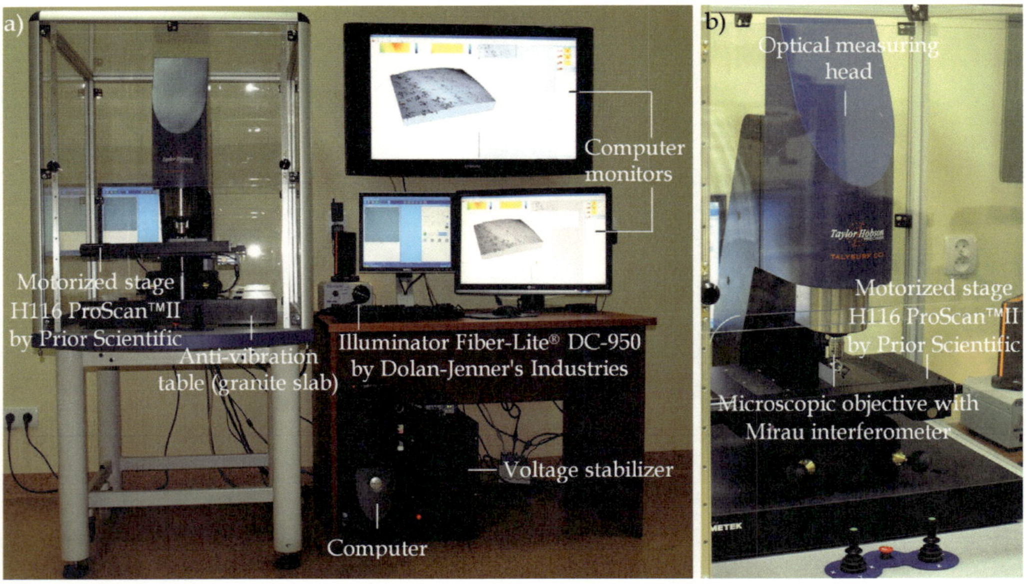

Fig. 7. White-light interference microscope Talysurf CCI 6000 produced by Taylor Hobson Ltd.: a) general view of the measuring system, b) near view of the instrument.

5.2 Pre-measurements procedure

The pre-measurements procedure for sample preparation usually involves:

- visual analysis of sample surface,
- cleaning the sample,
- overcoating of surface (optional, if necessary),
- fixturing the sample,
- orientation of sample.

The process requires a sample that must be clean and free from grease, smears, fingerprints, liquids and dust. If the samples are contaminated in some way before the measurement process begins they must be cleaned. Cleaning the sample would normally require the use of alcohol based liquids, or compressed air. Cleansing the sample surface is an extremely important issue, because the accumulated contamination can be generative source of measurement error.

In some cases, the reflective characteristics of a surface may hinder its measurement. This concern particularly affects those surfaces which are made of materials with relatively low reflectance. For such surfaces it may be desirable to overcoat with thin layers of vacuum coated gold or chromium.

The geometry and mass of the sample are the main factors affecting their method of fixturing. Use of non-permanent adhesives (e.g. adhesive tape, gels for fixturing etc.), may cause creep, whereas use of external force may cause sample distortion. Usually a range of table mounting points and accessories are used in the laboratory, including:

- magnetic and vacuum chucks/plates,
- disk fixtures
- tray holders.

The orientation of the sample may affect the results of the measurements. It is necessary to determine the proper orientation for all measured samples to obtain this same repeatability.

5.3 Measurements procedure

After the proper sample preparation the measuring procedure can begin. A typical measurement procedure (regardless of the CCI instrument) involves the following steps:

- power up the instrument and the computer,
- run the program (for control/measurements) on the computer,
- select the objective lens for the measurement,
- place the test element on the stage (adjusting the x and y axes),
- position the objective lens at its working distance from the element (adjusting the z axis),
- focus the microscope to obtain a sharp image,
- adjust stage level or tilt to optimize fringe contrast/appearance,
- adjust the test element position (x and y axes),
- minimize the number of fringes,
- adjust the light intensity level to an optimum value,
- check correct software-selectable items,
- initiate data acquisition.

A more detailed description of objective lens selection and image stitching procedures is presented below. Descriptions of other steps can be found also in (Leach et al., 2008; Petzing et al., 2010).

One of the important steps during the measuring process is selection of the objective lens. From a practical point of view it is most important to determine the main characteristics of the measured surface. It is on this basis that an appropriate objective lens is usually selected. General information concerning the selection of the objective lens is presented in Tab. 4, whereas general views of two types of microscope objective lenses with Mirau interferometer (used for surface roughness measurement) produced by Nikon (Japan) are show in Fig. 8.

Surface		Objective lens		
Area size	Surface element	Resolution	Magnification	Type of interference objective lens
Small	Steep slopes, rough surfaces	High	10×, 20×, 50×, 100×	Mirau
Large	Flat and smooth surfaces	Low	1×, 1.5×, 2×, 2.5×, 5×	Michelson
Very large			1×, 1.5×, 2×, 2.5×, 5× +image stitching procedure	

Table 4. General information on selection of the objective lens.

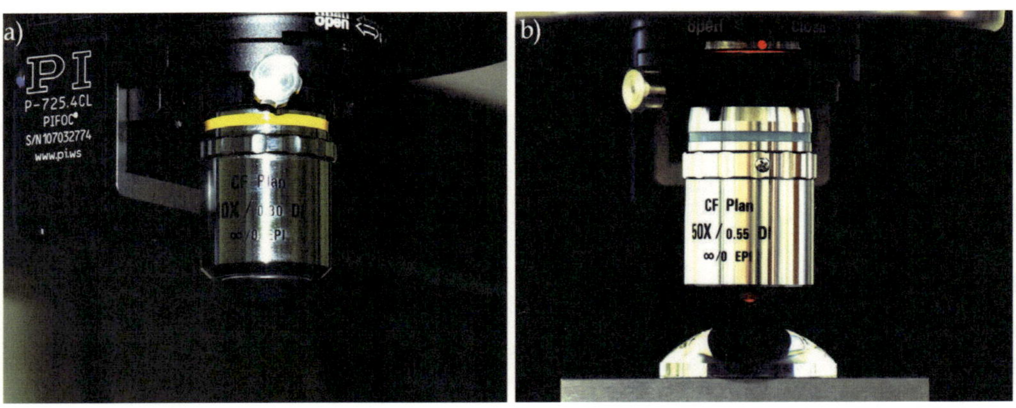

Fig. 8. Two types of the microscope objective lenses with Mirau interferometer used for surface roughness measurement: a) Nikon 10× CF Plan IC EPI (field of view 1.8 × 1.8 mm), b) Nikon 50× CF Plan IC EPI (field of view 0.36 × 0.36 mm).

In some cases there is a need for measurement of a larger area than that which would constitute the field of view of the objective lens. This is when the image stitching procedure, whereby a matrix of mages are joined, or stitched, together, can be utilised. Image stitching can be realized only if the CCI instrument is fitted with a motorized stage with horizontal translation, which allows the computer program to precisely adjust the surface position. Each acquired image is precisely positioned and there is a consistent overlap of images. This is important because it allows for a comparison of neighbouring images, and enables the precise correlation of lateral position and consistent vertical range values. The typical overlap between images ranges from 0 % to 25 %.

The operations of the image stitching procedure are described below using the example of the functions implemented in TalyMap Platinum software. This function is realized by

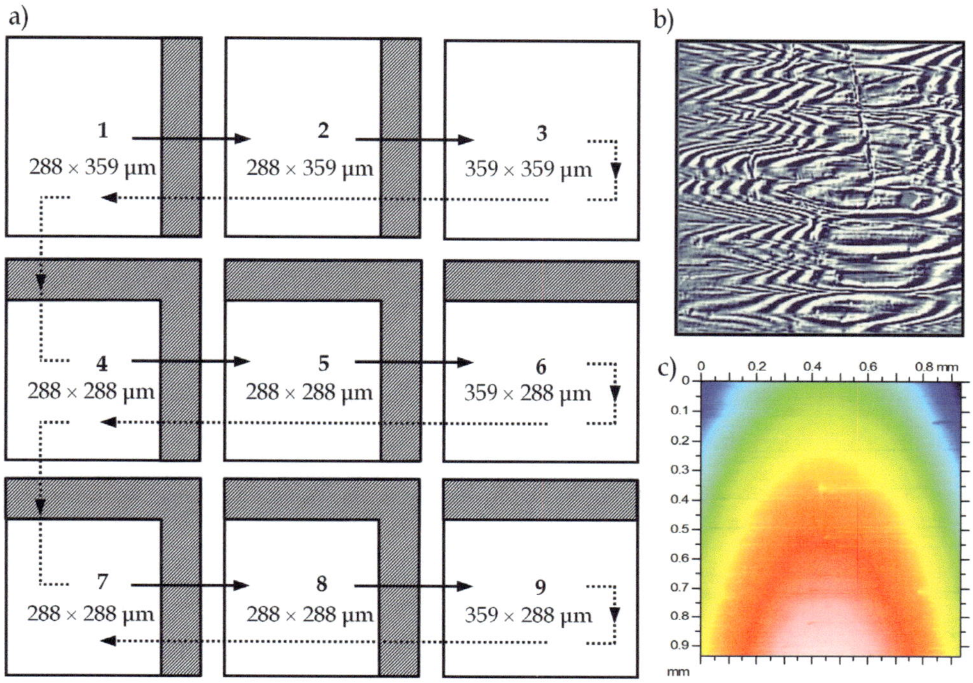

Fig. 9. The method of measurement of surface topographies with use of an image stitching function realized by TalyMap Platinum software: a) principle of image stitching procedure, b) single interferential image of measured surface with magnification 50×, c) output surface topography as a mosaic composed from a series of single surface topographies acquired as a 3 × 3 matrix.

Visual Basic script called Advances DataStiching v.1.2.1. In this case, the output surface topography is a mosaic composed of the 9 single surface topographies recorded as a 3 × 3 matrix.

A microscope objective lens with Mirau interferometer 50× CF Plan IC EPI, produced by Nikon, was used in the acquisition of the surface topographies The area measured for each of the topographies was the same 359 × 359 µm. During the stitching procedure, the area was often reduced by an overlap to a size 288 × 359 µm, 359 × 288 µm and 288 × 288 µm. The visual interpretations of the operations of the image stitching procedure are shown in Fig. 9.

5.4 Processing, analysis and interpretation of the results of measurements

Procedures relating to processing, analysis and interpretation (Petzing et al., 2010) are an important part of the measurement process following the acquisition of measurement data. In practice, these procedures are realized by special computer software. The popular commercial softwares for processing and analysis of measurement data include: Mountains Map by DigitalSurf (France), TrueMap by TrueGage (USA) and SPIP by Image Metrology A/S (Denmark). The pre-processing of measurement data usually includes (depending on software used):

- opening the file,
- levelling ,
- filling non-measured points.

After these procedures the file is ready to be processed, which will allow for the acquisition of:

- pseudo-colour surface maps,
- photo simulations,
- contour diagrams,
- mesh axonometrics,
- continuous axonometrics
- Abbot-Firestone curves,
- texture direction.

Above nly a few select procedures are given, in practice there may be many more.

They may also be more complicated and multi-threaded. Processing leads to the analysis, which typically involves:

- calculating the 2D parameters,
- calculating the 3D parameters.

As a result of analysis of the measurement data obtained, numerical values of the assessed paremeters characterizing the surface features are established.Based on these parameters and the machining parameters the whole machining process can be deduced. The interpretation of results in this case is very important, because it allows for the establishment of proper conditions within which the machining process will be optimally realised.

5.5 Examples of measurements and analysis

On the following pages some example results of analysis are shown, alongside collections of acquired surface topographies obtained by white-light interference microscope TalysurfCCI 6000 produced by Taylor Hobson Ltd., equipped with the TalyMap Platinum v.4.0.5.3985 software produced by DigitalSurf.

Fig. 10 presents an example analysis of a diffractive optical element (DOE) by Thorlabs, Inc. (USA).The DOEs are usually used in structured light generators. This sample has been manufactured from transparent polycarbonate. The value of light reflection coefficient for a polycarbonate surface was low, which made it a little awkward to analyse. The measurement was carried out on an area of approx. 4 sq. mm (Kaplonek & Tomkowski, 2009). Fig. 10 shows approx. 12% of the measured area (549 × 507 × 1.73 mm). The irregularities of this surface are periodical.

On Fig. 11 an example analysis of a 2.5″ HDD platter, by Toshiba Corp. (Japan), is shown.The topography has been measured on an area less than 1 sq. mm (980 × 979 × 0.06 μm). The irregularities of the surface are fairly insignificant and random. For these surfaces the values of 2D (*Ra,Rq, Rp, Rv, Rt, Rsk, Rku* and *Rz*) and 3D (*Sa, Sq, Sp, Sv, St, Ssk, Sku* and *Sz*) paremeters (Lukianowicz & Tomkowki, 2009) were also calculated. The collections of

Sample characteristics	Measuring instrument
Element: Diffractive optical element (DOE) Producer: Thorlabs, Inc. (USA) Material: polycarbonate	Instrument: Talysurf CCI 6000 Producer: Taylor Hobson Ltd., Software: TalyMap Platinum v. 4.0.5.3985
2D view: pseudo-colour surface map	2D view: photo-simulation of surface

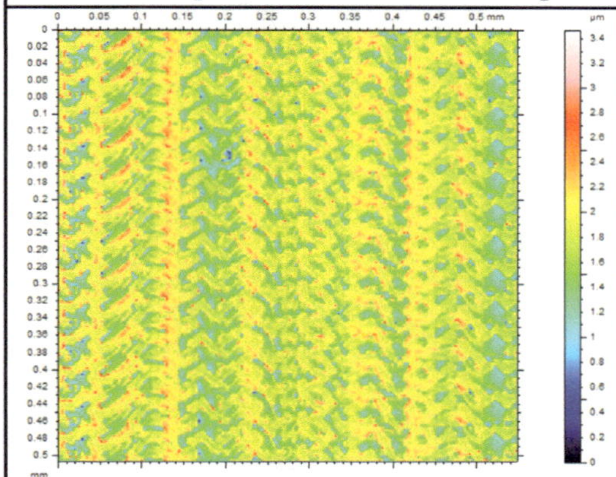

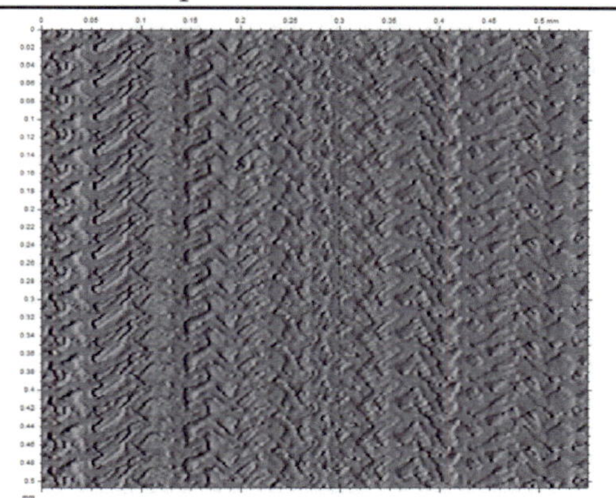

2D view: surface profile and parameters

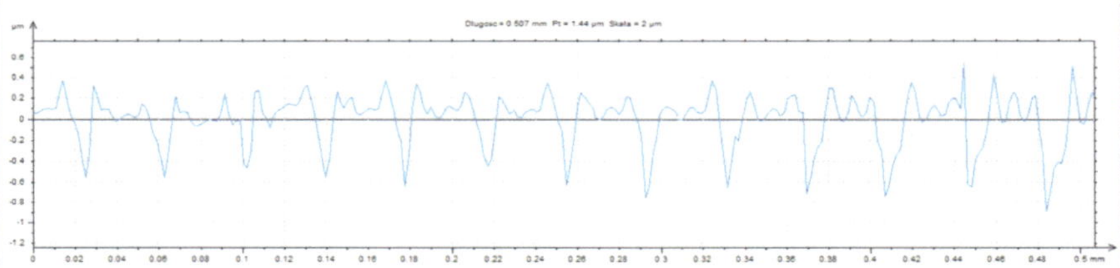

Ra = 0.153 µm
Rq = 0.205 µm
Rp = 0.298 µm
Rv = 0.562 µm
Rt = 0.999 µm
Rsk = -1.28
Rku = 4.18
Rz = 0.86 µm

3D view: surface topography and parameters

Surface area (x, y, z): 549 × 507 × 1.73 µm

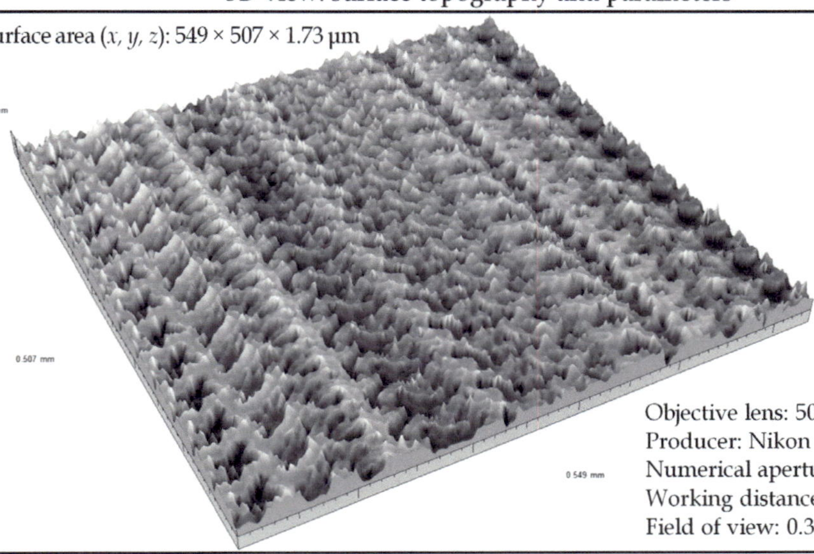

Sa = 0.26 µm
Sq = 0.324 µm
Sp = 1.66 µm
Sv = 1.8 µm
St = 3.46 µm
Ssk = -0.245
Sku = 3.27
Sz = 3.46 µm

Objective lens: 50× CF Plan IC EPI
Producer: Nikon
Numerical aperture: 0.55
Working distance: 3.4 mm
Field of view: 0.36 × 0.36 mm

Fig. 10. Exemplary analysis of a DOE sample for measurements obtained by white-light interference microscope Talysurf CCI 6000 produced by Taylor Hobson Ltd.

Sample characteristics	Measuring instrument
Element: 2.5″ HDD platter Producer: Toshiba Corp. (Japan) Material: aluminium substrate	Instrument: Talysurf CCI 6000 Producer: Taylor Hobson Ltd., Software: TalyMap Platinum v. 4.0.5.3985
2D view: pseudo-colour surface map	2D view: photo-simulation of surface

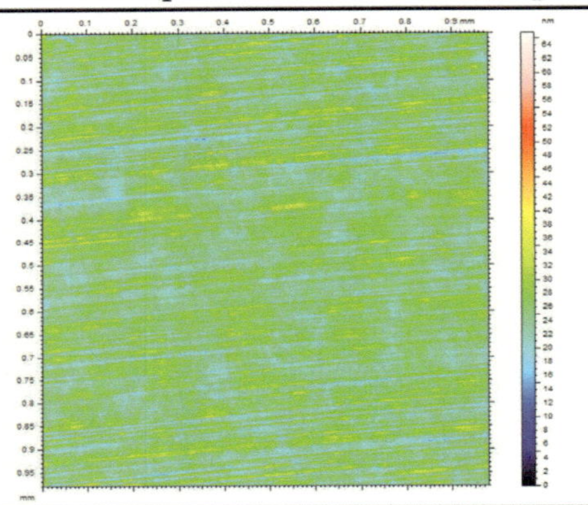

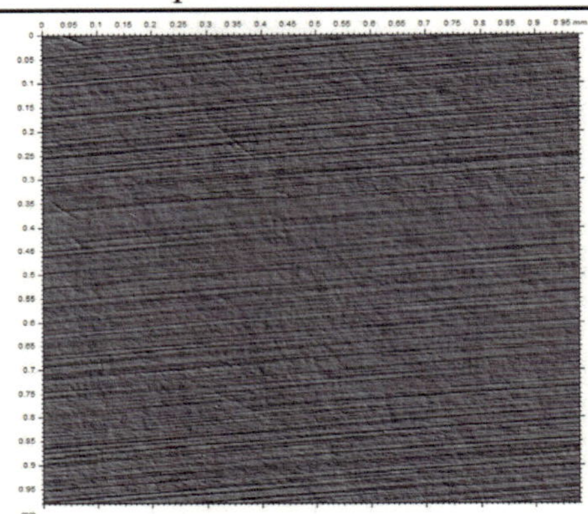

2D view: surface profile and parameters

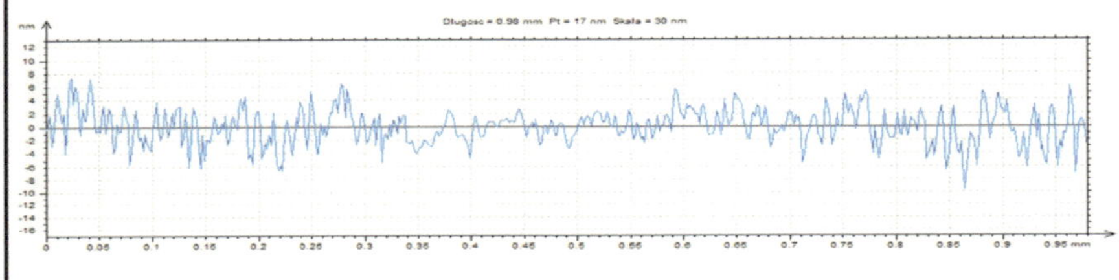

Ra = 1.42 *nm*
Rq = 1.73 *nm*
Rp = 3.33 *nm*
Rv = 3.8 *nm*
Rt = 10.9 *nm*
Rsk = -0.251
Rku = 3.55
Rz = 7.13 *nm*

3D view: surface topography and parameters

Surface area (x, y, z): 980 × 979 × 0.06 µm

Sa = 2.23 *nm*
Sq = 2.81 *nm*
Sp = 40.4 *nm*
Sv = 25.4 *nm*
St = 65.8 *nm*
Ssk = -0.11
Sku = 3.53
Sz = 65.8 *nm*

Objective lens: 50× CF Plan IC EPI
Producer: Nikon
Numerical aperture: 0.55
Working distance: 3.4 mm
Field of view: 0.36 × 0.36 mm

Fig. 11. Exemplary analysis of a 2.5″ HDD platter sample for measurements obtained by white-light interference microscope Talysurf CCI 6000 produced by Taylor Hobson Ltd.

Burnished surface (surface area 19.2 × 0.546 × 1.05 mm, Sa = 0.17 µm, Sq = 0.20 µm)

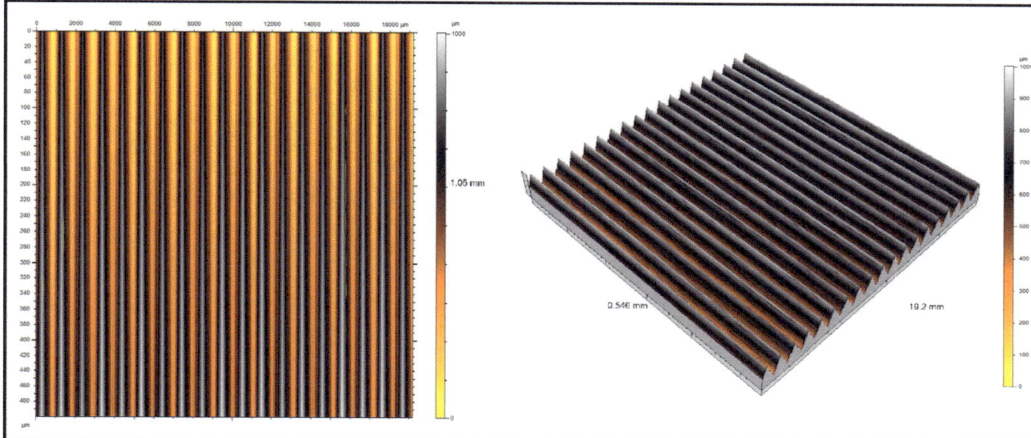

Milled surface (surface area 366 × 366 × 10.9 µm, Sa = 0.94 µm, Sq = 1.10 µm)

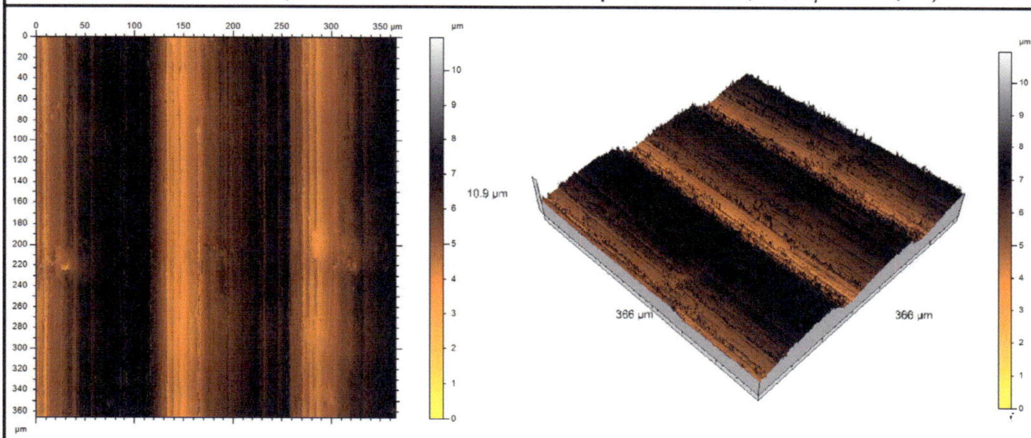

Ground surface with scratches (surface area 1.83 × 5.43 × 0.07 mm, Sa = 5.35 µm, Sq = 6.40 µm)

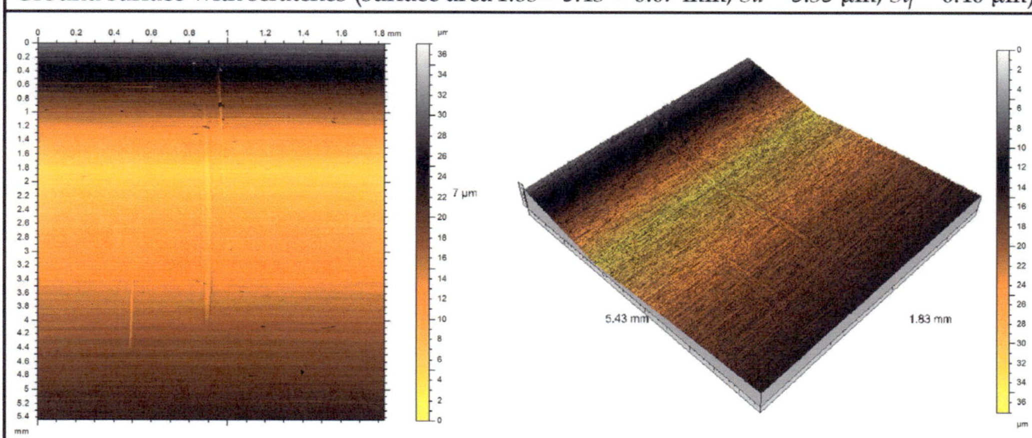

Fig. 12. Collection of surface topographies obtained by white-light interference microscope Talysurf CCI 6000 produced by Taylor Hobson Ltd.

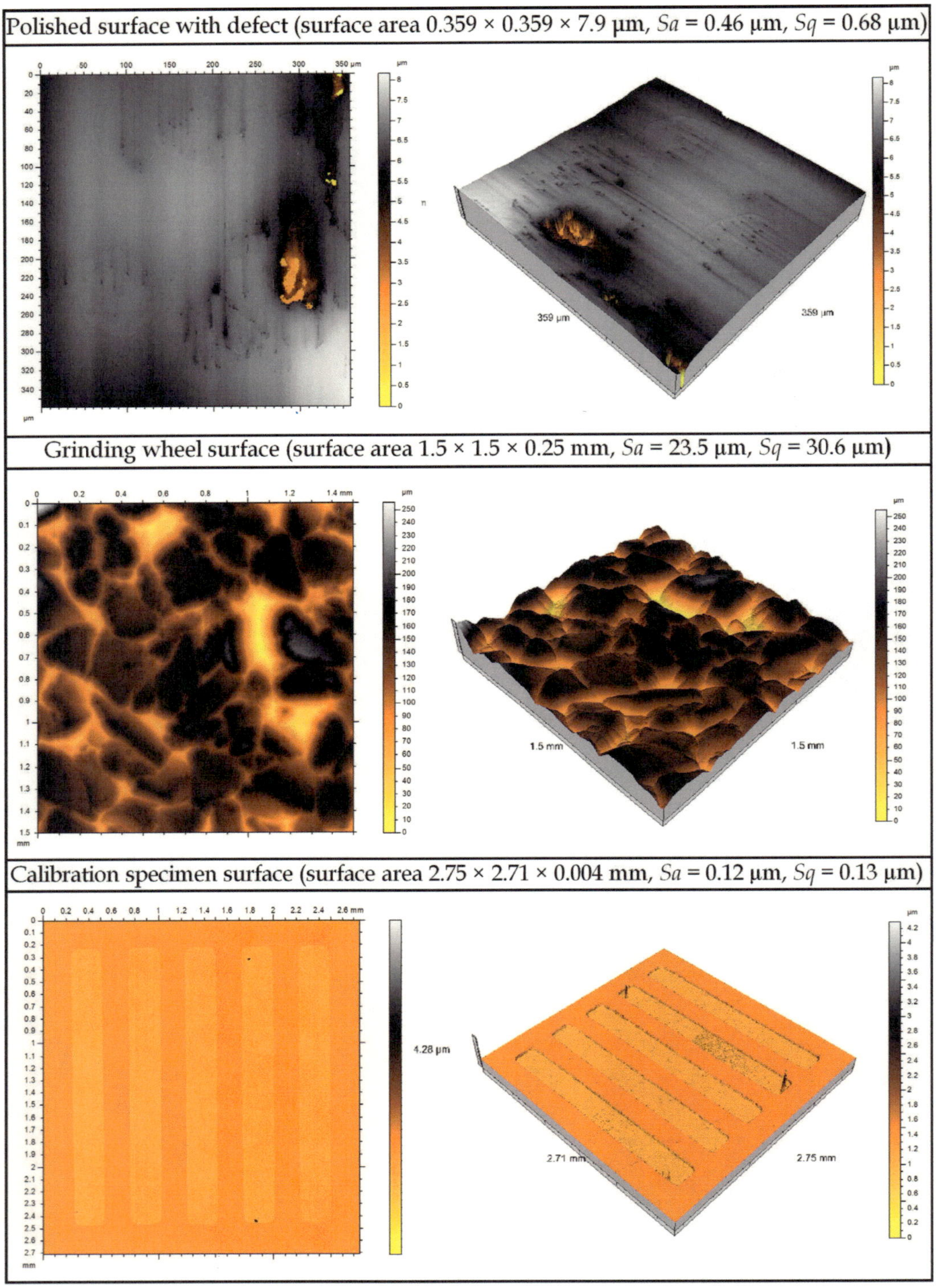

Fig. 13. Collection of surface topographies obtained by white-light interference microscope Talysurf CCI 6000 produced by Taylor Hobson Ltd.

topographies for different surfaces are shown in Fig. 12. and Fig. 13 (Kaplonek & Lukianowicz, 2010; Lukianowicz, 2010).

The first collection (Fig. 12) contains topographies of an engineering surface after the burnishing process. The value of the x axis is very high and amounts to 19.2 mm. This topography was generated using the image stitching procedure. In this same collection is a topography of a typically milled surface (measured area 366 × 366 × 10.9 μm) and a topography of a ground surface (measured area 1.83 × 5.43 × 0.07 mm) with defects in the form of scratches. These defects are typical for the surfaces of machine parts, which operate in industrial conditions.

In the second collection (Fig. 13) a polished surface is shown (measured area 0.359 × 0.359 × 7.9 μm) with a different defect, in this case taking the form of pitting. The next topography shows the surface of a grinding wheel 1-20×20×10 SG 80 M 8 V (measured area is 1.5 × 1.5 × 0.25 mm).

On the surface different sized abrasive grains are clearly visible. The last topography of this collection shows the surface of a calibration specimen, made for glass. Calibration specimens are usually used for routine calibrating of the stylus profilometers (measured area is 2.75 × 2.71 × 0.004 mm).

6. Limitations of CCI

During the interference measurements of surface topography some errors and limitations may be generated, for a number of different reasons. The sources of these errors are most often concerned with the measuring instrument and its software, as well as with the properties of the measured surface and the conditions in which the measurement was made. The limitations of CCI, and some factors affecting the accuracy of CCI based surface topography measurements, are discussed in this section.

6.1 Surface properties

One of the major contributing factors, which favor the generation of errors (Harasaki & Wyant, 2000, Pförtner & Schwider, 2001; Gao et al., 2008) in interference measurements of surface topography, are large gradients upon the surface analyzed. For such surfaces irregularities of slope and incline are significant.Slopes are usually described mathematically using the angles of slope tangent, surface derivatives or function gradient. In many cases, a considerable gradient causes the measurement errors due to limitations caused by the properties of the measuring instrument and its software. Fig. 14 shows three typical situations that may cause limitations in the measurement of surface topography:

- large slope angles of inequality,
- step height of surface irregularities,
- sloping recess.

No signal in the interference microscope caused by irregularities of slope occurs when, the slope angles of inequality θ exceed the maximum value θ_{max} given by the numerical aperture of the lens (Petzing et al., 2010):

$$\theta_{max} = \alpha_o = \arc\sin\left(\frac{N_A}{n}\right), \tag{6}$$

where: $N_A = n \sin\alpha_o$ - numerical aperture of interference microscope objective lens, n - refractive index of medium surrounding the objective lens, α_o - half the maximum value of the cone vertex angle formed by the rays from the microscope objective lens (Fig. 14a).

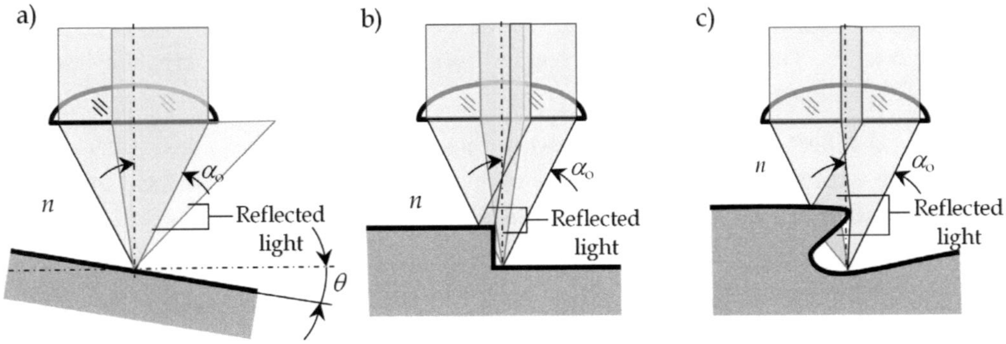

Fig. 14. Simplified diagram which showing the reflection of light focused by interference objective lens: a) from a sloped flat surface, b) from the surface with a step height, c) from the surface with sloping recess

Strong interference and no interference signal frequently reveal those surface areas, where there are significant step heights in surface irregularities or a sloping recess. During the measurements of these surfaces the self-shadowing of some areas, or partial self-shadowing of the light beam reflected from the surface, occurs (Fig. 14b and 14c).

For an analysis of the measurement errors of surface topography by interference methods, it is necessary to consider two models of surface irregularities. The first of these, is inequality, where the lateral dimensions are larger than the transverse resolving power d of an interference microscope objective lens given by the equation:

$$d = \frac{\lambda}{2N_A}, \tag{7}$$

where: N_A - numerical aperture of interference microscope objective lens, λ - average wavelength of light used to illuminate the surface. The slopes of these inequalities can be regarded as smooth.

An interference image at a location corresponding to the slope will depend upon the shape of the surface. This image can be observed, if the slope angles do not exceed the specified maximum angle θ_{max} given by equation (6). However, if the slope angle of inequality as well as the slop of the surface are excessive, then the reflected radiation can be directed out of the microscope objective lens (Fig. 15a). In the second model, the transverse dimensions of irregularity are smaller than the transverse resolving power d of the interference microscope objective lens. Such surfaces can be considered as rough surfaces, which scatter the incidence light in a more or less diffuse way.In this case a part of the light scattered by the surface will be directed at the optical system of interfence microscope (Fig. 15b).

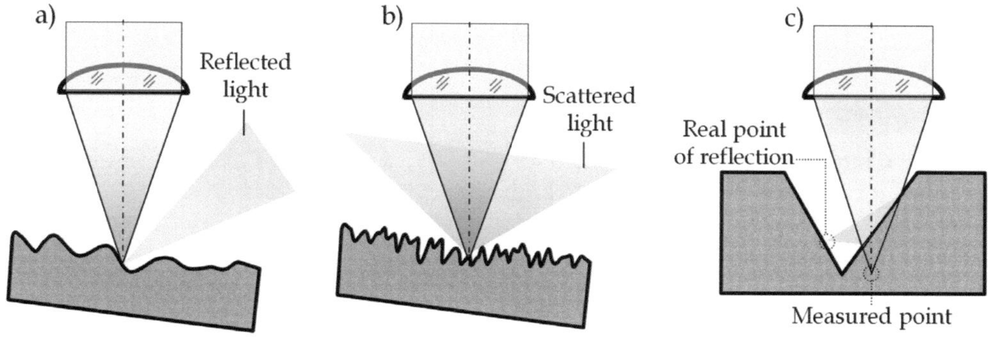

Fig. 15. Double-reflection of light focused by interference objective lens on the surface of the measured wedge groove.

6.2 Instrumental and measuring environment limitations

The significant instrumental limitations of CCI are limited resolution of the lens aperture and the interference microscope objective lens, as well as the limitations concerned with the source of light and the scanning system.These limitations can cause inaccurate representation of the surface topography. The reason for the additional errors may be improper calibration of the measuring system, and imperfect software. Among the limitations related to the environment in which the measurement is carried vibrations and temperature changes are particularly noteworthy, as they cause random changes to occur in the measured surface topography. Limitations of CCI methods were discussed in more detail in (Petzing et al., 2010).

7. Conclusions

The results presented in this chapter show that coherence correlation interferometry is an advanced optical technique used for precise measurement of surface topography. The dynamic development of this technique means that it should play a dominant role in modern interference microscopy. CCI extends the scope of inferferometric techniques used in the measurement of surface topography. Especially those, which are characterized by complex structure or a relatively low reflectance (e.g. calibration specimens made from glass or transparent films).

The comparison between coherence correlation interferometry techniques and phase shifting methods shows that CCI has a better tolerance for variations in surface texture as well as greater capabilities for measuring surface features and structures, such as step heights. An important feature is also the fact that the point where the maximum interference occurs can be found for each single pixel of the detector. Additional advantages are concerned with large measurement ranges, high accuracy and repeatability as well as shortened times of measurement.

8. Acknowledgments

Part of this work was supported by National Science Center of Poland. The Authors would like to thank Mr. Robert Tomkowski MSc, BSc, from the Laboratory of Micro- and Nano

Engineering at Koszalin University of Technology, for measurements of engineering surfaces carried out by white-light interference microscope Talysurf CCI 6000.

9. References

Ailing, T.; Chunhui, W.; Zhuangde, J.; Hongjun, W. & Bingcai, L. (2008). Study on Key Algorithm for Scanning White-Light Interferometry, *Proceedings SPIE*, Vol. 7155, pp. 71552N-71552N, Ninth International Symposium on Laser Metrology, 30 June-2 July 2008, Singapore, ISBN 0819473987

Bankhead, A. D. & McDonnell I. (2008). Surface Profiling Apparatus, US Patent, No. 7385707 B2

Blunt, L. & Jiang, X. (Eds). (2003). *Advanced Techniques for Assessment Surface Topography*, Kogan Page Science, ISBN 1903996112, London, UK, Sterling, VA, USA

Blunt, R. T. (2006). White Light Interferometry - a Production Worthy Technique for Measuring Surface Roughness on Semiconductor Wafers, *Proceedings of CS MANTECH Conference*, pp. 59–62, Vancouver, British Columbia, Canada, April 2006

Buchta, Z.; Mikel, B.; Lazar, J. & Čip, O. (2011). White-Light Fringe Detection based on a Novel Light Source and Colour CCD Camera, *Measurement Science and Technology*, Vol. 22, 094031 (6 pp.), ISSN 0957-0233

Cincio, R.; Kacalak, W. & Lukianowicz, Cz. (2008). System Talysurf CCI 6000 - Methodic of Analysis Surface Feature with using TalyMap Platinium, *Measurement Automation and Monitoring*, Vol. 54, No. 4, pp. 187–191, ISSN 0032-4140 (in Polish)

Conroy, M. & Armstrong, J. (2005). A Comparison of Surface Metrology Techniques, *7th International Symposium on Measurement Technology and Intelligent Instruments*, Journal of Physics, Conference Series, Vol. 13, pp. 458–465, ISSN: 1742-6588, Huddersfield, UK, September 2005

Conroy, M. (2008). Advances in Thick and Thin Film Analysis using Interferometry, *Wear*, Vol. 266, Issue 5-6, pp. 502-506, ISSN 0043-1648

Creath, K. (1988). Phase-Measurement Interferometry Techniques, In: *Progress in Optics*, E. Wolf (Ed.), Vol. XXVI, pp. 349-393, North-Holland, ISBN 0444870962, Amsterdam, Netherlands

Creath, K. & Schmit, J. (2004). Phase Shifting Interferometry, In: *Encyclopedia of Modern Optics*, Guenther, R. D.; Steel, D. G. & Bayvel, L. (Eds), Vol. 2, pp. 364–74, Elsevier Academic Press, ISBN 0122276000, Amsterdam, Netherlands

de Grot, P. & Colonna de Lega, X. (2004). Signal Modeling for Low-Coherence Height-Scanning Interference Microscopy, *Applied Optics*, Vol. 43, No. 25, pp. 4821-4830, ISSN 0003-6935

Deck, L. & de Grot, P. (1994). High-Speed Noncontact Profiler Based on Scanning White-Light Interferometry, *Applied Optics*, Vol. 33, No. 31, pp. 7334-7338, ISSN 0003-6935

Delaunay, G. (1953). Microscope Interférentiel A. Mirau pour la Mesure du Fini des Surfaces. *Revue d'Optique Theorique et Instrumentale*, Vol. 32, pp. 610–614, ISSN 0035-2489

Gao, F.; Leach R.; Petzing J. & Coupland J. M. (2008). Surface Measurement Errors using Commercial Scanning White Light Interferometers, *Measurement Science and Technology*, Vol. 19, No. 1, pp. 015303 (13 pp.), ISSN 0957-0233

Harasaki, A. & Schmit, J. (2002). Bat-Wing Attenuation in White-Light Interferometry, United States Patent, No. US 6493093 B2

Harasaki, A.; Schmit, J. & Wyant J. C. (2000). Improved Vertical-Scanning Interferometry, *Applied Optics*, Vol. 39, No. 13, pp. 2107-2115, ISSN 0003-6935

Harasaki, A. & Wyant, J. C. (2000). Fringe Modulation Skewing Effect in White-Light Vertical Scanning Interferometry, *Applied Optics*, Vol. 39, No. 13, pp. 2101-2106, ISSN 0003-6935

Hariharan, P. (2007). *Basis of Interferometry* (2nd edition), Elsevier Academic Press, ISBN 9780123735898, Amsterdam, Netherlands

Hariharan, P. & Roy, M. (1995). White-Light Phase-Stepping Interferometry: Measurement of the Fractional Interference Order, *Journal of Modern Optics*, Vol. 42, pp. 2357-2360, ISSN 0950-0340

Hýbl, O.; Berger, A. & Häusler G. (2008). *Information Efficient White-Light Interferometry*, Deutschen Gesellschaft für angewandte Optik e.V. Proceedings, 109, A 29, Available from http://www.dgao-proceedings.de, ISSN: 1614-8436

Kaplonek, W. & Tomkowski, R. (2009). Analysis of the Surface Topography of Diffractive Optical Elements by White Light Interferometry, *Measurement Automation and Monitoring*, Vol. 55, No. 4, pp. 272–275, ISSN 0032-4140 (in Polish)

Kaplonek, W. & Lukianowicz, Cz. (2010). Use of Coherence Correlation Interferometry for Measurements of Surface Topography, *Electrical Review*, Vol. 86, No. 10, pp. 43–46, ISSN 0033-2097 (in Polish)

Kim, J.-H.; Yoon, S.-W.; Lee J.-H.; Ahn W.-J. & Pahk H.-J. (2008). New Algorithm of White-Light Phase Shifting Interferometry Pursing Higher Repeatability by using Numerical Phase Error Correction Schemes of Pre-Processor, Main Processor, and Post-Processor, *Optics and Lasers in Engineering*, Vol. 46, Issue 2, pp. 140-148, ISSN 0143-8166

Kino, G. S. & Chim, S. S. (1990). Mirau Correlation Microscope, *Applied Optics*, Vol. 29, No. 26, pp. 3775-3783, ISSN 0003-6935

Kujawinska, M. (1993). Spatial Phase Measurement Methods, In: *Interferogram Analysis: Digital Fringe Pattern Measurement Technique*, Robinson, D. W. & Reid, G. T. (Eds), pp. 141-193, IOP Publishing, ISBN 075030197X, Bristol and Philadelphia

Larkin, K. G. (1996). Efficient Nonlinear Algorithm for Envelope Detection in White Light Interferometry, *Journal of the Optical Society of America A*, Vol. 13, No. 4, pp. 832–843, ISSN 1084-7529

Leach, R.; Brown, L.; Jiang, X.; Blunt R.; Conroy, M. & Mauger D. (2008). Guide to the Measurement of Smooth Surface Topography using Coherence Scanning Interferometry, In: *NPL Good Practice Guide*, No. 108, n.d., Available from http://www.npl.co.uk

Leach, R. (Ed.). (2011). *Optical Measurement of Surface Topography*, Springer-Verlag, ISBN 9783642120114, Berlin, Heidelberg, Germany

Linnik, V. P. (1933). A Simple Interferometer for the Investigation of Optical Systems, *Doklady Akademii Nauk SSSR (Proceedings of the USSR Academy of Sciences)*, No. 1, pp. 208-210

Lukianowicz, Cz. (2010). Use of White Light Scanning Interferometry for Assessment of Surface Topography, *Measurement Automation and Monitoring*, Vol. 56, No. 9, pp. 1055-1058, ISSN 0032-4140 (in Polish)

Lukianowicz, Cz. & Tomkowki, R. (2009). Analysis of Topography of Engineering Surfaces by White Light Scanning Interferometry, *Proceedings of 12th International Conference on Metrology and Properties of Engineering*(Pawlus P.; Blunt L.; Rosen B.-G.; Thomas T.; Wieczorowski M. & Zahouani H. Eds.), pp. 63–67, Rzeszow, Poland, ISBN 978-83-7199-545-3

Michelson, A. A. (1882). Interference Phenomena in a New Form of Refractometer, *The American Journal of Science*, Vol. 23, No. 3, pp. 120-129, ISSN 0002-9599

Niehues, J.; Lehmann, P. & Bobey, K. (2007). Dual-Wavelength Vertical Scanning Low-Coherence Interferometric Microscope, *Applied Optics*, Vol. 46, No. 29, pp. 7141-7148, ISSN 0003-6935

Park, M.-C. & Kim, S.-W. (2000). Direct Quadratic Polynomial Fitting for Fringe Peak Detection of White Light Scanning Interferograms, *Optical Engineering*, Vol. 39, No. 4, pp. 952-959, ISSN 0091-3286

Patorski, K.; Kujawińska, M. & Sałbut, L. (2005). *Laser Interferometry with Automatic Fringe Pattern Analysis* (1st edition), Patorski, K. (Ed.). Oficyna Wydawnicza Politechniki Warszawskiej, ISBN 9788372074911, Warszawa, Poland, (in Polish)

Pawłowski, M. E.; Sakano, Y.; Miyamoto, Y. & Takeda, M. (2006). Phase-Crossing Algorithm for White-Light Fringes Analysis, *Optics Communications*,Vol. 260, Issue 1, pp. 68–72, ISSN 0030-4018

Petzing, J.; Coupland, J.; & Leach R. (2010). The Measurement of Rough Surface Topography using Coherence Scanning Interferometry, In: *NPL Good Practice Guide*, No. 116, n. d., Available from http://www.npl.co.uk

Pförtner, A. & Schwider, J. (2001). Dispersion Error in White-Light Linnik Interferometers, *Applied Optics*, Vol. 40, No. 34, pp. 6223-6228, ISSN 0003-6935

Pluta, M. (1993). *Advanced Light Microscopy, Vol. 3, Measuring Techniques*, North Holland, ISBN 9780444988195, Amsterdam, Netherlands

Sato, S. & Ando, S. (2009). Weighted Integral Method in White-Light Interferometry: Envelope Estimation from Fraction of Interferogram, *Proceedings SPIE*, Vol. 7389, pp. 73892V-8, Optical Measurement Systems for Industrial Inspection VI, 15-18 June 2009, Munich, Germany, ISBN 9780819476722

Schmit, J. (2005). White Light Interferometry, In: *Encyclopedia of Modern Optics*, Guenther, R. D.; Steel, D. G. & Bayvel, L. (Eds), Vol. 2, pp. 375–387, Elsevier Academic Press, ISBN 0122276000, Amsterdam, Netherlands

Stahl, H. P. (1990). Review of Phase-Measuring Interferometry, *Proceedings of the SPIE*, Vol. 1332, Optical Testing and Metrology III: Recent Advances in Industrial Optical Inspection, pp. 704-719, ISBN 9780819403933, San Diego, California, USA, July 1990

Thomas, T. R. (1999). *Rough Surfaces* (2nd edition), Imperial College Press, ISBN 0750300396, London, UK

Veeco Instruments Inc., (2002). *Wyko NT1100 Optical Profiling System*. (Brochure), Available from http://ppewww.physics.gla.ac.uk/~williamc/NT1100_spec.pdf

Viotti, M. R.; Albertazzi, A.; Dal Pont, A. & Fantin, A. V. (2007). Evaluation of a Novel Algorithm to Align and Stitch Adjacent Measurements of Long Inner Cylindrical Surfaces with White ,Light Interferometry, *Optics and Lasers in Engineering*, Vol. 45, Issue 8, pp. 852-859, ISSN 0143-8166

Whitehouse, D. J. (1994). *Handbook of Surface Metrology* (1st edition), Institute of Physics Publishing, ISBN 0750300396, Bristol and Philadelphia

Whitehouse, D. J. (2003). *Handbook of Surface and Nanometrology*, Institute of Physics Publishing, ISBN 0750305835, Bristol and Philadelphia

Wieczorowski, M. (2009). *Using Topographic Analysis in Measurements of Surface Irregularities*, Wydawnictwo Politechniki Poznańskiej, ISBN 9788371438066, Poznan, Poland, (in Polish)

Zygo Corporation, (2010). *NewView™ 7300 Specifications* (Brochure), Available from: http://www.zygo.com/met/profilers/newview7000/nv7300spec.pdf

Zygo Corporation, (2007). *NewView™ 600 3D Optical Profiler* (Brochure), Available from: http://www.zygo.com/met/profilers/newview7000/nv7300spec.pdf

2

Surface Micro Topography Measurement Using Interferometry

Dahi Ghareab Abdelsalam
Engineering and Surface Metrology Laboratory,
National Institute for Standards,
Egypt

1. Introduction

Surface topography measurement plays an important role in many applications in engineering and science. The three dimensional (3D) shapes of objects need to be measured accurately to ensure manufacturing quality. Optical methods have been used as metrological tools for a long time. They are non-contacting, non-destructive and highly accurate. In combination with computers and other electronic devices, they have become faster, more reliable, more convenient and more robust. Among these optical methods, interferometry has received much interest for its shape measurement of optical and non-optical surfaces. Information about the surface under test can be obtained from interference fringes which characterize the surface. Two-beam interference fringes have been used to investigate the shape of optical and non-optical surfaces for a long time (Born et al., 1980). Phase distribution is encoded in an intensity distribution as a result of interference phenomena, and is displayed in the form of an interference pattern. Phase distribution should therefore be retrievable from the interference pattern. There are many methods of phase evaluation. In this chapter, we present recent developments in interferometry techniques carried out in our laboratory. In Sec. 2, the flat fielding method for coherent noise suppression is described. In Sec. 3, a combination of flat fielding and apodized apertures is presented. In Sec. 4, the phase retrieval using single-shot off-axis geometry, Zernike's polynomial fitting, Bünnagel's method, phase shifting interferometry, and two-wavelength interferometry are described. Section 5 describes a new off-axis interferometry configuration and the principle of single-shot, dual-wavelength interferometry for measuring a step height of 1.34 μm nominally. In Sec. 6, a multiple-beam interferometry technique at reflection for measuring the microtopography of an optical flat nominally of $\lambda / 20$ is described. Section 7 gives concluding discussions and remarks.

2. Flat fielding

The use of digital detectors in digital holography for recording a series of iterferograms is usually accompanied by dark current (thermal noise), shot noise, and scattering noise due to some dust particles, scratches, etc., on optical elements. Such coherent noise can affect the quality of the reconstruction of the object wave (amplitude and phase). Subtracting the dark current (thermal noise) clears the camera of any accumulated charge and reads out the cleared CCD. Figure.1(a) shows a three-dimensional (3D) thermal noise captured when there is no illumination through the optical system. Figure.1(b) shows a 3D influence of the non-uniformity of illumination which is mainly caused by the non-uniform Gaussian intensity distribution, instability of the laser source used, and the vignetting of lenses used in the system. Some factors which cause coherent noise have been reduced drastically by application of the flat fielding method.

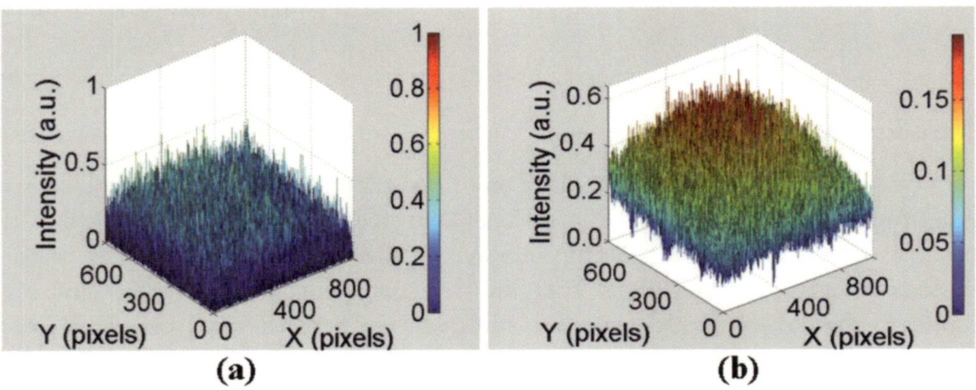

Fig. 1. Three-dimensional (3D) intensity distribution shows the effect of (a) dark frame (thermal noise), (b) influence of the non-uniformity of illumination (flat frame).

A CCD camera is calibrated by a process known as "Flat fielding" or "Shading correction". Flat fielding can be illustrated in the following formula (Abdelsalam et al., 2010; Steve, 2006; Abdelsalam et al., 2010),

$$\bar{I}_C = [M(\bar{I}_R - \bar{I}_B)] / (\bar{I}_F - \bar{I}_B), \tag{1}$$

where $\bar{I}_C$ is the average of calibrated captured interferograms, $\bar{I}_R$ is the average of non-calibrated captured interferograms, $\bar{I}_B$ is the average of dark frames, M is the average pixel value of the corrected flat field frame, and $\bar{I}_F$ is the average of flat field frames. Dark frames have been taken in advance and stored in the computer when there is no illumination through the optical system. Subtracting the dark frame clears the camera of any accumulated charge and reads out the cleared CCD. Flat field frames measure the response of each pixel in the CCD array to illumination and is used to correct any variation in illumination over the CCD sensor. The non-uniformity of illumination is mainly caused by the non-uniform Gaussian intensity distribution, as depicted in Fig.2(a). And also, the vignetting of lenses used in the system, which decrease the light intensity towards the image periphery or the dust particles on optical components like a glass window in front of CCD as illustrated in Fig.2(b), can be some inherent sources of errors in practical experiments. The flat fielding process corrects the uneven illumination.

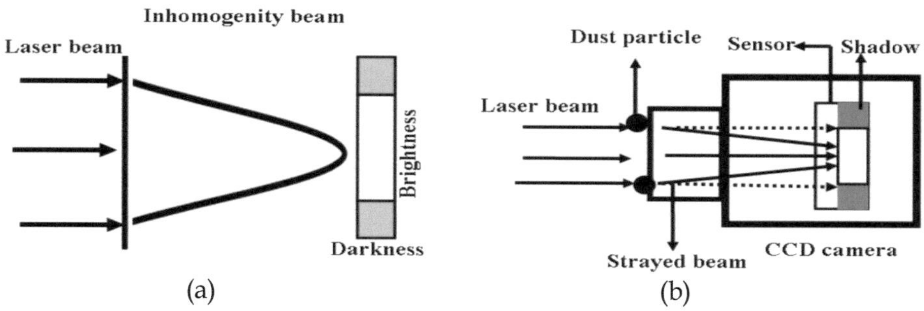

Fig. 2. Uneven illumination which produces darkness at the edges of the image (a) inhomogenity of the laser beam, (b) shadow detection of the dust particles hanging at the CCD camera aperture.

Flat field frames are also captured in advance and stored in the computer by blocking the wave from the object arm. Note that flat field frames are captured when laser source is switched on. The captured single off-axis interferograms have been calibrated by using the stored dark and flat field frames. The specimen is a sample of 200 μm step height. Figure 3(a) shows an average of 50 off-axis interferograms I_R captured when the sample is adjusted as an object in the Mach-Zender interferometer. In order to subtract the thermal noise, fifty dark frames I_B have been captured when there is no illumination through the optical system. Shot noise (random arrival of photons), fixed pattern noise (pixel to pixel sensitivity variation), and scattering noise due to some dust particles or scratches on optical elements have been suppressed drastically when the flat fielding Eq.5 is applied. Where I_F is the average of 50 off-axis interferograms captured by blocking the wave from the object arm when laser source is switched on. Figure 3(b) shows the correction of Fig.3(a) after application of the flat fielding method.

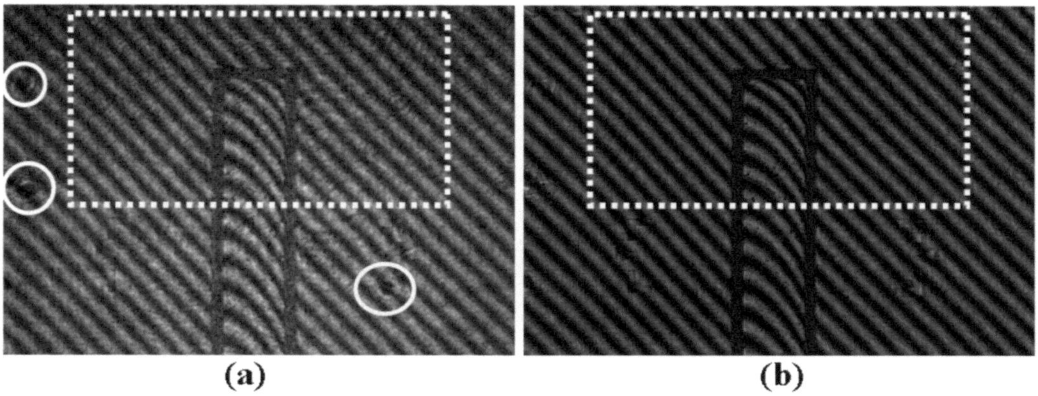

Fig. 3. Average of 50 off-axis interferograms of a specimen of 200 μm step height. (a) Before correction with flat fielding. (b) After correction with flat fielding.

It is shown apparently from Fig.3(b) that most of random coherent noise (i.e. circulated noise which may come from dust particles and scratches) which distorts the holographic fringe pattern is reduced drastically by application of the flat fielding method.

3. Flat fielding with apodized aperture

The coherent noise is further reduced when the apodization with cubic spline interpolation technique (Cuche et al., 2000) is applied to the calibrated interferogram with flat fielding (Abdelsalam et al., 2011). Apodization of the interferogram could be achieved experimentally by inserting an apodized aperture in front of the CCD. However, as a digital image of the interferogram is acquired, it is more practical to perform this operation digitally by multiplying the digitized interferogram with a 2D function representing the transmission of the apodized aperture (this is explained in more detail in ((Cuche et al., 2000)). A profile of this function is presented in Fig.4(a). The aperture is completely transparent (transmission equal to unity) in the large central part of the profile. At the edges, the transmission varies from zero to unity following a curve defined by a cubic spline interpolation. After application of the apodized aperture technique to the corrected interferogram with flat fielding (Fig.3(b)), the obtained off-axis interferogram is numerically processed (Gabor, 1948; Kühn et al., 2007) to obtain the object wave (amplitude and phase). Figure 4(b) shows the reconstructed amplitude-contrast image after application of the flat fielding method with the apodized aperture technique.

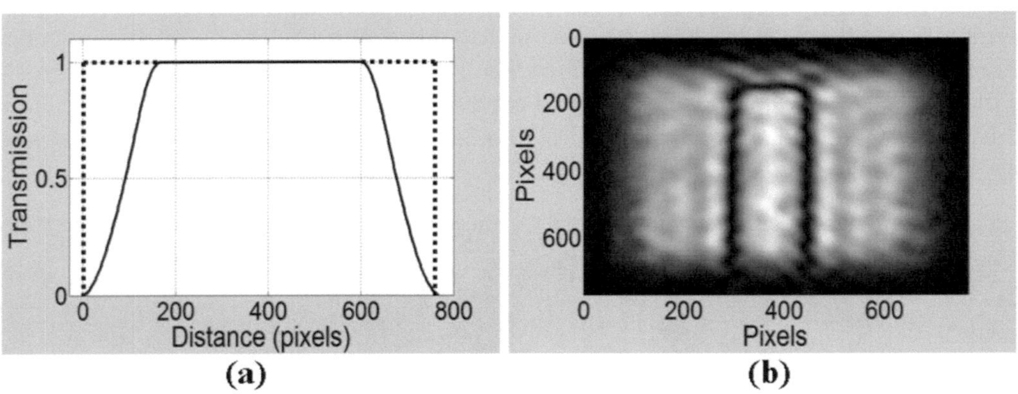

Fig. 4. Transmission profile of the apodized aperture. (a) The transmission from 0 to 1 at the edge of the aperture follows a cubic spline interpolation; the dashed line indicates the transmission of the unapodized aperture. (b) Numerically reconstructed amplitude-contrast image of Fig.3(b).

Figure 5(a) and Fig.5(b) show the reconstructed phase of the original off-axis interferogram (before correction) and after application of the proposed method (a combination of the flat fielding method with the apodized aperture technique), respectively. It can be noticed that the quality of the reconstructed phase-contrast image is improved as shown clearly in Fig.5(b) and Fig.5(d). The reconstructed phase images shown in Fig.5(a) and Fig.5(b) are exactly in the same size inside the white rectangles in Fig.3(a) and Fig.3(b), respectively.

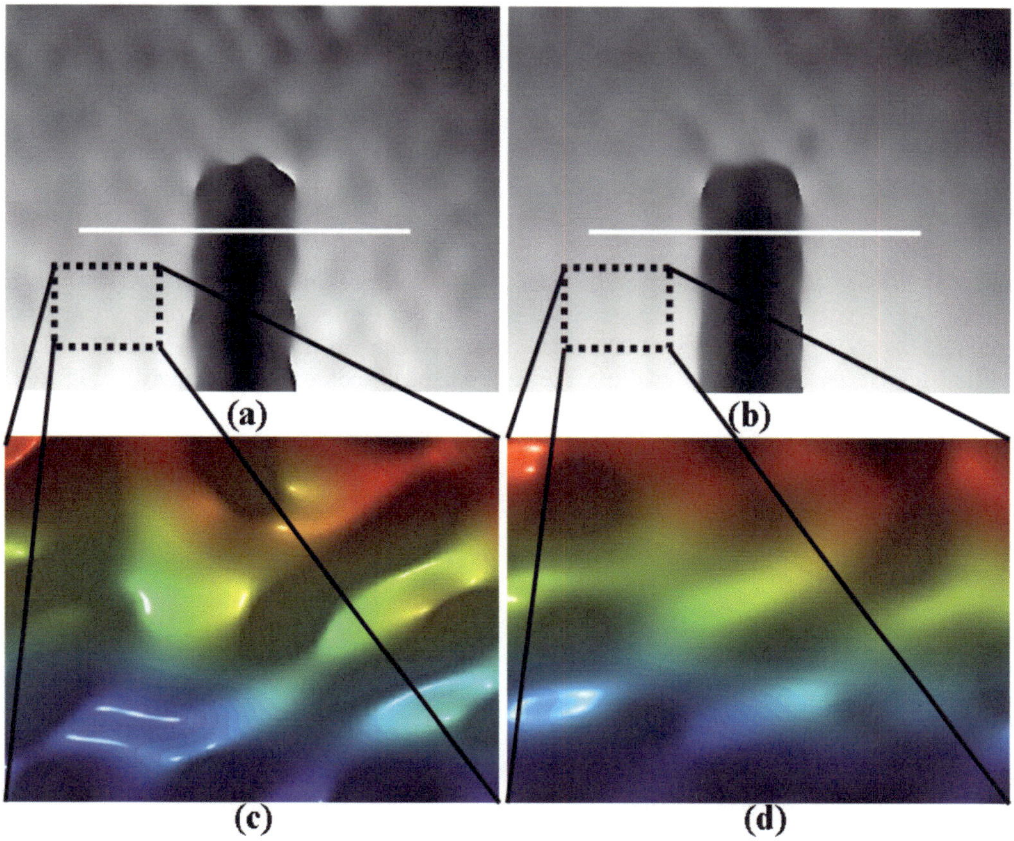

Fig. 5. Reconstructed phase images after converting to height from the off-axis interferograms. (a) Original (before correction with the proposed method). (b) After correction with the proposed method. (c) 3D of the selected rectangle of (a). (d) 3D of the selected rectangle of (b).

4. Phase retrieval

4.1 Off-axis geometry

In this section, the phase of single-shot circular fringes is extracted by using the off-axis geometry. Figure 6 shows the schematic diagram of the optical setup of the Fizeau interferometer (Abdelsalam et al., 2010). The tested smooth spherical surface of 25.4 mm in size was mounted as an object in the interferometer. A laser diode beam passes through a collimating lens and expands. This expansion is necessary to illuminate a greater area of the surface to be imaged and to reduce the error measurement due to the inhomogeneity in the Gaussian beam. The collimated beam of the laser light ($\lambda = 635nm$) falls upon the beam splitter, which transmits one half and reflects the other half of the incident light. The reflected collimated beam is then incident on the interferometer, which changes the path length of the light inside it due to the irregularities of the surface of the interferometer. When the object and the reference ($\lambda / 20$ flatness) are mounted close and parallel, two types of circular reflection fringes are seen; one, called the insensitive fringes, due to the interference from the two interfaces of the object surfaces, and the second, called the sensitive fringes, due to the interference of the reference interfaces and the object interfaces,. When 2D-FFT was applied for the inteferogram that had the two types of fringes (insensitive and sensitive) as shown in Fig. 7(a), six spectra were produced as shown in Fig. 7(b): three spectra produced from the insensitive fringes and the others produced from the sensitive fringes. In this case, the complex fringe amplitude becomes difficult to determine because the Fourier spectra of the interferogram that contains the insensitive and the sensitive fringes are not separated completely. The problem of the insensitive fringes was solved by adjusting the object so that it became parallel to the reference (in-line scheme). Therefore, the insensitive circular reflection fringes were seen in the center of the field of view. These circular fringes were transformed to a background spectrum (DC term) when 2D-FFT was applied (Cuche et al., 1999). By tilting the reference (off-axis case) as shown in Fig. 6, the sensitive circular reflection fringes were displaced, and nearly curved fringes at reflection with higher spatial frequency were seen, as shown in Fig.7(c). Only three spectra were obtained as shown in Fig.7(d) from Fig. 7(c) when 2D-FFT was implemented.

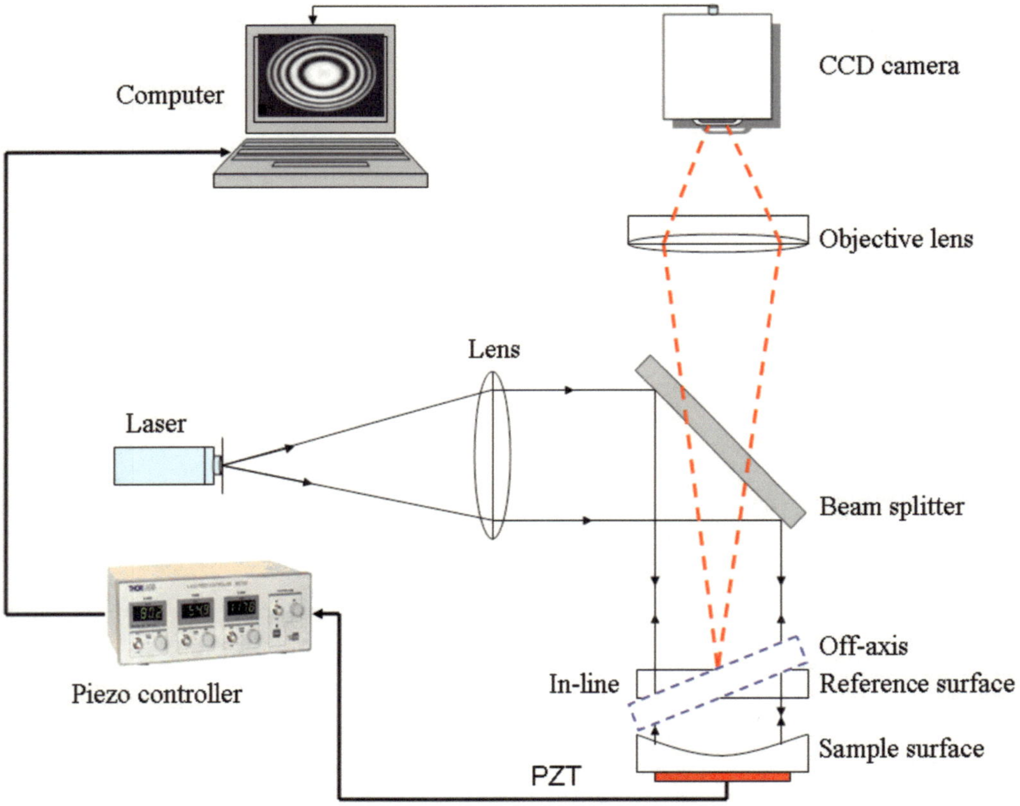

Fig. 6. Schematic diagram of the optical setup.

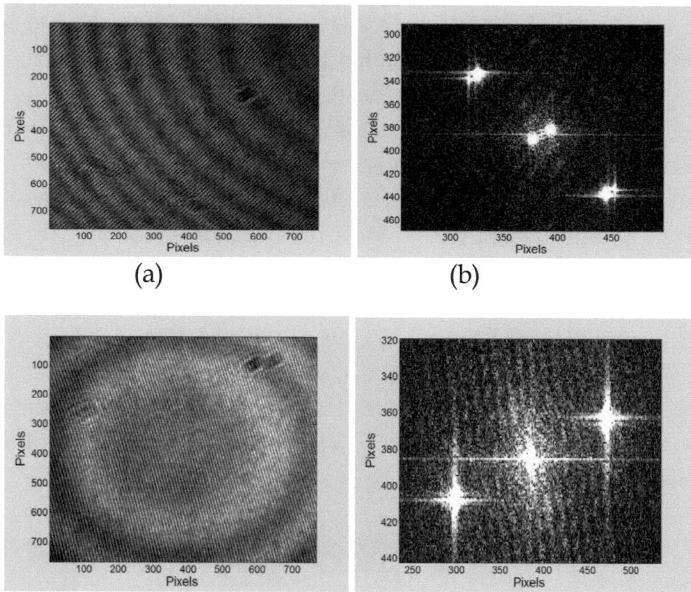

(a) (b)

(c) (d)

Fig. 7. Captured interferograms and their spectra using 2D-FFT. (a) Off-axis insensitive and sensitive fringes. (b) Spectra of (a). (c) In-line insensitive fringes and off-axis sensitive fringes. (d) Spectra of (c).

The interferogram was captured by the CCD camera of 1024×768 pixels with pixel size $\Delta x = \Delta y = 6.4\mu m$. Assume that the coordinate system of the interferogram plane is the mn plane. When waves from both the object and reference of the interferometer meet to interfere, the intensity of the interferogram is given as follows:

$$I(m,n) = \Psi = R^{*}O \cdot \tag{2}$$

Here, Ψ represents the intensity of the recorded interferogram, O is the object beam, R is the reference beam,* denotes the complex conjugate and m, n are integers. The reconstructed wave front is an array of complex numbers. An amplitude-contrast image and a phase-contrast image can be obtained by using the following intensity $[\mathrm{Re}(\Psi)^2 + \mathrm{Im}(\Psi)^2]$ and the argument $\tan^{-1}[\mathrm{Re}(\Psi) / \mathrm{Im}(\Psi)]$, respectively. In the reconstruction process, the intensity of the interferogram is multiplied by the amplitude of the original reference wave called a *digital reference wave* ($R_D(m,n)$). If we assume that a perfect plane wave is used as the reference for interferogram recording, the computed replica of the reference wave $R_D(m,n)$ can be calculated as follows:

$$R_D(m,n) = A_R \exp[i(2\pi / \lambda)(k_x m\Delta x + k_y n\Delta y)] \cdot \tag{3}$$

Where, A_R is the amplitude, λ is the wavelength of the laser source, and k_x and k_y are the two components of the wave vector that must be adjusted such that the propagation direction of $R_D(m,n)$ matches as closely as possible with that of the experimental reference wave. By using this digital reference wave concept, we can obtain an object wave which is reconstructed in the central region of the observation plane. The captured interferogram of the surface object was processed using Matlab codes to obtain a reconstructed object wave (amplitude and phase). Figure 8 shows the detail numerical reconstruction process of a single shot off-axis hologram of a surface object. As depicted in Fig. 8(a) through 8(d), 2D-FFTs were implemented for the spatial filtering approach. The inverse 2D-FFT was applied after filtering out the undesired two terms, and the complex object wave depicted in Fig. 8(d) and 8(e) in the interferogram plane was extracted. After the spatial filtering step, the object wave in the interferogram plane was multiplied by the digital reference wave $R_D(m,n)$. The final reconstructed object wave (amplitude and phase) as demonstrated in Fig. 8(g) and 8(h) was recorded by selecting appropriate values for the two components of the wave vector $k_x = 0.002955mm^{-1}$ and $k_y = 0.01143mm^{-1}$. The reconstructed phase shown in Fig. 8(h) is non-ambiguous and shows the results wrapped onto the range $-\pi$ to π. In order to retrieve the continuous form of the phase map ϕ, an unwrapping step has to be added to the phase retrieval process (Ghiglia et al., 2010).

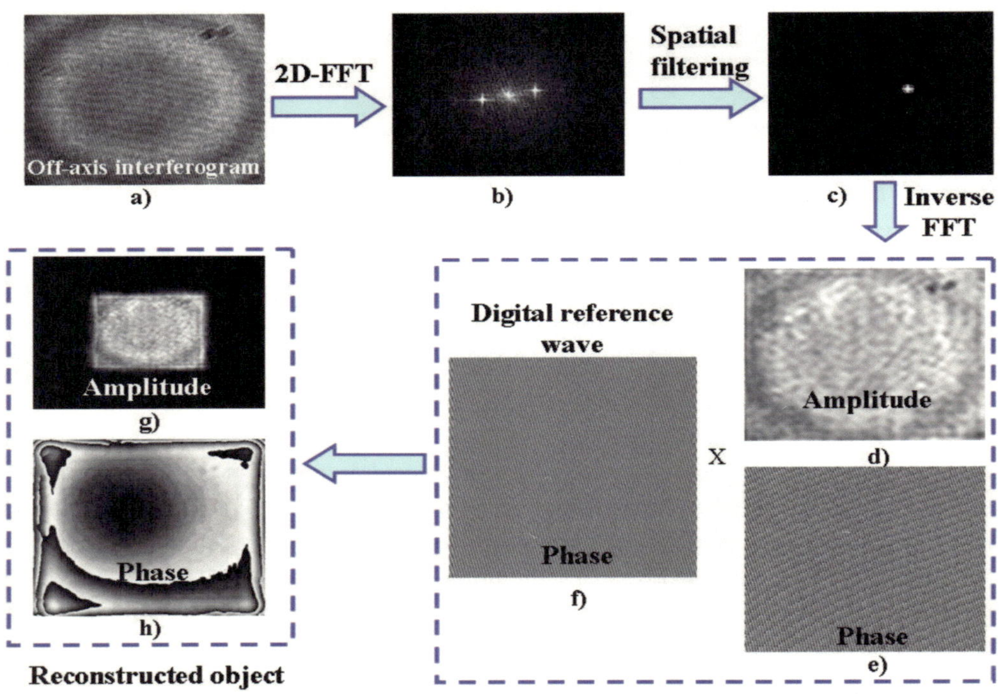

Fig. 8. Reconstruction steps of the conventional spatial filtering based phase contrast off-axis interferometry: a) Off-axis interferogram, b) Fourier transformed spatial frequency domain data, c) Spatially filtered domain data, d) - e) Inversely Fourier transformed data, f) Phase map of the digital reference wave, g) - h) Reconstructed object wave.

The two-dimensional (2D) surface profile height h can be calculated directly as follows:

$$h = \frac{\phi}{4\pi}\lambda . \tag{4}$$

Figure 9(a) shows the 480×480 pixels unwrapped phase map for the wrapped phase map in Fig. 8(h). The 3D view of the unwrapped phase map is shown in Fig. 9(b).

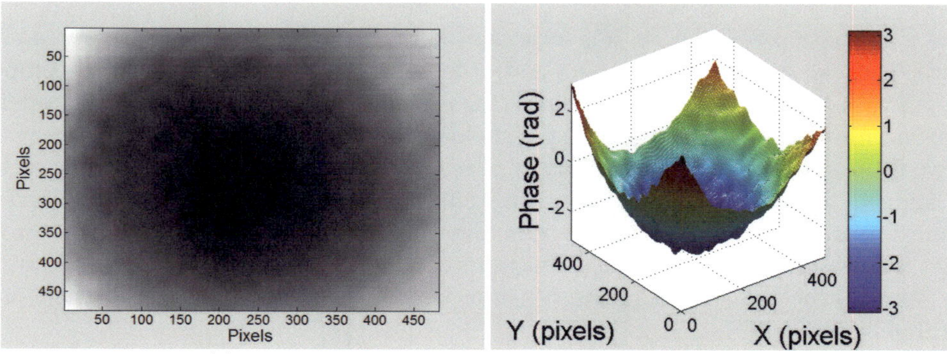

(a) (b)

Fig. 9. (a) 480 x 480 pixels unwrapped phase map for the wrapped phase map in Fig.4(h). (b) Three-dimensional view of the unwrapped phase map of (a).

The phase information shown in Fig. 9(b) was converted to metrical 3D surface height information as shown in Fig.10(a). Figure 10(b) presents the measured profile curve along 450 pixels in the x-direction and its cubic fitting.

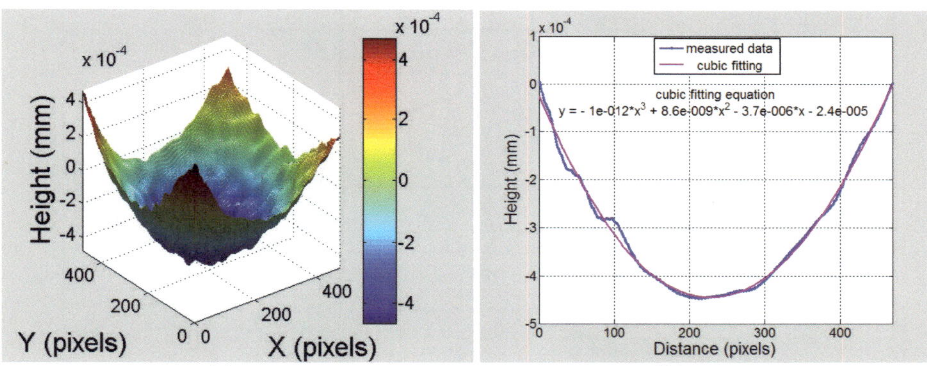

(a) (b)

Fig. 10. (a) Three-dimensional surface height resulting from the unwrapped phase shown in Fig. 9(b). (b) Two-dimensional surface height along 450 pixels in the x-direction.

The peak to valley value calculated from Fig. 10(b) was of the order of $0.45x10^{-3}mm$.

4.2 Zernike's polynomial fitting

Zernike polynomials have been used extensively to fit wavefront data for the interferometry to obtain an accurate representation of the surface. The characteristics of the surface are described by the calculation of the weighting coefficients of the Zernike polynomials. The surface height function $Z_r(x_r, y_r)$ can be represented by a linear combination of M polynomials $F(x_r, y_r)$ and their weighting coefficients G (David, et al., 1993 ; Wang, et al., 1980) is expressed as follows:

$$Z_r(x_r, y_r) = \sum_{j=1}^{M} F_j(x, y) G_j \,, \tag{5}$$

where r is the sample index so, it is important to calculate the coefficients to represent the surface. This can be done by thinning and ordering the fringes, respectively. A computer program with Matlab was written to obtain the thinning of the fringes. The fringe pattern is digitized into the computer and then thresholded to yield a binary gray-level image. The median filter was used to delete salt and pepper noise in the image. The flat fielding method was applied to the image to remove the effect of the offset of the camera and the inhomogeneity of the collimated laser beam intensity. The x-y scan suitable for the circular fringes was applied to the corrected image to obtain the fringe centers. After the fringes were thinned, the assignment of the fringe orders was determined. The flowchart of the algorithm is shown in Fig.11.

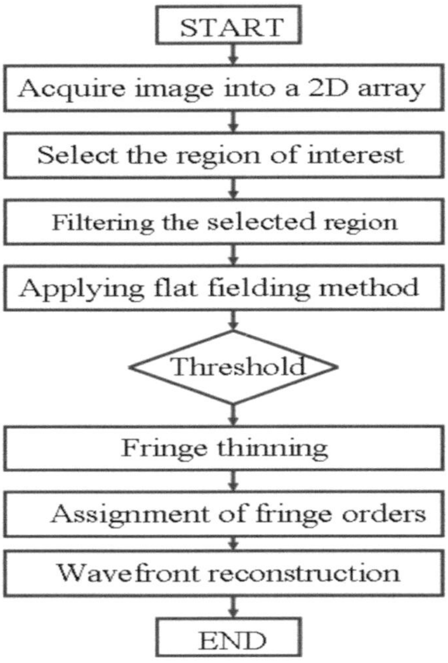

Fig. 11. Flow chart for the fringe pattern processing.

The data were fitted to a polynomial set. This is typically accomplished with a least-squares method, where S is defined as the sum of the square of the difference between the data points and the fitted polynomials is expressed as follows:

$$S = \sum_{r=1}^{N} \left[Z_r - \sum_{j=1}^{M} G_j F_j(x_r, y_r) \right]^2 . \tag{6}$$

If a perfect fit were possible, S would be equal to zero, because there would be no difference between the measured values and the representing polynomials evaluated at the corresponding points. However, there are always differences between the real surface and its representation. Therefore S is nonzero. The function of a least-squares fit is to find the coefficients for a given set of polynomials, which minimizes S. The coefficients are found by taking the derivative of S with respect to each coefficient and setting the result equal to zero. The result of minimizing S for all coefficients can be expressed by using summations in matrix form as follows:

$$\begin{bmatrix} \sum_{r=1}^{N} Z_r F_1(x_r,y_r) \\ \cdot \\ \cdot \\ \sum_{r=1}^{N} Z_r F_M(x_r,y_r) \end{bmatrix} = \begin{bmatrix} \sum_{r=1}^{N} F_1(x_r,y_r)F_1(x_r,y_r) \ldots\ldots \sum_{r=1}^{N} F_M(x_r,y_r)F_1(x_r,y_r) \\ \cdot \\ \cdot \\ \sum_{r=1}^{N} F_1(x_r,y_r)F_M(x_r,y_r) \ldots\ldots \sum_{r=1}^{N} F_M(x_r,y_r)F_M(x_r,y_r) \end{bmatrix} \begin{bmatrix} G_1 \\ \cdot \\ \cdot \\ G_M \end{bmatrix}. \quad (7)$$

The equations were solved numerically and G_j were obtained. The reconstruction of the surface was obtained automatically after substituting the coefficients in Eq.5. The same sample used in off-axis geometry has been adjusted in the in-line case of the setup shown in Fig.6. The captured interferogram is shown in Fig.12(a). The corrected fringe pattern with flat fielding shown in Fig.12(b) was processed by the automatic processing technique (see the flow chart in Fig.11) and the 3D surface profile of the surface under test was obtained automatically.

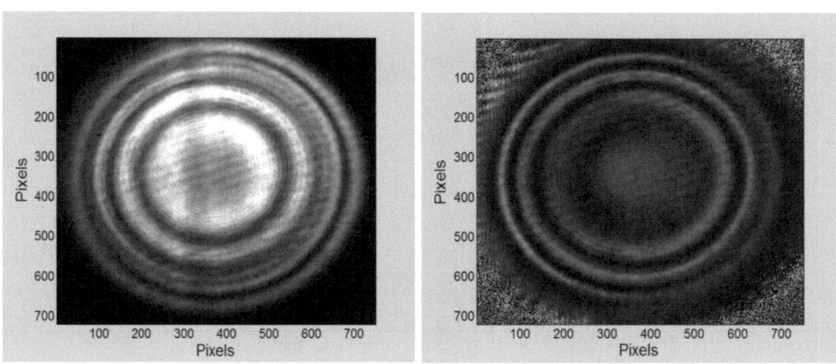

(a) (b)

Fig. 12. (a) Circular fringe pattern before correction with flat fielding. (b) After correction.

The 3D reconstruction surface of the circular fringe pattern of Fig.12(b) is shown in Fig.13(a). The 2D surface profile along 450 pixels in the x-direction and its cubic fitting is shown in Fig.13(b). The surface form of the tested spherical surface was estimated from Fig.13(b) to be of the order of $0.44x10^{-3}$ mm.

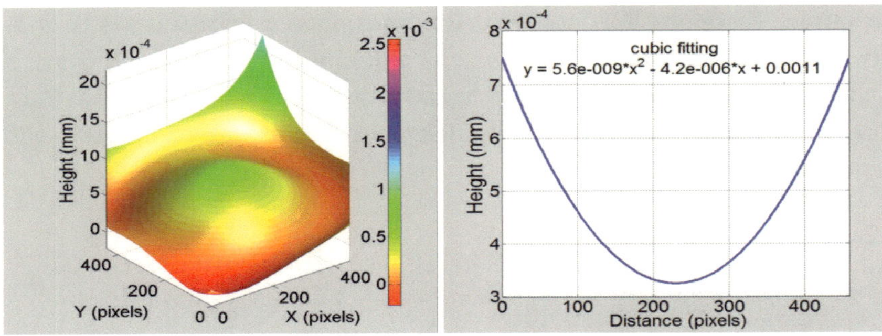

(a) (b)

Fig. 13. Processing results of the circle interference fringe pattern. (a) Three-dimensional reconstruction surface by using Zernike polynomial fitting method. (b) Two-dimensional surface profile.

4.3 Bünnagel's method

The Bünnagel method (Bünnagel, 1956) was used in this paper to analyze the interferogram of the tested smooth spherical surface. This method has been used to analyze multiple-beam interferograms with very sharp fringes manually. Nowadays, with the invention of high speed microprocessors and high speed CCD cameras, the Bünnagel method can be applied to two-beam interferograms. The feature height obtained from the Bünnagel method depends only on the fringe distance as shown in Eq.(11). From the schematic diagram shown in fig. 5, the topographical height related to fringe distance is calculated as follows:

$$h_2 = q\cos(\alpha),\ \tan(\alpha) = s / d_2,\ s + q = 2\lambda / 2, \tag{8}$$

when α is taken very small, so that $\cos(\alpha) = 1$. The height can be written as follows:

$$h_2 = 2\lambda / 2 - d_2 \tan(\alpha). \tag{9}$$

Also from fig.4, one can write $\tan(\alpha) = \dfrac{4\lambda / 2}{d_4}$ or

$$h_2 = \frac{\lambda}{2}\left(2 - \frac{d_2 / d_4}{4}\right), \tag{10}$$

or for general case

$$h_m = \frac{\lambda}{2}\left(m - \frac{d_m / d_n}{n}\right), \tag{11}$$

where h_m is the topographical height at fringe m.

d_m is the distance from fringe m to selected points AC and d_n is the distance from fringe n (at selected point B) to AC.

The same sample used in the off-axis geometry method has been adjusted in the in-line case of the setup shown in Fig.6. The captured interferogram is shown in Fig.12(a). The corrected fringe pattern with flat fielding is shown in Fig.12(b). Three fringes from the center of the circular fringes shown in Fig.12(b) have been dealt with. Since the topographical height is related to the fringe distance as expressed in Eq.10, a line profile as shown in Fig.15(a) has been taken over the 3 fringes from the center of Fig.12(b). The source used in the experiment is a laser diode of $\lambda = 635nm$. The surface height at the center of the 3 fringes has been calculated by using Eq.11. The 2D surface profile along 450 pixels in the x-direction and its cubic fitting is shown in Fig.15(b). The surface form of the tested spherical surface was estimated from Fig.13(b) to be of the order of $0.46x10^{-3}$ mm.

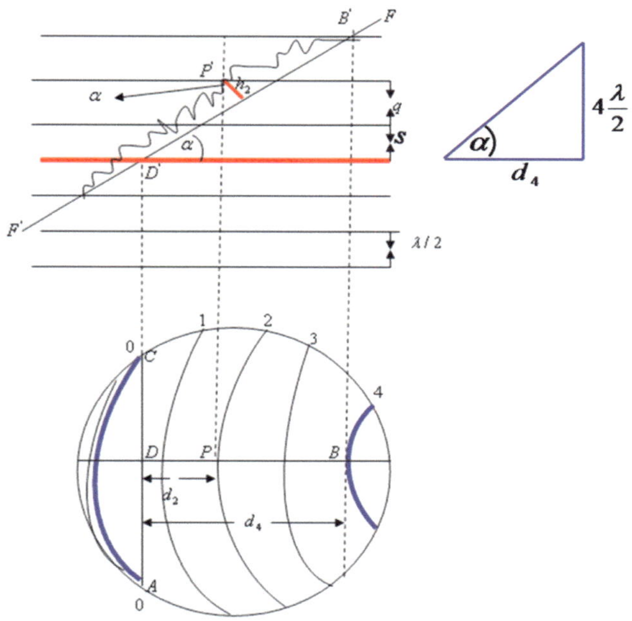

Fig. 14. Schematic diagram of the interferogram and height description.

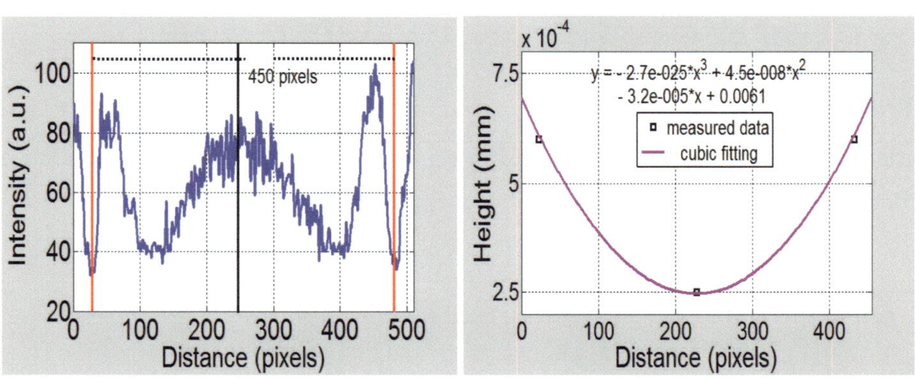

Fig. 15. (a) Intensity profile taken in the middle of Fig.12(b). (b) Two-dimensional surface profile.

4.4 Phase shifting algorithms

Phase shifting is a very robust technique for the analysis of fringe patterns. Phase shifting measures the phase of a pixel depending on the values of that pixel in different images. The intensity distribution of the phase-shifted frames for the initial state of the object can be expressed as follows:

$$I(x,y) = I_O + I_R + 2\sqrt{I_O I_R}\cos(\varphi + j\pi / 2), \tag{12}$$

where I_O and I_R are the intensities of the object and the reference waves, respectively, φ is the phase, and j is the number of the phase-shifted frames. Several phase-shifting algorithms have been developed with a view to reduce the effect of the phase shift and other errors on phase calculation (Kumar, et al., 2009). The evaluated phase is wrapped between $-\pi$ and π due to arctangent function. In this section we present a comparative study of several of these algorithms in fringe analysis.

4.4.1 Three-step algorithm

Using the first stored phase-shifted frames ($j = 0 - 2$), the phase distribution of the object can be expressed as follows (Kumar, et al., 2009):

$$\phi_j = \tan^{-1}\left(\frac{I_1 - 2I_2 + I_3)}{I_1 - I_3}\right), \tag{13}$$

4.4.2 Four-step algorithm

It uses the first four stored phase-shifted frames ($j = 0 - 3$). The phase distribution can be expressed as follows (Kumar, et al., 2009):

$$\phi_j = \tan^{-1}\left(\frac{(I_2 - I_4)}{-I_1 + I_3}\right). \tag{14}$$

4.4.3 Five-step algorithm

From the first five stored phase-shifted frames ($j = 0 - 4$). The phase distribution can be written as follows (Kumar, et al., 2009):

$$\phi_j = \tan^{-1}\left(\frac{2(I_2 - I_4)}{-I_1 + 2I_3 - I_5}\right). \tag{15}$$

4.4.4 Six-step algorithm

If we use the first six stored phase-shifted frames ($j = 0 - 5$). The phase distribution can be expressed as follows (Kumar, et al., 2009):

$$\phi_j = \tan^{-1}\left(\frac{I_1 - I_2 - 6I_3 + 6I_4 + I_5 - I_6)}{4(I_2 - I_3 - I_4 + I_5)}\right) - \frac{\pi}{4}. \tag{16}$$

4.4.5 Seven-step algorithm

Using all the seven stored phase-shifted frames (j=0–6), we have phase distribution as follows (Kumar, et al., 2009):

$$\phi_j = \tan^{-1}\left(\frac{4I_2 - 8I_4 + 4I_6)}{-I_1 + 7I_3 - 7I_5 + I_7}\right). \tag{17}$$

The tested surface (same sample used in the off-axis geometry method) was mounted as an object in the interferometer and adjusted very closely to the reference. The distance of the cavity between the object and the reference was changed very slightly by using a PZT varied by voltage. The four-step algorithm explained in section 4.4.2 is used to extract the phase. Four different interferograms with $0, \pi/2, \pi$ and $3\pi/2$ radian phase shifts were captured and corrected with the flat fielding method as shown in Fig.16.

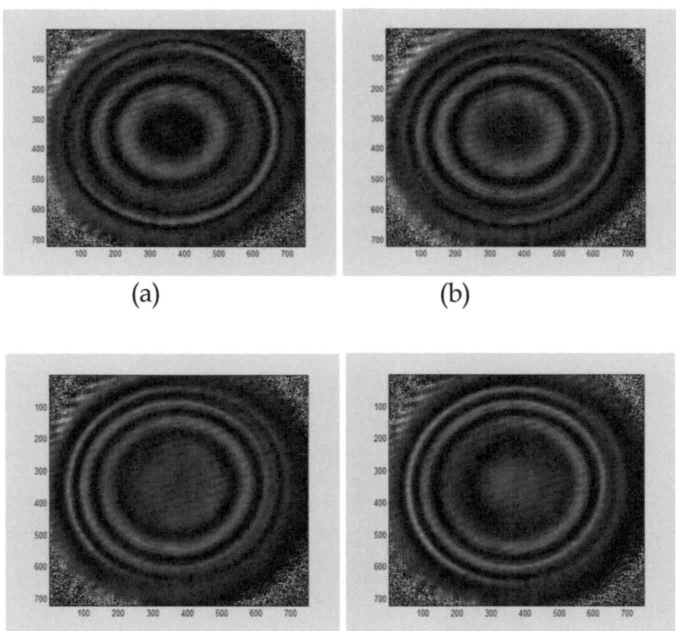

(a) (b) (c) (d)

Fig. 16. Four different interferograms (x-axis and y-axis are pixels) with (a) 0, (b) $\pi/2$, (c) π, and (d) $3\pi/2$ radian phase shifts after correction with the flat fielding method.

The wrapped phase map (Eq.14 is used) from the corrected interferograms is shown in Fig.17(a). The wrapped phase map is then unwrapped to remove the 2π ambiguity and the unwrapped phase map is shown in Fig.17(b). Figure.17(c) shows 480 x 370 pixels unwrapped phase map at the middle of Fig.17(b). The phase information shown in Fig.17(c) was converted to metrical 3D surface height information by using Eq.4 as shown in Fig.17(d).

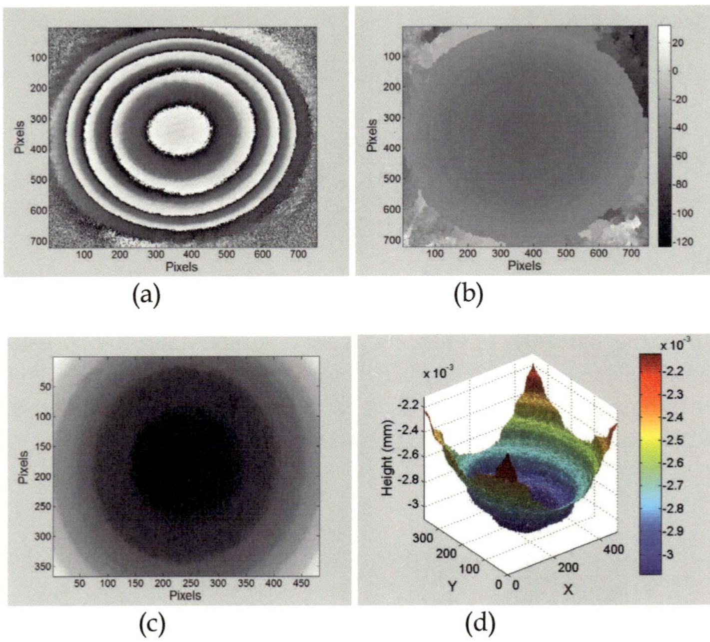

Fig. 17. (a) Wrapped phase map resulted from the four-frames. (b) Unwrapped phase information in 2D grey-scale. (c) 480 x 370 pixels unwrapped from the middle of (b). (d) Three-dimensional surface height of (c).

Figure.18 presents the measured profile curve along 450 pixels in the x-direction and its cubic fitting. The peak to valley value calculated from Fig.18 was of the order of $0.47x10^{-3}mm$.

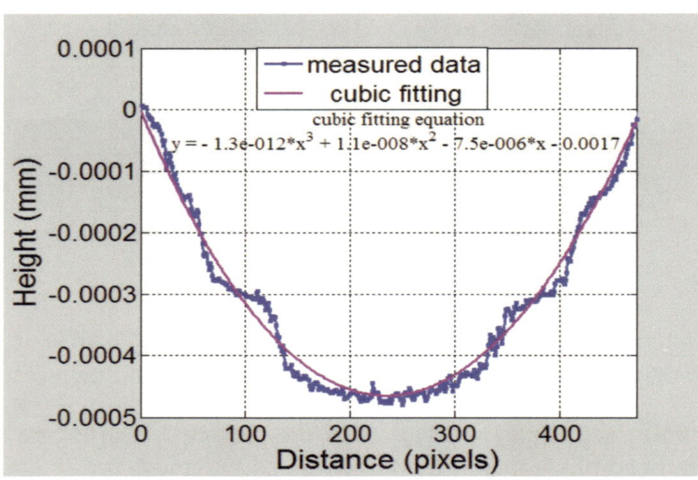

Fig. 18. Two-dimensional surface height along 480 pixels at x-direction.

As shown from the results and the cubic fitting equations, the measured value measured with off-axis geometry is very close to the value measured with the Zernike's polynomial fitting method, the Bünnagel's method, and the four-frame phase shifting technique. The little deviation may be due to the vibration because the phase shifting algorithm is more sensitive to vibration than single shot off-axis geometry.

4.5 Two-wavelength interferometry

The conventional interferometric surface profilers using a single wavelength (the phase takes the form of a sawtooth wave, as seen in Fig. 19) have a serious limitation: they can only handle smooth profiles and step heights of less than half a wavelength. The approaches adopted to overcome the problem of the small dynamic range are based on two-wavelength interferometry, multi-wavelength interferometry and white-light interferometry. In the two-wavelength method, the system can use long-range measurements with careful choice of wavelengths. If the phase of the surface under test is measured at two wavelengths, λ_1 and λ_2, then the difference between the two corresponds to the modulo-2π phase map, which could have been generated with a longer wavelength (synthetic beat-wavelength) expressed as follows:

$$\Lambda = \frac{\lambda_1 \lambda_2}{\lambda_2 - \lambda_1} . \tag{18}$$

We can see that the smaller the difference between the two wavelengths, the larger the synthetic wavelength, typically within the range of micrometers to millimeters. The resulting graph of a discontinuity in phase for two wavelengths before subtraction is shown in Fig.20. The algorithm for the unwrapping of phase discontinuities is simple; one wavelength is simply subtracted from the other, as seen in Fig. 21. The result appears to be discontinuous, but by merely adding 2π wherever the phase map is negative (addition modulo 2π), the result is a continuous slope that accurately recreates the original object.

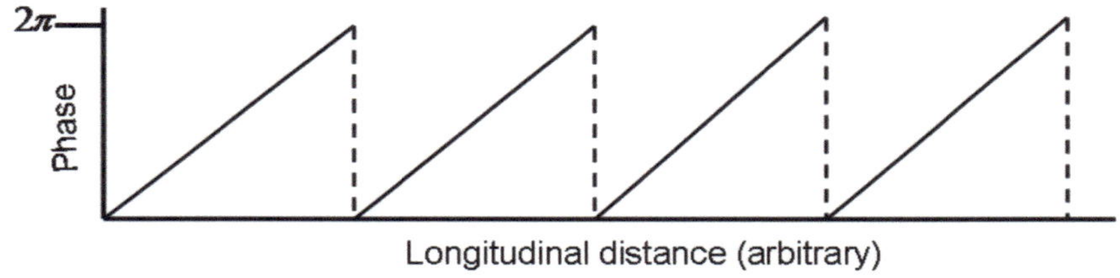

Fig. 19. Phase map of using one wavelength.

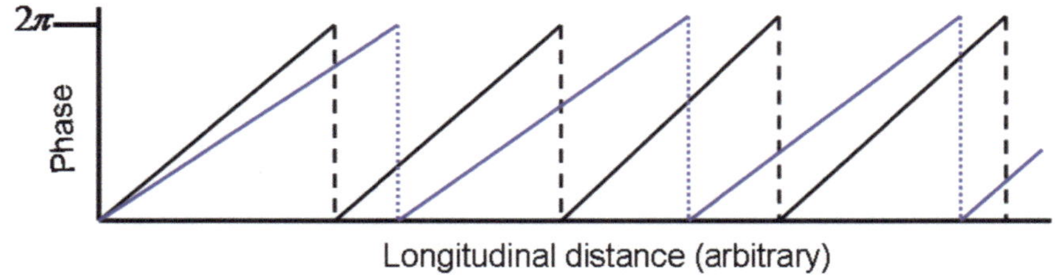

Fig. 20. Phase maps for two wavelengths.

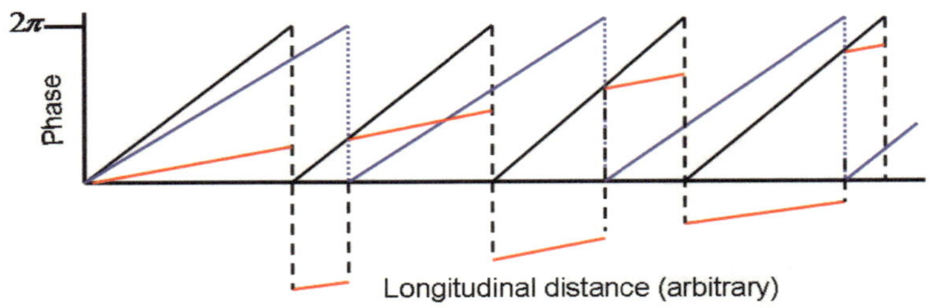

Fig. 21. Subtraction of phase maps.

5. Single-shot, dual-wavelength interferometry

5.1 Principle

The principle is based on the acquisition of an interferogram by a CCD camera with two object beams O_1 and O_2 of two different wavelengths λ_1 and λ_2 that interfere with two counterpart reference beams R_1 and R_2, emitted by the same pair of laser sources, in an off-axis configuration (Abdelsalam et al., 2011). The expression for the intensity pattern, which results from both interferomgrams at λ_1 and λ_2, can be written as follows:

$$I(k,l) = |O_1|^2 + |R_1|^2 + |O_2|^2 + |R_2|^2 + R_1^*O_1 + R_1O_1^* + R_2^*O_2 + R_2O_2^* . \quad (19)$$

Here, I is the interferogram intensity, k and l are integers, and * denote the complex conjugate. In Eq. (1), the first four intensity terms are of zero order, which can be directly filtered in the Fourier domain, and the last four are the interference terms with the object wave O_i (the virtual images) or their conjugate O_i^* (the real images), being modulated by the spatial carrier frequency in the spatial frequency domain. The carrier frequencies are dependent upon the k-vectors of R_1 and R_2. Now, if the two reference waves have different propagation directions, especially in the configuration where their k-vector projections on the CCD plane are orthogonal, each interference term occupies a different position in the Fourier domain. Provided that there is no overlap between the interference terms, it is straightforward to isolate each frequency component by the spatial filtering approach (Cuche et al., 2000). After filtering out the object and conjugate terms, $R_1^*O_1$ and $R_2^*O_2$, the filtered spectrum data in the spatial frequency domain turn back to the spatial domain by using inverse 2D FFT separately for each wavelength. Let us define these two filtered complex waves Ψ_1 and Ψ_2 as follows:

$$\Psi_i = R_i^*O_i , \quad (20)$$

where $i = 1,2$. The above filtered complex wave is an array of complex numbers. An amplitude-contrast image and a phase-contrast image can be obtained by using the following intensity $[\mathrm{Re}(\Psi_i)^2 + \mathrm{Im}(\Psi_i)^2]$ and the argument $\arctan[\mathrm{Re}(\Psi_i)/\mathrm{Im}(\Psi_i)]$, respectively. Finally, in order to obtain the object information, the Ψ_i needs to be multiplied by the original reference wave called a digital reference wave $(R_{D_i}(m,n))$. Here, m and n are integers. If we assume that a perfect plane wave is used as a reference for interferogram recording, the computed replica of the reference wave R_{D_i} can be represented as follows:

$$R_{D_i}(m,n) = A_{R_i} \exp[i(2\pi / \lambda_i)(k_{x_i} m\Delta x + k_{y_i} n\Delta y)] , \tag{21}$$

where, A_{R_i} is the amplitude, λ_i is the wavelength of the laser source, $\Delta x = \Delta y$ are the pixel sizes, and k_{x_i} and k_{y_i} are the two components of the wave vector that must be adjusted such that the propagation direction of R_{D_i} matches as closely as possible with that of the experimental reference wave. By using this digital reference wave concept, we can obtain an object wave which is reconstructed in the central region of the observation plane for each wavelength separately. As stated above, the reconstructed complex wave fronts O_1 and O_2 in the reconstruction plane contain both the amplitude and the phase information. However, both complex wave fronts suffer from phase ambiguity for abrupt height varying specimens. In order to overcome this limitation of the single wavelength approach, we can use a synthetic beat-wavelength that can be defined as follows:

$$\Phi = \arg(O_1 O_2^*) = \phi_1 - \phi_2 = 2\pi x \left(\frac{\lambda_2 - \lambda_1}{\lambda_1 \lambda_2} \right) = 2\pi \frac{x}{\Lambda} , \tag{22}$$

where, x is OPD (Optical Path Difference), which means twice of the topography for reflection scheme. ϕ_i is the reconstructed phase for wavelength λ_i and Λ is the synthetic beat wavelength expressed in Eq.18. The two-dimensional (2D) surface profile height h can be calculated directly as follows:

$$h = \frac{\Phi}{4\pi} \Lambda . \tag{23}$$

Figure 22 shows the flow chart of the algorithm that has been used to analyze the off-axis interferogram containing both wavelengths.

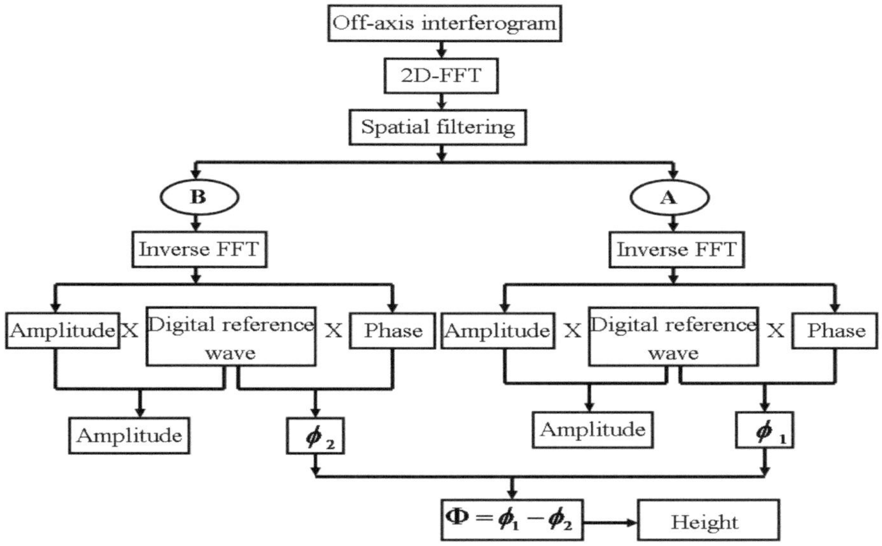

Fig. 22. Flowchart of the algorithm that has been used to analyze the off-axis interferogram containing both wavelengths.

5.2 Experimental results

The proposed new scheme on the dual-wavelength off-axis system which is basically based on a Mach-Zehnder interferometer is shown in Fig. 23. The laser sources used in the experiment are laser diodes with $\lambda_1 = 635nm$ and $\lambda_2 = 675nm$, yielding a synthetic wavelength $\Lambda = 10.72\mu m$. The key concept of the proposed scheme is to separate each wave wavelength beam pair in different reference arms, while combining them in an object arm. We employ two PBSs (Polarizing Beam Splitters) and two HWPs. Each laser beam is expanded by using a spatial filter and collimated by a collimating lens. An iris in front of the lens allows the adjustment of the size of the collimated beam. For the wavelength λ_1 , the laser beam is divided into object and reference beams through the PBS 1. After passing through the PBS 1, the transmitted beam becomes p-polarized (parallel to the optical plane) while the reflected beam gets s-polarized (perpendicular to the optical plane). The transmitted beam acts as the reference beam of the wavelength λ_1 and it travels to the CCD sensor maintaining the same linear p-polarization by the triangular prism and mirror 1. Meanwhile, the reflected one which is entitled to be the object beam meets a HWP which is designed for the wavelength λ_1 . By setting the optic axis of the HWP for rotation by 45 degrees from the s-polarization axis, we can rotate the linear polarization state by 90 degrees. The polarization state of the object wave for λ_1 is changed from s-polarization to p-polarization. After that, it passes through the PBS 2 and BS 2 to illuminate the surface of the object and travels to the CCD sensor. Likewise, for the wavelength λ_2 , the laser beam is divided into object and reference beams through PBS 2. After passing through the PBS 2, the transmitted beam becomes p-polarized (parallel to the optical plane) while the reflected beam gets s-polarized (perpendicular to the optical plane) exactly as done for λ_1 .

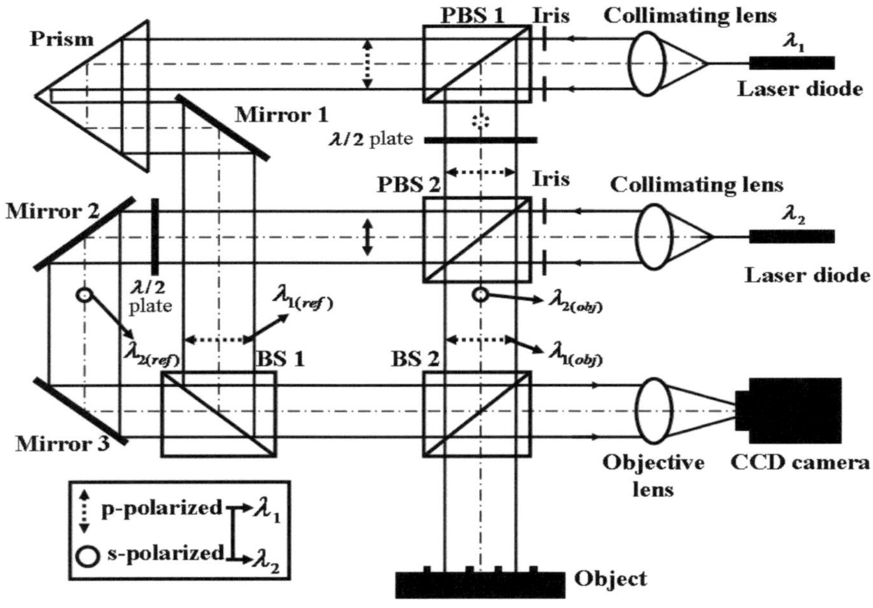

Fig. 23. Schematic diagram of the proposed single shot dual-wavelength interferometric system.

The transmitted beam corresponds to the reference beam of the wavelength λ_2. But it meets a HWP which is designed for the wavelength λ_2, and consequently, the polarization state is changed from p-polarization to s-polarization state when the optic axis of the HWP designed for the wavelength λ_2 is set for rotation by 45 degrees from the s-polarization axis, in the exactly same manner as in the case of λ_1.Then, it travels to the CCD sensor maintaining the same linear s-polarization state. Meanwhile, the reflected beam, which is in the linear s-polarization state, remains unchanged and illuminates the surface of the object and then travels to the CCD sensor. Here, the object beam for both wavelengths illuminates the object, and the reflected object beam propagates on-axis with the light diffracted from the 3-D object. To obtain the off-axis interferogram, mirrors that reflect the reference waves R_1 and R_2 are tilted such that the reference wave reaches the CCD with an incidence angle while the object wave propagates perpendicular to the CCD. Finally, the CCD camera records the interferogram that result from the interference between the object wave O_1 and the reference wave R_1. And also, the interference between the object wave O_2 and the reference wave R_2 is recorded simultaneously. Figure 24 shows the experimental results obtained through the proposed dual-wavelength off-axis interferometric scheme. The surface under test is an object with a nominal height 1.34 μm. The general procedure which is needed for measuring the complex object wave in dual-wavelength off-axis interferometry is shown in Fig.24. Both amplitude and phase information of the object for the two wavelengths can be obtained with a single interferogram. Figure 24 represents the experimental results of application of the 2D-FFT based spatial filtering method for each wavelength separately. After the spatial filtering step, the digital reference wave R_{D_i} is used for the centering process. Then, the final reconstructed object is obtained by adjusting the values of k_x and k_y for each wavelength separately.

The digital reference wave used in the calculation process should match as close as possible to the experimental reference wave. This has been done in this paper by selecting the appropriate values of the two components of the wave vector k_x = 0.00819 mm^{-1} and k_y = -0.00654 mm^{-1} for the interferogram captured at $\lambda_1 = 635nm$ and k_x = -0.09944 mm^{-1} and k_y = 0.00674 mm^{-1} for the interferogram captured at $\lambda_2 = 675nm$. Once the two phase maps are obtained for each wavelength, a phase map can be calculated on the synthetic beat-wavelength. Figure 24(a) shows the investigated sample of a nominal step height of 1.34 μm. The single-shot dual-wavelength off-axis interferogram of the investigated sample is shown in Fig. 24(b). The filtering widows A and B as shown in Fig. 24(c) for $\lambda_1 = 635nm$ and $\lambda_2 = 675nm$, respectively, have been chosen carefully. The reconstructed amplitude and phase map for $\lambda_1 = 635nm$ and for $\lambda_2 = 675nm$ are shown in Fig. 24(d-e) and Fig. 24(e-f), respectively. The phase map on the synthetic beat-wavelength is shown in Fig. 24(h). The two dimensional (2D) surface profile along the selected line of Fig.24(h) is shown in Fig.24(i) after converting to height by using Eq.(23). The noise shown in Fig.24(i) is inevitable noise may come from the spatial filtering window and 2D-FFT processes. The noise vibration shown in Fig.24(i) is not coherent but varies sinusoidally; thus, it is more logical to express the height error as the root mean square (rms) value of the disturbed height (Malacara, et al., 2005). The rms height can be expressed as (Pan, et al., 2011):

$$h_{rms} = \sqrt{\left[h - \bar{h}\right]^2 / (N-1)}, \tag{24}$$

where h denotes the height distribution, $\bar{h}$ denotes the mean values, N represents the pixel number of row of calculation region. The modified height profile of Fig.24(i) after applying the rms for the distributed height is shown in Fig.25. The proposed rms method can provide a satisfied solution especially for measuring objects having high abrupt height difference. Based on the measured height in Fig.25, the average step height has been estimated to be around 1.30 μm .

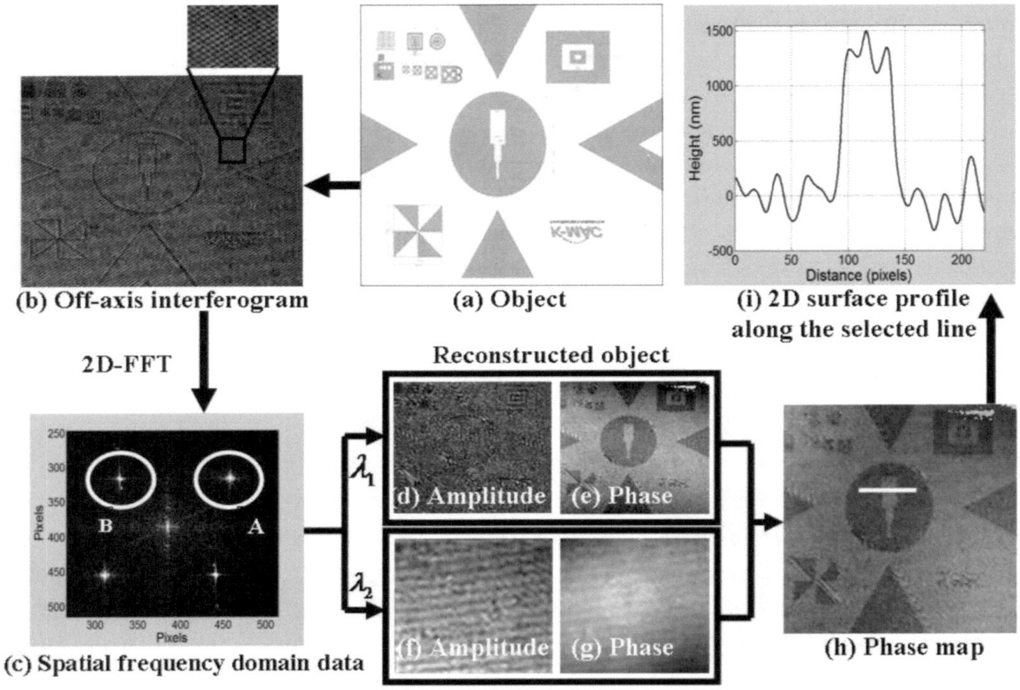

Fig. 24. Sequential reconstruction steps of the spatial filtering based phase contrast dual-wavelength off-axis interferometry (a) original phase object, (b) off-axis interferogram, (c) Fourier transformed spatial frequency domain data, (d)-(e) reconstructed amplitude and phase map for $\lambda_1 = 635nm$, (f)-(g) amplitude and phase map for $\lambda_2 = 675nm$, (h) object phase map on the synthethic beat-wavelength, and (i) two dimensional (2D) surface profile along the selected line of (h).

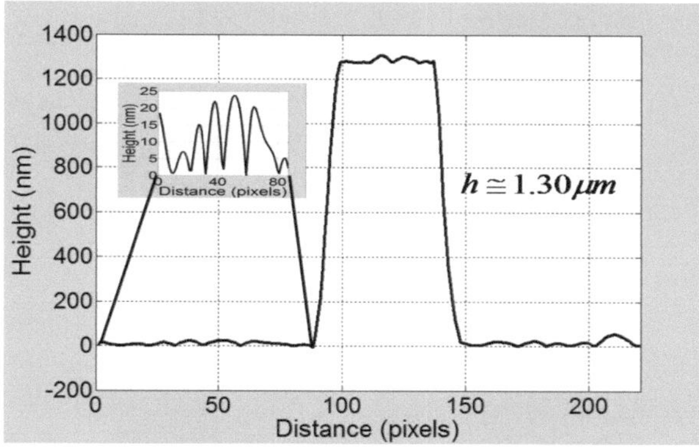

Fig. 25. Two-dimensional surface profile of Fig.24(i) after applying the rms method for the distributed height.

6. Multiple-beam interferometry

Multiple-beams interferometry means the generation of interference between reflected wavefronts from two surfaces, employing a succession of coherent beams. Such beams can combine and produce highly sharpened fringes. The earliest multiple-beam interferometers were developed by Fabry and Perot for the accurate measurement of wavelengths and the standard meter. Tolansky (Tolansky, 1960) studied the microtopography of different types of surfaces by using multiple-beam wedge interferometry (Fizeau-Tolansky type). However, the fact that the Fizeau-Tolansky wedge had an angle at the edge introduced many sources of difficulties which were the subject of many studies in the period (1945-1975). Sharp fringes are obtained when the surfaces forming the cavity are coated with higher reflectivity film (silver film is used here). The fringes obtained are called multiple-beam fringes. The theory of the intensity distribution of Fabry-Perot fringes at reflection from an infinite number of beams collected was dealt with by Holden (Holden, 1949). In the use of multiple beam interferometry several differences are encountered to the more conventional two-beam systems. In the two-beam case, the recorded intensity variation follows the Cos^2 law, which achieves an error estimation of ($\lambda/2$) for simple fringe counting. In contrast, multiple beam fringes are extremely sharp and simple measurements with such fringes can reveal surface micro topography close to ($\lambda/500$). The narrowness of the multiple-beam reflected fringes as shown in Fig.26 reflects directly on the accuracy with which the position of the fringe can be determined.

Figure.27 shows a schematic representation of the Fabry-Perot (F-P) interferometer in reflection. It shows that the air gap between the two plates is constant. The coating layer of the lower component is opaque while the upper component facing the incident light is semi transparent. The reflected rays are of the following amplitudes $r_1, r_3t^2, r_2r_3^2t^2, r_2^2r_3^3t^2, ...$ When the series is taken to infinity, the intensity distribution in the reflected system (Holden, 1949) is given as follows:

$$I_{R(\infty)} = r_1^2 + \frac{[t^4 r_3^2 + 2t^2 r_1 r_3 \cos(F+\delta) - 2t^2 r_1 r_2 r_3^2 \cos(F)]}{[1 + r_2^2 r_3^2 - 2r_2 r_3 \cos(\delta)]}. \quad (25)$$

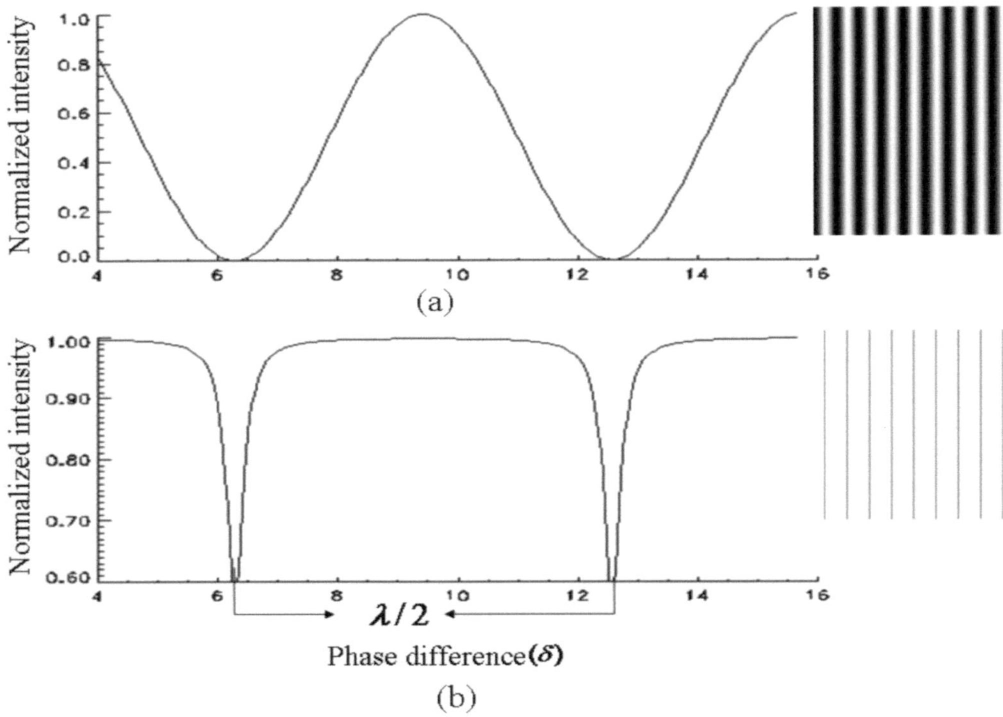

Fig. 26. (a) Two-beam cosine fringes. (b) Multiple beam Fabry-Perot fringes.

Where $r_1^2 = R_1, r_2^2 = R_2$ are the fractions of the light intensity reflected from the glass/metallic layer and air/metallic layer of the upper component, respectively. Also, $r_3^2 = R_3$ refers to the reflected intensity at air/metallic layer coating from the lower component, and $t^2 = T$ is the transmitted intensity through the metallic layer of the upper component. F is the combined phase function $F = 2\gamma - \beta_1 - \beta_2$. The phase difference δ between any two successive beams is given as follows:

$$\delta = \frac{2\pi}{\lambda}(2\mu t Cos(\theta)) \pm 2\beta \pm 2\gamma \,. \tag{26}$$

Where β_2 and β_3 are the change of phase at air/metallic layer reflection for the upper and lower components respectively and γ is the phase change in transmission. β_1 is the change of phase at glass/metallic layer reflection for the upper component of the interferometer when facing the incident light.

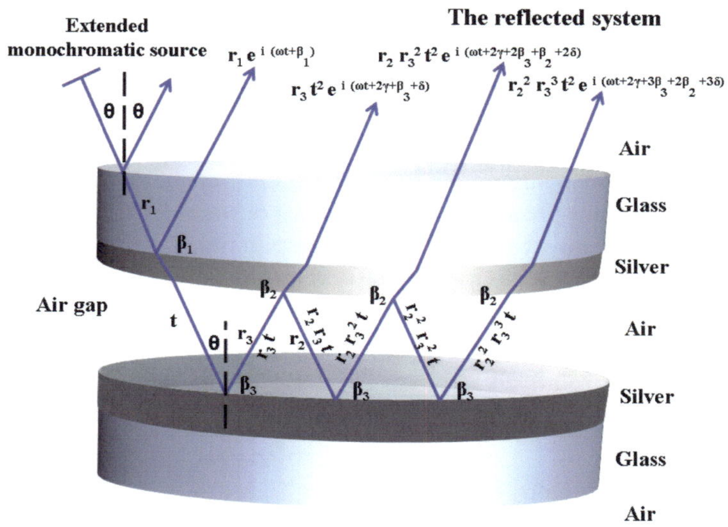

Fig. 27. Schematic representation of the Fabry-Perot (F-P) interferometer in reflection.

Figure.28(a) shows a schematic diagram of the Fizeau-Tolansky interferometer in reflection with coated inner surfaces producing fringes of equal thickness. Figure.28(b) shows a schematic diagram of the expected locations of the fringes in the Fizeau-Tolansky interferometer. Apparently, the fringe formation suffers from the existence of the wedge angle α causing a departure from the exact phase condition. Many authors discussed that effect (Gehrcke, 1906, Kinosita, 1953).The phase lag was estimated by Tolansky to be $\varepsilon = (4/3)\pi M^3\alpha^2 N$ (Tolansky, 1960). Where M is the order of the last beam of effective amplitude contributing to the point of peak intensity, α is the wedge angle and N is the order of interference. In this section, a multiple-beam Fizeau-Tolansky interferometer in reflection for measuring the microtopography of an optical flat nominally of $\lambda / 20$ is described (Abdelsalam et al., 2010). The method used for the fringe analysis is the Bünnagel's method (see section 4.3).

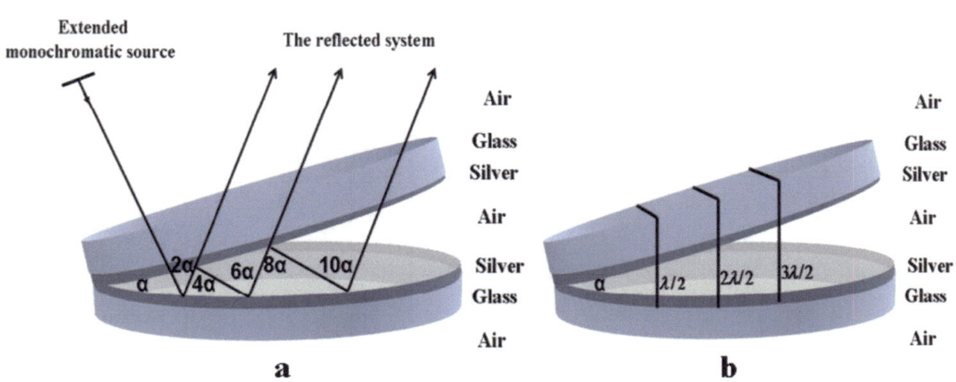

Fig. 28. The wedge interferometer (a) a schematic diagram of the Fizeau-Tolansky interferometer in reflection and (b) a schematic diagram of the expected locations of the fringes in the Fizeau-Tolansky interferometer.

The schematic diagram of the optical set up is illustrated in fig.29. The He-Ne laser source of 632.8 nm wavelength is allowed to pass through a beam expander of pinhole diaphragm diameter 20μm, where the beam diameter is expanded by a factor of 15. This is necessary for uniformly illuminating a greater area of the surface to be imaged and to reduce the error in measurement due to inhomogeneity in the Gaussian beam. The collimated beam of laser light falls upon the beam splitter which transmits one half and reflects one half of the incident light with constant phase change. The plates of the interferometer are coated with silver film of reflectivity nearly 90% which corresponding to 37 nm thickness and are mounted close together, with separation t. The reflected collimated beam is then incident on the interferometer which changes the path length of the light inside it due to the irregularities of the surface of the interferometer. When the reflected beams are combined, the intensity of the fringes is given by Eq.25. The interferogram was captured by the CCD camera and then refined by using flat fielding.

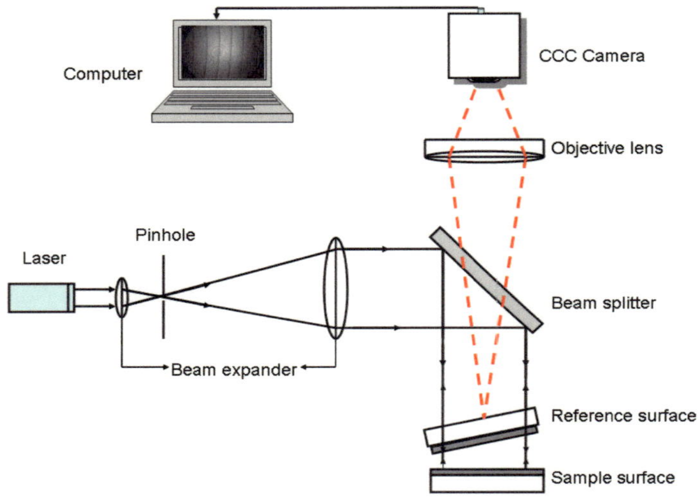

Fig. 29. Schematic diagram of the optical setup.

The average of 100 images captured by CCD camera is shown in Fig. 30(a). The corrected interferogram of Fig. 30(a) using flat fielding is shown in Fig. 30(b).

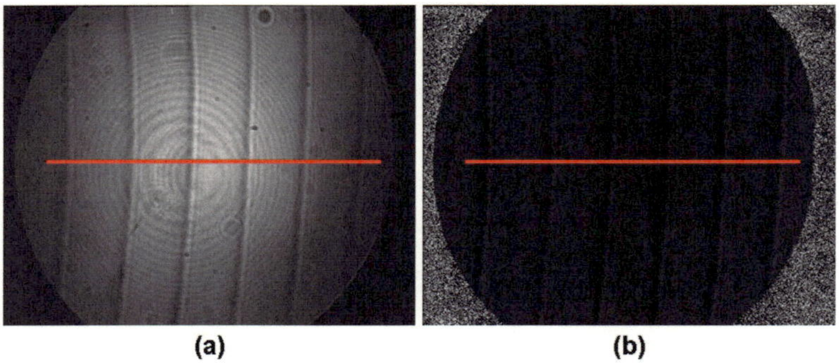

Fig. 30. (a) Average of one hundred images taken from the setup. (b) The corrected interferogram of (a).

Fig. 31(a) shows the profile taken at the middle of fig.30(a) as immersed in the background due to effect of the offset of the camera and the inhomogeneity of the collimated laser beam intensity. Fig.31(b) shows the profile taken at the middle of fig.30(b) after correction when the harmful background was removed i.e. was subtracted from the obtained distribution.

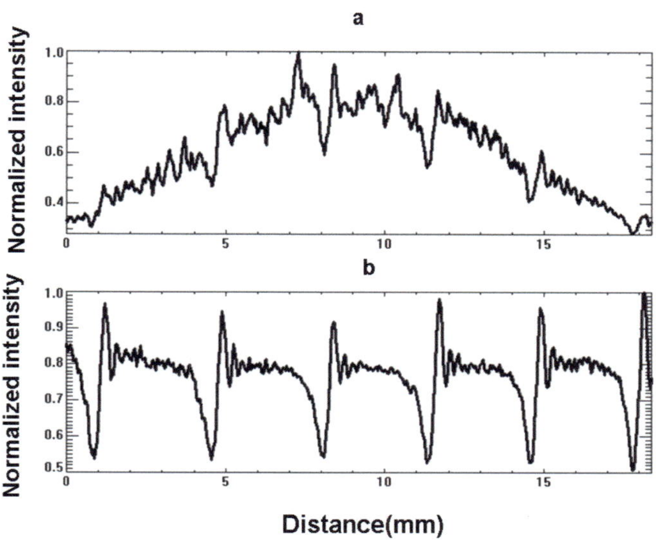

Fig. 31. Normalized intensity against distance along profile for average of one hundred images.(a) Before correction with flat fielding.(b) After correction with flat fielding.

The object surface with flatness $\lambda / 20$ was tested using a calibrated reference of $\lambda / 50$ flatness. Equation (11) has been applied into the corrected image (Fig.30 (b)) to get the topographical height as shown in fig.32. It was found that the peak to valley height of the optical flat equals $4.03x10^{-5}mm$ i.e. $\cong \lambda / 16$. To obtain the form with high accuracy, ten profiles were calculated every 30 pixel in y axis and it was found that the mean of 10 values of the height for 10 different cross sections in the corrected interferogram was calculated as:

Total height (mm) = (4.03+3.62+3.88+4.25 + 3.08+3.49+3.54+3.98+4.15+4.59)x10^{-5}/10 = 3.86 x 10^{-5}(mm) $\cong \lambda / 17$.

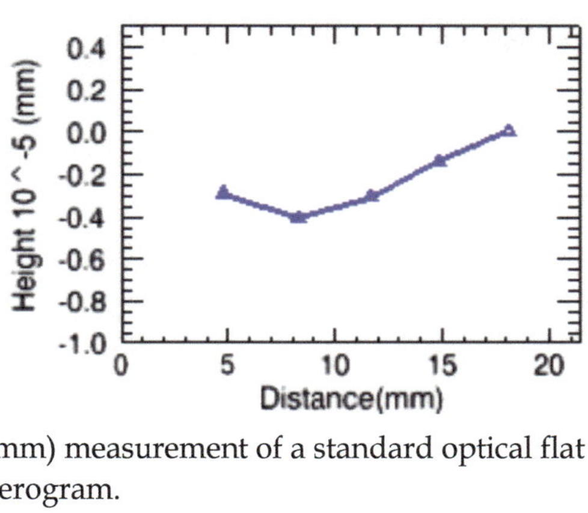

Fig. 32. Surface height (mm) measurement of a standard optical flat of nominally λ / 20 resulted from one interferogram.

Because of there is no data between any two successive fringes in multiple beam case, phase shifting technique in multiple-beam is used to span the distances between the fringes. The scanning here is very important for obtaining good results and high measurement accuracy by increasing the number of points.

6.1 Phase shifting technique in multiple-beam interferometry

The basic idea of phase shifting technique is that, if the amplitude difference between the interfering beams is made to vary in some known manner such as changing in discrete steps (stepping), the intensities of the fringes are changed and hence the fringes are shifted. The most common way to vary the intensity distribution is to mount the object surface on a piezoelectric transducer (PZT) and change the voltage to the PZT. The calculations are based on an assumption that the PZT is linear in its motion. The distance of the cavity has been changed by varying the voltage and the displacement was measured precisely using laser interferometer. The voltage was applied and was varied four times as shown in the table 1. The distance inside the cavity is changed and then the fringe positions are changed. For each displacement one hundred images were captured using CCD camera and then corrected with flat fielding as show in Fig.33.

Voltage (V)	10	15	20	25
Distance (nm)	40	140	240	340

Table 1.Voltage versus distance.

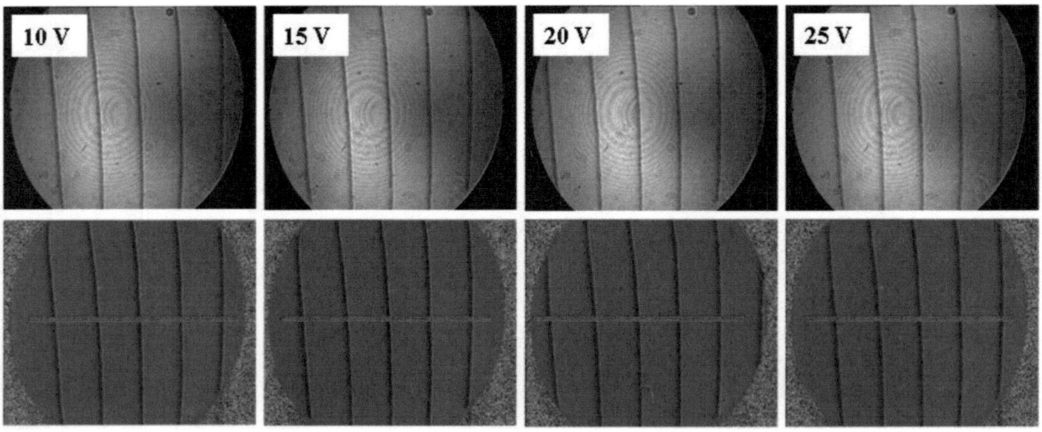

Fig. 33. Average of one hundred images and their corresponding corrections with flat fielding for 10V, 15V, 20V, and 25V.

The profiles of one cross section from the middle of the four interferograms have been illustrated in fig.34. The feature height of the optical flat was calculated for the four interferograms using Eq. (11). It was observed that there is a fluctuation in the regulations of the data which may be produced from the reading of the laser interferometer. This may be overcome using a programming process to get exactly the positions of the fringes. The data were arranged and sorted to obtain the feature height in one profile as shown in fig.35. The peak to valley value was calculated from fig.35 to be of the order of $\lambda/20$.

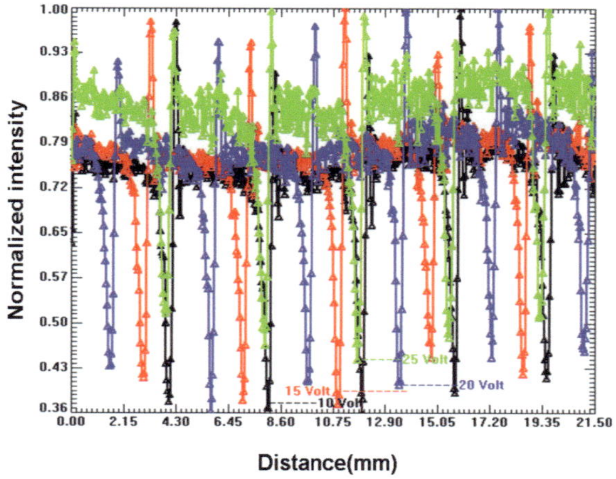

Fig. 34. The profiles of one cross section from the middle of the four interferograms for 10V,15V,20V, and 25V.

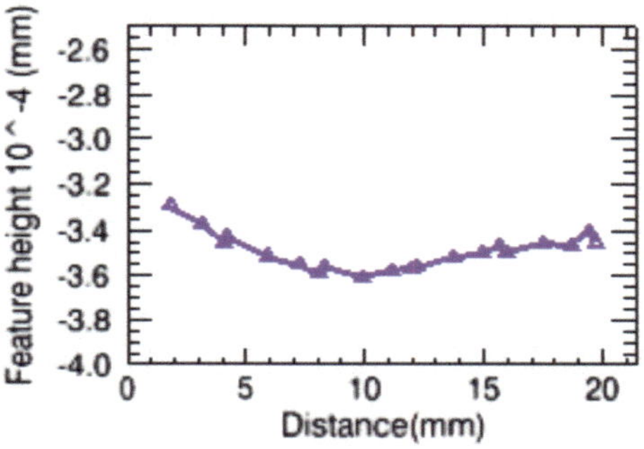

Fig. 35. Surface topography measurement of a standard optical flat of nominally $\lambda / 20$ resulted from the four interferograms.

7. Conclusion

In conclusion, we have presented the recent developments of interferometry techniques for surface micro topography measurement. In this chapter, a new approach to reduce coherent noise in interferometry phase-contrast image is presented. It is accomplished by combining the flat fielding method with the apodized apertures technique. In this chapter, different numerical reconstruction algorithms are reviewed and compared. Finally, we have described the principle of multiple-beam interferometry and its ability to feature very small height objects. Experimental results are presented to verify the principles.

8. Acknowledgment

Most of the work has been done in NIS (Egypt), Chonbuk National University (South Korea), and PTB (Germany). The Author would like to thank Dr. Daesuk Kim, Mechanical System Engineering Division, Chonbuk National University, South Korea and Dr. Michael Schulz, head of the Working Group "Form and Wavefront Metrology", PTB, Germany for data analysis.

9. References

Abdelsalam, D. ; Magnusson, R.; & Kim, D. (2011). Single-shot dual wavelength digital holography based on polarizing separation, Applied Optics, 50, pp. 3360-3368.

Abdelsalam, D.; & Kim, D. (2011). Coherent noise suppression in digital holography based on flat fielding with apodized apertures, Optics Express, 19, pp. 17951-17959.

Abdelsalam, D.; Baek, B.; Cho, Y.; & Kim, D. (2010). Surface form measurement using single-shot off-axis Fizeau interferometer, J. Opt. Soc. Korea, 14, pp. 409-414.

Abdelsalam, D.; Shaalan, M.; & Eloker, M. (2010). Surface microtopography measurement of a standard flat surface by multiple-beam interference fringes at reflection, Optics and Lasers in Engineering, Vol. 48, pp. 543-547.

Abdelsalam, D.; Shaalan, M.; Eloker, M.; & Kim, D. (2010). Radius of curvature measurement of spherical smooth surfaces by multiple-beam interferometry in reflection, Optics and Lasers in Engineering, Vol. 48, pp. 643-649.

Born, M.; Wolf, E. (1980). Principles of Optics, Cambridge University Press, pp 459-490.

Bünnagel, R. (1956). Einfaches verfahren zur topographischen darstellung einer optischen planflache. Opt Acta, 3, pp. 81-85.

Cuche, E. ; Marquet, P.; & Depeursinge, D. (2000). Spatial filtering for zero-order and twin image elimination in digital off-axis holography, Applied Optics, 39, pp. 4070-4075.

Cuche, E.; Bevilacque, F.; & Depeursinge, C. (1999). Digital holography for quantitative phase contrast imaging, Optics. Letters., 24, pp. 291-293.

Cuche, E.; Marquet, P.; & Depeursinge, C. (2000). Aperture apodization using cubic spline interpolation :application in digital holographic microscopy, Optics communications, 182, pp. 59-69.

David, F.; John, T.; Lopez, R.; & Stahl, H. (1993). Vector formulation for interferogram surface fitting, Applied Optics, 32, pp. 4738-43.

Gabor, D. (1948). A new microscopic principle, Nature, 161, pp. 777-778.

Gehrcke, E. (1906). Interferenzen, pp 39, Germany.

Ghiglia, D. ; & Pritt, M. (1998). Two-dimensional Phase Unwrapping : Theory, Algororithm, and Software, Wiley, New York , USA.

Holden, J. (1949). Multiple-beam interferometer-intensity distribution in the reflected system, Proc. Phys. Soc., 62, pp. 405-417.

Kinosita, K. (1953). Numerical evaluation of the intensity curve of a multiple-beam Fizeau fringe, J. Phys. Soc. Japan, 8, pp. 219-225.

Kühn, J.; Colomb, T.; Montfort, F.; Emery, Y.; Cuche, E.; & Depeursinge, C. (2007). Real-time dual-wavelength digital holographic microscopy with a single hologram acquisition, Optics Express, 15, pp. 7231-42.

Kumar, U. ; Bhaduri, B. ; Kothiyal, M. ; & Mohan Krishna. (2009). Two wavelength micro-interferometry for 3-D surface profiling, Optics and Lasers in Engineering, 47, pp. 223-229.

Malacara, D. ; Servin, M.; & Malacara, Z. (2005). Interferogram analysis for optical testing, United States of America : Taylor and Francis Group, pp. 384-385.

Pan, F.; Xiao, W.; Liu, S. ; Wang, F., & Li, R. (2011). Coherent noise reduction in digital holographic phase contrast microscopy by slightly shifting object, Optics Express, 19, pp. 3863-3869.

Steve, B. (2006). Hand book of CCD astronomy, Cambridge, UK.

Tolansky, S. (1960). Surface microtopography, London : Longmans.

Wang, JY.; & Silva, DE. (1980). Wave-front interpretation with Zernike polynomials. Applied Optics, 19, pp. 1510-8.

3

Advanced Multitemporal Phase Unwrapping Techniques for DInSAR Analyses

Antonio Pepe
Istituto per il Rilevamento Elettromagnetico dell'Ambiente (IREA), Research National Council (CNR)
Italy

1. Introduction

Differential Synthetic Aperture Radar Interferometry (DInSAR) represents nowadays a well-established remote sensing technique to generate spatially dense surface deformation maps of large areas on Earth (Massonnet & Feigl, 1999; Bürgmann et. al., 2000). To achieve this task, the phase difference (known as interferogram) between SAR data pairs related to temporally-separated observations is exploited, retrieving a measure of the ground displacement projection along the radar line of sight (LOS) with centimeter to millimeter accuracy (Gabriel et. al., 1989).

Altough DinSAR methodology was originally applied to analyze single deformation episodes (Massonnet et. al., 1993; Goldstein et. a., 1993; Massonnet et. al., 1995; Peltzer et. al., 1995), it has recently applied with success to investigate the temporal evolution of the detected deformation phenomena through the generation of displacement time-series (Ferretti et. al., 2001; Berardino et. al., 2002) . In this case, the analysis is based on the computation of deformation time-series via the inversion of a properly chosen set of interferograms, produced from a sequence of temporally-separated SAR acquisitions relevant to the investigated area. In this context, two main categories of advanced DInSAR techniques for deformation time-series generation have been proposed in literature, often referred to as Persistent Scatterers (PS) (Ferretti et al., 2001; Hooper et. al., 2004), and Small Baseline (SB) (Berardino et. al., 2002; Mora et. al., 2003) techniques, respectively. The PS algorithms select all the interferometric data pairs with reference to a single common master image, without any constraint on the temporal and spatial separation (baseline) among the orbits. In this case, the analysis is carried out at the full resolution spatial scale, and is focused on the pixels containing a single dominant scatterer thus ensuring very limited temporal and spatial decorrelation phenomena (Zebker & Villasenor, 1992). Instead, for what concerns the SB techniques, the interferograms are generated by considering multiple master images, because this allows having interferometric data pairs with small temporal and spatial baselines. Accordingly, distributed targets can be also investigated, and the analysis may exploit both single-look and multi-look interferograms. Within the SB framework, a popular approach is the one referred to as Small BAseline Subset (SBAS), which was originally developed for analyzing multi-look interferograms (Berardino et. al., 2002) and subsequently adapted to the full resolution ones; in the latter case the low resolution results are properly extended to the full resolution scale, as discussed in (Lanari et. al., 2004).

Despite the differences among the various implementations of the PS and SB algorithms, which are outside the scope of this work, a common problem to be faced is the Phase Unwrapping (PhU) operation representing the retrieval process of the full phase signals from their (measured) modulo-2π restricted components, often referred to as "interferometric fringes". In the context of the advanced DInSAR techniques, the need of jointly analyzing sequences of multi-temporal interferograms has more recently promoted the development of new PhU approaches with improved performances with respect to those focused on processing single interferograms. In this chapter, following the description of the basic rationale of the most widely used PhU techniques, we present a short review of the most recent space-time PhU methodologies with a particular emphasis on the Extended Minimum Cost Flow (EMCF) and its recent improvements. Some experimental results obtained by applying these latter approaches will be shown to demonstrate the validity of the presented algorithms.

2. Phase unwrapping

We start by introducing the key ideas at the base of most of the Phase Unwrapping (PhU) techniques proposed up to now in literature. As stated before, the phase unwrapping problem concerns the retrieval of the full interferometric phase (unwrapped ones) from its computed modulo-2π restricted components. Within DInSAR processing codes, PhU operation actually represents one the most critical task to be successfully accomplished. The problem in reconstructing the true phase from the measured one arises in the presence of aliasing errors mainly caused by the phase noise caused by low coherence and undersampling phenomena due to locally high fringe rates (related to the deformation signals associated to the interferograms).

To clarify the key aspects of the phase unwrapping problem, let us firstly focus on the problem to unwrap single interferograms. To this aim, let $\phi(a_z, r_g)$ be the interferometric phase measured in correspondence to the pixel of SAR coordinates (a_z, r_g). The unwrapping problem can be stated as the searching of the multiple 2π-integer that must be added to the wrapped phases to retrieve the full interferometric phase $\psi(a_z, r_g)$. Indeed:

$$\psi(a_z, r_g) = \phi(a_z, r_g) + 2\pi H(a_z, r_g) \tag{1}$$

This represents an ill-posed problem admitting an infinite set of acceptable solutions that can benefit from the knowledge of external information. In the absence of such information, the unwrapped field is typically reconstructed by computing the phase spatial gradients and integrating them over a "consistent" path. Accordingly, PhU algorithms can be grouped in two main categories depending on the fact that the unwrapping solution is path-dependent or not. To efficiently solve the problem, it is convenient to represent it in terms of discrete coordinates (x, y) being:

$$\begin{aligned} a_z &= a_{z0} + x\Delta a \\ r_g &= r_{g0} + y\Delta r \end{aligned} \tag{2}$$

wherein Δa and Δr are the azimuth and range pixel spacing, respectively and a_{z0} and r_{g0} the corresponding azimuth and range first lines. The discrete counterpart of the partial derivatives are then computed by using the popular wrapped-differences-of-wrapped-phases estimator $\langle \Delta\varphi \rangle_{-\pi,\pi} = arctg\left(\sin(\Delta\varphi)/\cos(\Delta\varphi)\right)$. Accordingly, the phase gradient vector can be defined as follows:

$$\hat{\nabla}\boldsymbol{\psi} = \langle \Delta_x \psi \rangle_{-\pi,\pi} \hat{\mathbf{x}} + \langle \Delta_y \psi \rangle_{-\pi,\pi} \hat{\mathbf{y}} \tag{3}$$

whose components with respect to the two spatial axis $\hat{x}$ and $\hat{y}$ are:

$$\begin{aligned} \langle \Delta_x \psi(x,y) \rangle_{-\pi,\pi} &= \langle \psi(x+1,y) - \psi(x,y) \rangle_{-\pi,\pi} = \langle \phi(x+1,y) - \phi(x,y) \rangle_{-\pi,\pi} \\ \langle \Delta_y \psi(x,y) \rangle_{-\pi,\pi} &= \langle \psi(x,y+1) - \psi(x,y) \rangle_{-\pi,\pi} = \langle \phi(x,y+1) - \phi(x,y) \rangle_{-\pi,\pi} \end{aligned} \tag{4}$$

Accordingly, the phase gradient vector is estimated by wrapping possible phase differences greater than +/-π in the $[-\pi,\pi]$ interval by adding the correct multiples of 2π and by implicitly assuming that, in a properly sampled interferogram, the phase differences of adjacent samples are likely to be restricted to the $[-\pi,\pi]$ interval. Moreover, the probability that the phase difference exceed $\pm\pi$ depends both on the noise level (i.e. the lower is the coherence value the more likely the derivatives exceed $\pm\pi$). For this reason, the introduced operator is not conservative, that is:

$$rot\left(\hat{\nabla}\psi\right) = \nabla \times \left(\hat{\nabla}\psi\right) \neq 0 \tag{5}$$

Consequently, the integration of such phase gradients depends on the chosen integration path. This suggests us that a way to unwrap interferograms is to correct the phase gradients in such a way that the unwrapped phase can be recovered independently of the integration direction. Moreover, we can observe that the phase gradient estimate has the advantage that its errors are localized and come in integer multiples of 2π so that its curl (hereafter referred to as residue field) can be profitably used to reconstruct the full phase terms. The residue field expression (see Figure 1) is the following:

$$\begin{aligned} r(x,y) &= \nabla \times \left(\hat{\nabla}\psi(x,y)\right) = \Delta_x\left[\langle \Delta_y \psi(x,y) \rangle_{-\pi,\pi}\right] - \Delta_y\left[\langle \Delta_x \psi(x,y) \rangle_{-\pi,\pi}\right] = \\ &= \langle \Delta_x \psi(x,y) \rangle_{-\pi,\pi} + \langle \Delta_y \psi(x+1,y) \rangle_{-\pi,\pi} - \langle \Delta_x \psi(x,y+1) \rangle_{-\pi,\pi} - \langle \Delta_y \psi(x,y) \rangle_{-\pi,\pi} \end{aligned} \tag{6}$$

Its values are either zero (no residues) or +/- 2π (positive or negative residue, respectively). Therefore, the goal of the phase unwrapping procedure is to eliminate potential integration paths enclosing unequal numbers of positive and negative residues. Residue derives from two sources in the radar measurements. The former is related to the true discontinuities in the data: the fringe spacing may be so fine on certain topographic slopes or, from large inter-observation displacement, such to exceed the Nyquist criterion of half-cycle spacing. The latter is the noise present in the data set, whatever from thermal and other noise sources or from decorrelation due to baseline length and temporal changes in the scene. However, residues from whatever source require compensation in the phase unwrapping procedure.

One of the major PhU algorithm exploits the fact that residues mark the endpoints of lines in the interferogram along which the true phase gradient exceed π/sample, these lines are commonly referred to as "branch-cuts" or "ghost-lines". Most of the algorithm up to now proposed are based on a proper compensation of these residues and, among these, we will concentrate in the following on the residue-cut and the least-square approaches (Zebker & Lu, 1998).

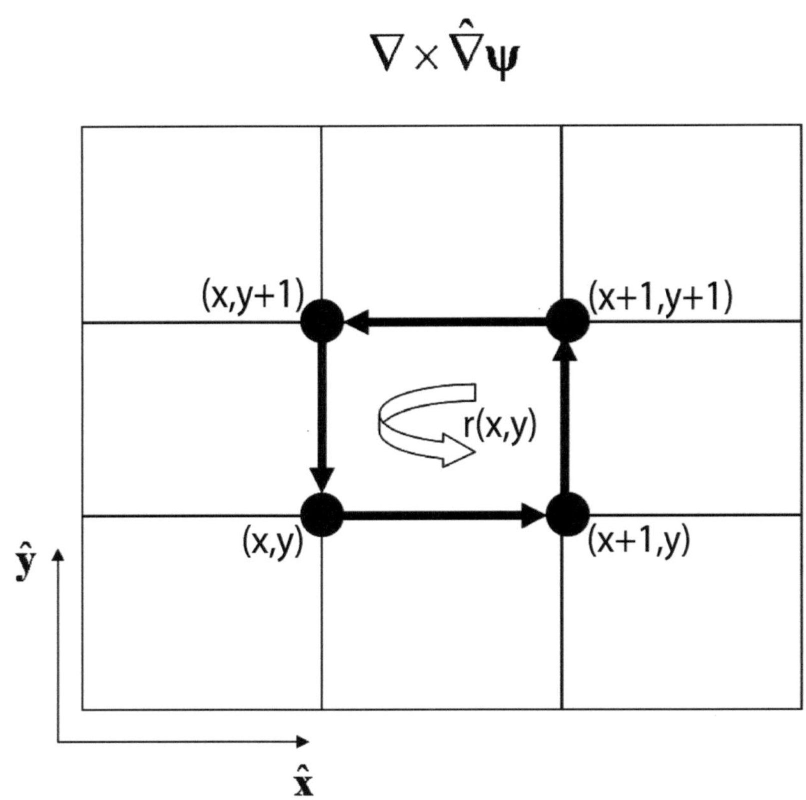

Fig. 1. Representation of the residue field.

2.1 Branch-cut algorithms

Within this class of algorithms we will point out our interest in particular on the so-called "residue-cut tree algorithm". The initial residue-cut phase unwrapping procedure proposed in (Golstein et. al., 1998) is implemented by first of all identifying the locations of all residues in an interferogram, and then connecting them with branch-cuts, so as to prevent the existence of integration paths that can encircle unbalanced numbers of positive and negative residues. The tree algorithm is a relatively conservative algorithm that tends to grow rather dense networks of trees in residue-rich regions. The algorithm initially connects closely spaced, oppositely charged, pairs of residues with cuts that prevent integration paths between them and, if all permitted integration paths enclose equal numbers of positive and negative residues, the consistency is assured. Progressively longer trees are permitted until all residues are connected to, at least one, other residue and until the next charge on each tree is zero. Networks on small trees are used to prevent any single branch from becoming too long and isolating large sub-areas from the rest of the image. A consequence of the indiscriminate branch growth until charge neutrality is achieved from all trees, and all residues accounted for, is that in residue-rich regions the tree growth is so dense that the region is isolated from the remainder of the image and no unwrapped phase estimate can be obtained. By concluding, this conservative approach nearly eliminates mistakes but, at the expense of providing an incomplete unwrapping result.

2.2 Least-squares algorithms

The second major class of Phase Unwrapping algorithms, commonly in use today, was presented by (Ghiglia & Romero, 1994), who applied a mathematical formalism first developed by Hunt [69] to the radar interferometry phase unwrapping problem. Hunt developed a matrix formulation suitable for general phase reconstruction problems; Ghiglia found that a discrete cosine transform technique permits accurate and efficient least-squares inversion, even for the very large matrices encountered in the radar interferometry special case. In particular, the un-weighted LS method performs the following minimization problem

$$\min_{\tilde{\boldsymbol{\psi}}}\left\{\sum_i\sum_j\left|\nabla\tilde{\boldsymbol{\psi}}(x,y)-\hat{\nabla}\boldsymbol{\psi}(x,y)\right|^2\right\} \tag{7}$$

where $\tilde{\boldsymbol{\psi}}$ is the unknown unwrapped phase field. We may stress that, with respect to the branch-cut approaches, in these cases the solution will be no longer congruent with the original interferometric phase.

Equation (7) represents a variational problem, whose Euler equation is the Poisson one:

$$\nabla^2\tilde{\psi}(x,y)=\nabla\cdot\nabla\tilde{\psi}(x,y)=\Delta_x\left(\left\langle\Delta_x\psi(x,y)\right\rangle_{-\pi,\pi}\right)+\Delta_y\left(\left\langle\Delta_y\psi(x,y)\right\rangle_{-\pi,\pi}\right) \tag{8}$$

under the Neumann boundary condition, which can be finally solved by using simple cosine or Fourier transform filtering (Fornaro et. al., 1996).

One major difference between the residue-cut and least-squares solutions is that in the residue-cut approach only integral numbers of cycles are added to the measurements to produce the result. Conversely, in the least-squares approach, any value may be added to ensure smoothness and continuity in the solution, thus the spatial error distribution may differ between the approaches, and the relative merits of each method must be determined depending on the application.

Least-squares methods are very computationally efficient when they make use of fast Fourier transform techniques but the resulting unwrapping is not very accurate, because they tend to spread the errors instead of concentrating them on a limited set of points. To overcome this problem a weighting of the wrapped phase can be useful. However, the proposed, weighted least squares algorithms are iterative and not as efficient as the un-weighted ones, and the result accuracy strongly depends on the used weighting mask.

Several other approaches, which can be found in the bibliography at the end of this work, have also been investigated but, in the following, we will address the capability of one of these, known as the minimum cost flow phase unwrapping technique, developed for the two-dimensional case by (Costantini, 1998), and belonging to the branch-cuts PhU algorithm class.

2.3 Minimum cost flow algorithms

Within the branch-cuts phase unwrapping algorithms an easy and fast algorithm is based on a solution of an equivalent minimum cost flow network and will be here addressed.

Branch-cut methods are based on the integration of the estimated neighbouring pixel differences of the unwrapped phase along conservative paths, thus avoiding the regions where these estimated differences are inconsistent. The problem of building cuts delimiting these regions is very difficult and the resulting phase unwrapping algorithm is very computationally expensive. However, we may exploit the fact that the neighbouring pixel differences of the unwrapped phases are estimated with possibly an error that is an integer multiple of 2π. This circumstance allows the formulation of the phase unwrapping problem as the one of minimizing the weighted deviations between the estimated and the unknown neighbouring pixel differences of the unwrapped phases with the constraint that the deviations must be integer multiple of 2π. With this constraint, the unwrapping results will not depend critically on the weighting mask we used, and errors are prevented to spread.

Minimization problems with integer variables are usually computationally very complex. However, recognizing the network structure underlying the phase unwrapping problem, it makes possible to employ very efficient strategies for its solution. In fact, the problem can be equated to the one of finding the minimum cost flow on a network, for the solution of which there are very efficient algorithms. To explain its basic principles and clarify how it can be performed, we refer to the unknown, unwrapped phase field and we impose that the result is consistent, thus requiring the irrotational property of this field:

$$\begin{aligned}\nabla\times\nabla\psi(x,y)&=\Delta_x\left(\Delta_y\psi(x,y)\right)-\Delta_y\left(\Delta_x\psi(x,y)\right)=\\&=\Delta_x\psi(x,y)+\Delta_y\psi(x+1,y)-\Delta_x\psi(x,y+1)-\Delta_y\psi(x,y)=0\end{aligned}\tag{9}$$

Obviously, we can also express each term of (9) with respect to the wrapped phase derivates by introducing, for each phase term, a corresponding, unknown 2π-multiple term, as follows:

$$\begin{aligned}\Delta_x\psi(x,y)&=\left\langle\Delta_x\psi(x,y)\right\rangle_{-\pi,\pi}+2\pi K_x(x,y)\\\Delta_y\psi(x+1,y)&=\left\langle\Delta_y\psi(x+1,y)\right\rangle_{-\pi,\pi}+2\pi K_y(x+1,y)\\\Delta_x\psi(x,y+1)&=\left\langle\Delta_x\psi(x,y+1)\right\rangle_{-\pi,\pi}+2\pi K_x(x,y+1)\\\Delta_y\psi(x,y)&=\left\langle\Delta_y\psi(x,y)\right\rangle_{-\pi,\pi}+2\pi K_y(x,y)\end{aligned}\tag{10}$$

These relations eventually lead to the following equation

$$K_x(x,y)+K_y(x+1,y)-K_x(x,y+1)-K_y(x,y)=-\frac{r(x,y)}{2\pi} \tag{11}$$

that relates the integer unknowns to the measurable residues. At this stage, the phase unwrapping problem can be formulated as the searching of the K terms that satisfy the constraints (11) and solve the following minimization problem:

$$\min_{K_x K_y}\left\{\sum_x\sum_y c_x(x,y)\left|K_x(x,y)\right|+\sum_x\sum_y c_y(x,y)\left|K_y(x,y)\right|\right\} \tag{12}$$

being c( ) the so-called cost functions allowing us to individuate areas where the location of branch-cuts is likely or unlikely. It is easy to verify that, whether the costs were chosen constant, the problem (12) would be equivalent to search for the minimum total cut-line length. Cost functions are essentially expressed as a function of the estimated local interferogram quality (by exploiting the spatial coherence, or the phase gradient density or other properly identified quality factors). The problem given in (12) is a non-linear minimization problem with integer variables, and, if the following change of variables is considered

$$\begin{aligned}
K_x^-(x,y)&=\min\left[0,K_x(x,y)\right]\\
K_x^+(x,y)&=\max\left[0,K_x(x,y)\right]\\
K_y^-(x,y)&=\min\left[0,K_y(x,y)\right]\\
K_y^+(x,y)&=\max\left[0,K_y(x,y)\right]
\end{aligned} \tag{13}$$

it can be re-formulated via two different linear problem, as follows:

$$\min_{K_x K_y}\left\{\sum_x\sum_y \begin{matrix}\left[c_x(x,y)K_x^+(x,y)+c_y(x,y)K_y^+(x,y)\right]\\ +\left[c_x(x,y)K_x^-(x,y)+c_y(x,y)K_y^-(x,y)\right]\end{matrix}\right\} \tag{14}$$

It can be seen that the problem stated in (14) can be transformed to define a minimum cost flow problem on a network (see Figure 2), with the new variables representing the net flow running along the network arcs. Once the network has been solved, the solutions in terms of the 2π-multiple integer functions will be expressed as follows:

$$\begin{aligned}
K_x(x,y)&=K_x^+(x,y)-K_x^-(x,y)\\
K_y(x,y)&=K_y^+(x,y)-K_y^-(x,y)
\end{aligned} \tag{15}$$

2.4 Extended Minimum Cost Flow (EMCF) algorithm

This section is focused on the presentation of the rationale at the base of the so-called Extended Minimum Cost Flow (EMCF) PhU technique that represents a way to efficiently incorporate temporal information in the phase unwrapping problem of sequence of multi-temporal differential interferograms. Before discussing the characteristics of the proposed approach, some considerations about the MCF technique are first in order. The original minimum cost flow algorithm, thought to be applied to a regular spatial grid, was improved in order to deal with sparse data (Costantini, 1998). In particular, the grid of the investigated samples is typically chosen to be relevant to the coherent pixels present into the DInSAR interferograms, while the Delaunay triangulation is used to define the neighbouring points and the elementary cycles in the set of the identified coherent sparse pixels. The possibility to extend this approach to the three-dimensional case, in order to simultaneously unwrap an interferometric sequence, has already been investigated in (Costantini et. al., 2002). However, in this case, the problem cannot be formulated in terms of network minimum cost flow procedures. Accordingly, no computationally efficient codes are available and, therefore, the overall unwrapping process can be extremely time-consuming, particularly for long interferogram sequences.

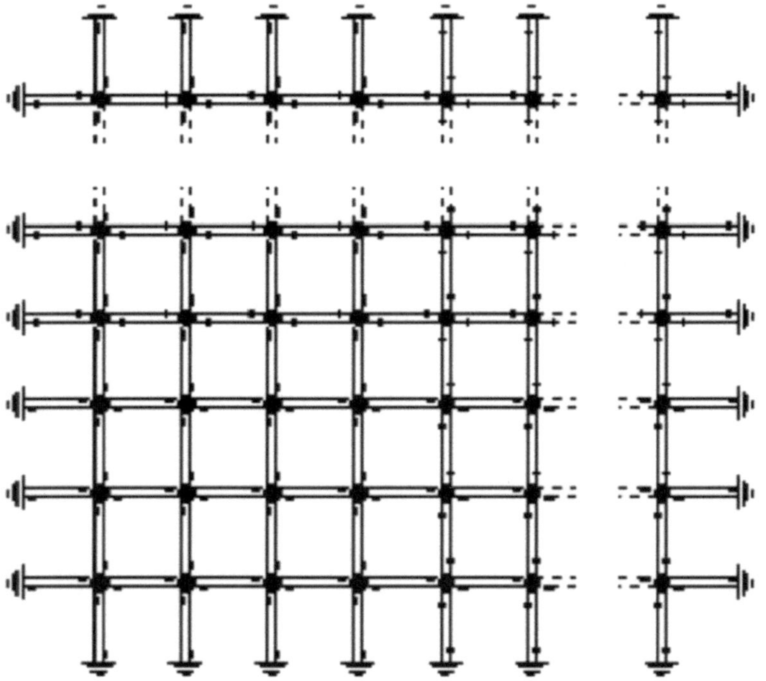

Fig. 2. A pictorial representation of the equivalent network to be solved for retrieving the 2π integer multiples needed to reconstruct a conservative phase field. The network nodes are associated to each residue-cut and the bidirectional arcs are related to the phase differences arcs.

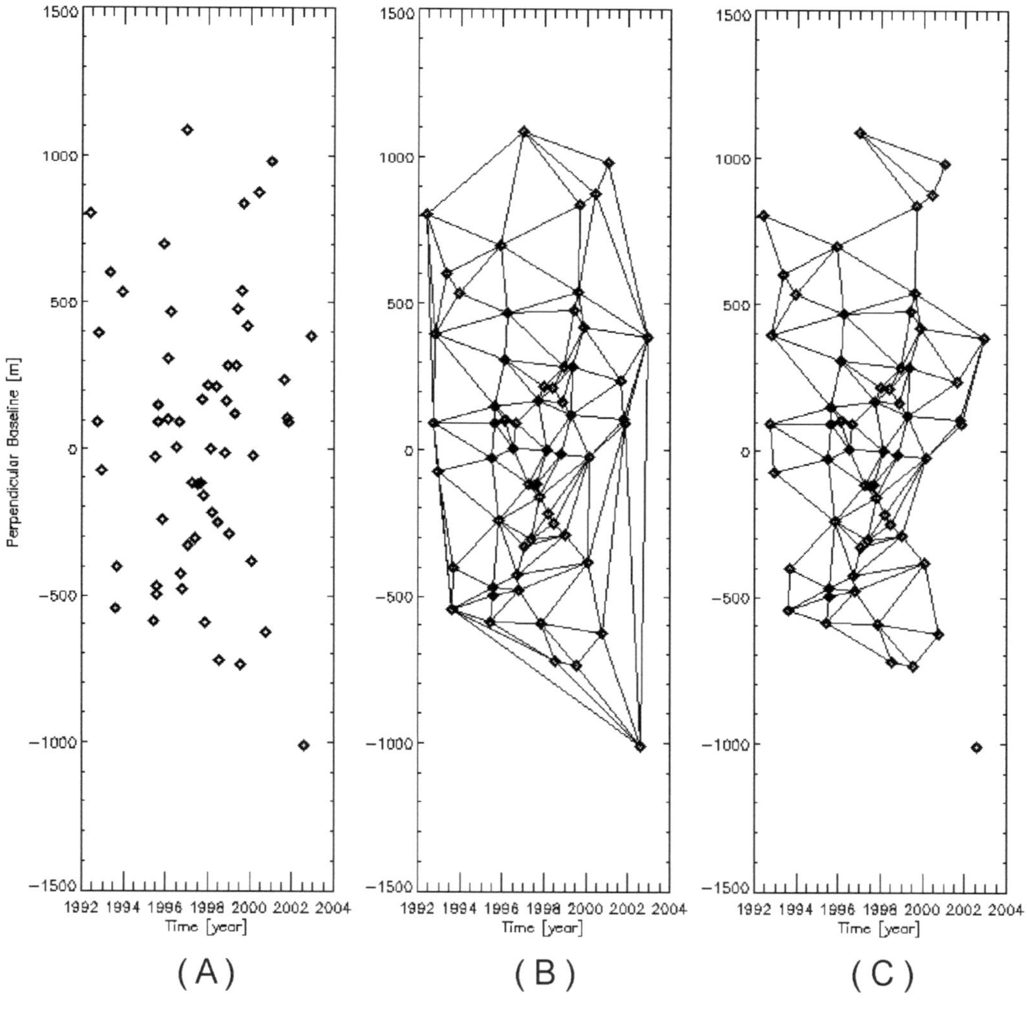

Fig. 3. SAR data representation in the temporal/perpendicular baseline plane for the ERS SAR data analyzed in the following experiments. (A) SAR image distribution. (B) Delaunay triangulation. (C) Triangulation after removal of triangles with sides characterized by spatial and/or temporal baseline values exceeding the selected thresholds (corresponding in our experiments to 300 m and 1500 days, respectively).

The proposed unwrapping solution, in addition to the spatial characteristics of each DInSAR interferogram, also exploits the temporal relationships among a properly selected interferogram sequence, thus allowing the performance improvement of the MCF technique.

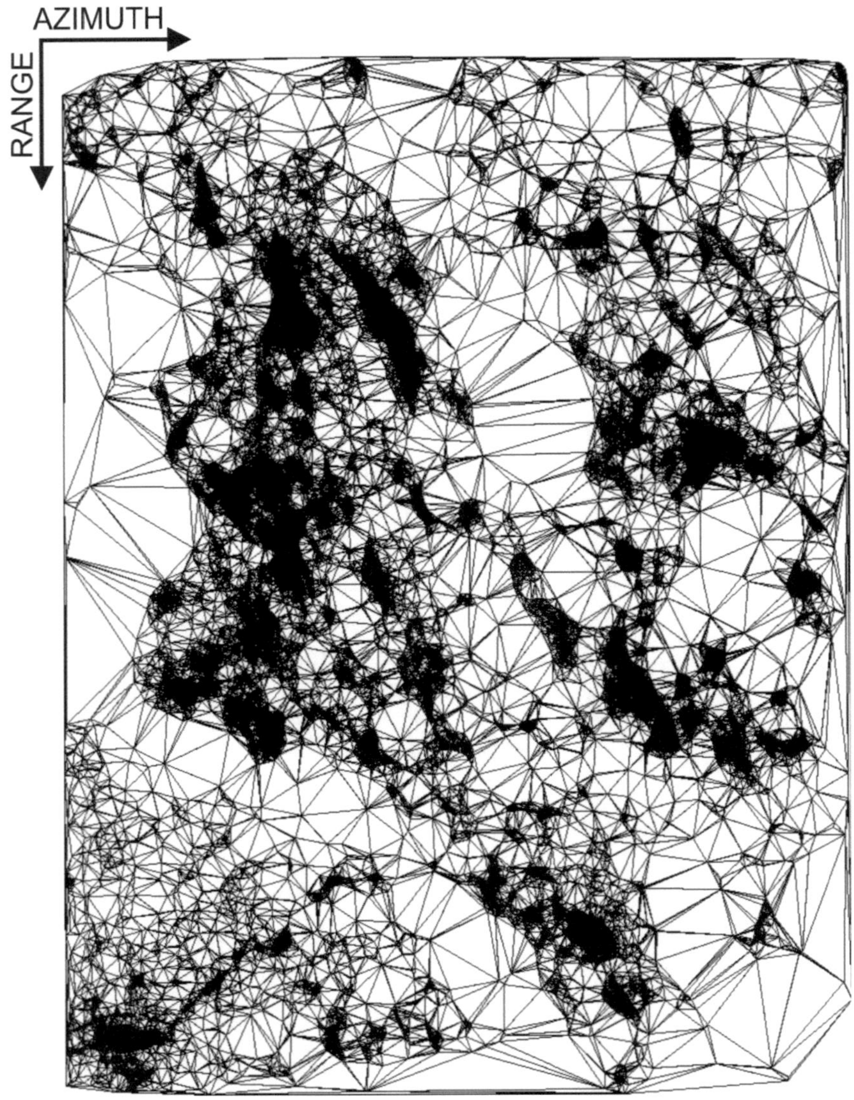

Fig. 4. Delaunay triangulation in the azimuth/range plane involving the set of spatially coherent pixels (in black).

The starting point of our approach, mostly oriented to deformation time-series generation, is the computation of two Delaunay triangulations. The former is relevant to the SAR data acquisitions distribution in the so-called "Temporal/Perpendicular baseline" plane and allows us to identify the DInSAR interferograms sequence to be computed; the latter is carried out in the Azimuth/Range plane and involves the coherent pixels common to the interferograms within the sequence. The unwrapping operation of the overall data set is then performed via a two-step processing procedure: first of all, we identify all the segments of the computed "spatial" triangles and, for each of these, we carry out a "temporal" PhU step by applying the MCF technique to the grid relevant to the temporal/perpendicular baseline plane. The second step uses, for each arc, the previously computed unwrapped phases as a starting point for the subsequent spatial unwrapping operation, performed on each single interferogram; again the standard MCF network programming technique is applied and, in this case, we use the "costs" of the previous solutions to set the weights associated with the single arcs involved in the spatial unwrapping. Overall, the basic rationale of the procedure is quite simple: to exploit the temporal relationships among the computed multi-temporal interferograms to bootstrap the subsequent spatial unwrapping operation. We remark that the presented approach is focused on multilook interferograms. Moreover, the possibility to apply the MCF network programming algorithm to both the temporal and spatial unwrapping steps leads to a computationally efficient procedure. Let us start our analysis by investigating the generation process of the interferograms needed for the algorithm implementation discussed in the following sections. Accordingly, we consider a set of N+1 independent SAR images of the same area, which are co-registered with respect to a reference one, with respect to which we may compute the temporal and spatial (perpendicular) baseline components. Accordingly, each SAR image can be represented by a point in the temporal/perpendicular baseline plane (see Figure 3A) where we may also compute a Delaunay triangulation (see Figure 3B). Despite the constraints on the maximum allowed baseline extensions, we remark that the obtained triangulation may involve arcs relevant to data pairs whose baselines exceed the assumed maxima, thus potentially leading to generate interferograms strongly decorrelated. To avoid these effects without loosing our triangulation representation, we may remove all the triangles involving arcs with too large baselines, as shown in Figure 3C. Equivalently, we may also remove the triangles corresponding to interferograms including data pairs with large Doppler centroid differences; this is often the case for interferograms involving ERS data acquired after 2000, i.e., following the gyroscopes failure events (Miranda et. al., 2008). Notice that this triangles removal step may lead to discarding some acquisitions and/or to the generation of more than one independent subset of triangles, i.e., to a data representation consistent with the one described in the D-InSAR Small Baseline Subset (SBAS) procedure. Therefore, the compatibility between this data organization and the one exploited in the SBAS technique is clearly evident. Following the identification of the final triangulation, the computation of the DInSAR interferogram sequence is performed. At this stage, we can generate the "mask" of the pixels in the spatial plane that are considered coherent within the generated sequence. This mask can be obtained, for instance, by considering those pixels with an estimated coherence value greater than a selected threshold, which are common to a part or even to the entire interferogram sequence. Accordingly, as a final step, we compute a second Delaunay triangulation which involves the arcs connecting the neighbouring pixels of the computed mask.

The presented PhU procedure is based on a two-step processing approach that benefits of the information available from both the grids. In particular, the key idea is to carry out first, for each arc connecting neighbouring pixels, a "temporal" unwrapping operation which implies the basic MCF approach. The second step relies on the use of these results as a starting point for the "spatial" unwrapping performed on each single interferogram. The key issues of these two processing steps are described in the following analysis which is focused on the use of multilook interferograms. The temporal PhU method benefits from the relationships existing between the phase differences of pixel pairs relevant to the wrapped and unwrapped signals. In particular, if we consider the spatial arc connecting the generic A and B pixel pair in the spatial plane, the unknown, unwrapped phase difference can be expressed as follows

$$\Delta\tilde{\psi}_{AB} = \tilde{\psi}(A) - \tilde{\psi}(B) = \langle\tilde{\varphi}(A) - \tilde{\varphi}(B)\rangle_{-\pi,\pi} + 2\pi H_{AB} = \\ = \Delta\tilde{\varphi}_{AB} + 2\pi H_{AB} \tag{16}$$

Therefore, we may compute a Delaunay triangulation shown in Figure 4. In order to define a set of elementary cycles relevant to the coherent pixels only, and we may impose the irrotational property for the phase gradient, in a discrete space. At this stage, the PhU problem can be solved, as done for the original case, in a very efficient way by recognizing that a network structure underlines it and, by searching for the relevant network minimum cost flow solution, as fully described in (Pepe & Lanari, 2006). The unwrapped DInSAR phase differences computed via the previous temporal PhU step are then finally used as a starting point for the spatial unwrapping operation. This second step is carried out on the single interferograms through the application of the basic MCF unwrapping technique.

3. EMCF-based region growing technique

A different class of PhU algorithms is based on the general concept of the "region growing" allowing us to considerably increase the spatial density of the areas with correct unwrapped results. In particular, we present a space-time region growing (RG) technique whose core is represented by the EMCF algorithm described in section 2. Basic idea of such approaches is to retrieve the unwrapped phase of a set of investigated pixels that are in the neighbourhood of an already unwrapped region. In particular, the unwrapped phase of the analyzed pixels is computed by exploiting the phase values of neighbouring unwrapped points (Xu & Cumming, 1999). Notice that with respect to classic region growing approaches here we apply a method that allows to handle with non-linear deformation trends: this is beneficial to analyze areas characterized by strongly non-linear displacements behaviours. Some experimental results will be also presented to demonstrate the effectiveness of the proposed PhU methodology. In particular, the implemented RG-EMCF algorithm has been applied to sequences of multilook differential interferograms, relevant to the Central Nevada region, characterized by large non-linear deformation phenomena, as well as to a test site located in the Gardanne area (France), affected by strong decorrelation phenomena.

The method provides a prediction of the (unwrapped) phase values at a pixel in the neighborhood of a region of pixels that have already been unwrapped. Following the preliminary PhU operation, we identify two sets of pixels. The former consists of Q high quality points $\mathbf{S} = (S_0, S_1, \ldots, S_{Q-1})$, hereinafter referred to as seed points, that are characterized by temporal coherence values greater than a given threshold (typically set to 0,7 for DInSAR analyses). Notice that the temporal coherence is a measure of the correctness of the obtained unwrapping results originally proposed in (Pepe & Lanari, 2006) that is not fully discussed here for the sake of brevity. The region-growing procedure starts by analyzing the candidate pixels in the proximity of the DInSAR reference point location, and proceeds along a path (for instance, a spiral) including all the previously selected candidate pixels.

For the sake of generalization, let us describe the *p-th* iteration of the algorithm, aimed at analyzing the multitemporal unwrapped phases at the generic candidate pixel C_p. This is done by exploiting the knowledge of the unwrapped phase values $\mathbf{\Psi}(S_{pi})$ at the seed pixels S_{pi}, located within a rectangular area centered around the location of the candidate pixel [see Fig. 5(a)]. Thereby, we obtain different predictions of the (unwrapped) phases at the

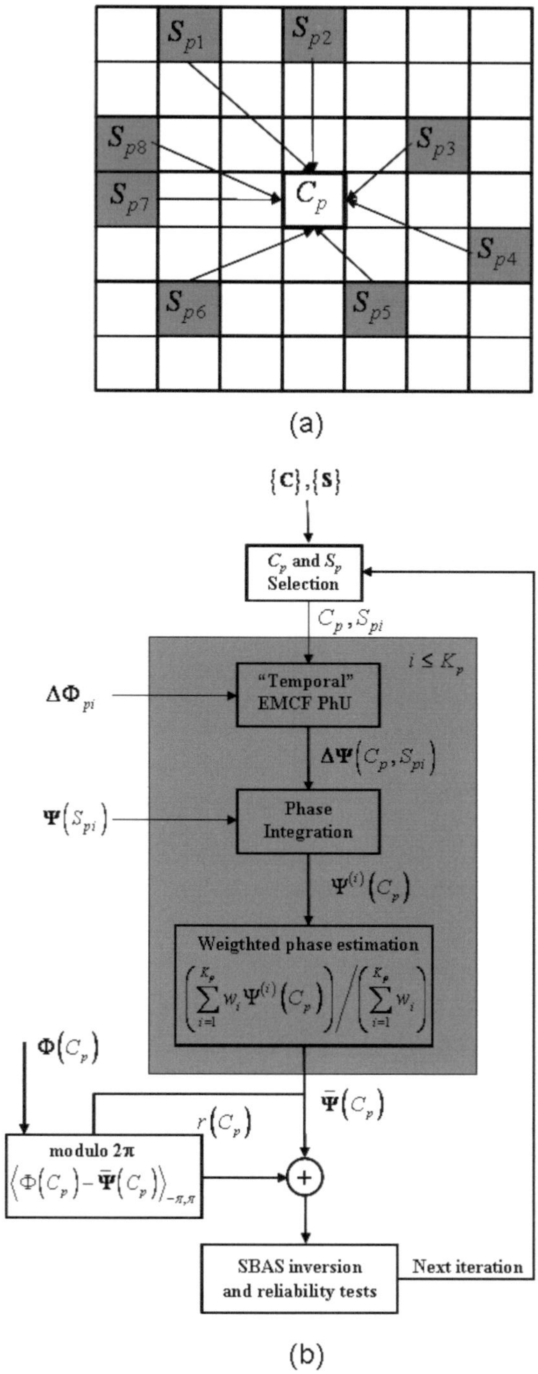

Fig. 5. EMCF-based region-growing algorithm. (a) Pictorial representation of the *p-th* iteration of the algorithm: the phase values at the candidate pixel (located at the center) are predicted by the phases of the eight-seed-pixels (gray boxes) located in its neighborhood. (b) Diagram block of the algorithm.

selected candidate pixel, namely $\boldsymbol{\Psi}^{(i)}\left(C_p\right)$, computed by integrating the phase differences along the paths connecting the candidate and the i-th seed pixel:

$$\boldsymbol{\Psi}^{(i)}\left(C_p\right)=\boldsymbol{\Psi}\left(S_{pi}\right)+\Delta\boldsymbol{\Psi}\left(C_p,S_{pi}\right) \quad i=1,\ldots,K_p \tag{17}$$

where the estimate of the (unwrapped) phase difference $\Delta\boldsymbol{\Psi}\left(C_p,S_{pi}\right)$ over the given candidate/seed arc is retrieved by applying the temporal PhU strategy detailed in (Pepe & Lanari, 2006). However, since different integration paths lead to independent phase predictions at the candidate pixel, we carry out a weighted average of such individuals predictions:

$$\bar{\boldsymbol{\Psi}}\left(C_p\right)=\left(\sum_{i=1}^{K_p} w_i \boldsymbol{\Psi}^{(i)}\left(C_p\right)\right)\Bigg/\left(\sum_{i=1}^{K_p} w_i\right) \tag{18}$$

where the weights {w} are set taking into account the (achieved) temporal network minimum costs (Pepe & Lanari, 2006). The average prediction in (18) is eventually used to compute the (unwrapped) phase vector at the selected candidate pixel.

$$\boldsymbol{\Psi}\left(C_p\right)=\bar{\boldsymbol{\Psi}}\left(C_p\right)+\mathbf{r}\left(C_p\right) \quad i=1,\ldots,K_p \tag{19}$$

where the $\mathbf{r}\left(C_p\right)=\left\langle\boldsymbol{\Phi}\left(C_p\right)-\bar{\boldsymbol{\Psi}}\left(C_p\right)\right\rangle_{-\pi,\pi}$ term on the right-hand side of (19) guarantees that unwrapped and wrapped phase vectors differ by 2π-integer multiples.

A check whether the prediction in (19) is correct is finally performed: if this is the case, the selected candidate pixel is added to the set of seed points, otherwise it is discarded from the following analyses. To the purpose, two different reliability tests are exploited.

Test 1- We invert (19) by the SBAS strategy and measure the related temporal coherence value (Pepe et al. 2006). The test is passed when the measured temporal coherence is greater than a given threshold.

Test 2- It relies on the assumption the larger the values of the residual phase vector the larger the probability to retrieve incorrect phase estimates. Accordingly, we analyze the residual phase dispersion through:

$$\Lambda\left(C_p\right)=\frac{\left|\sum_{k=0}^{M-1}\exp\left[jr\left(C_p\right)\right]\right|}{M} \tag{20}$$

This second test is passed when this additional coherence value is greater than a selected threshold.

The block diagram of the algorithm is sketched in Fig. 5b.

3.1 Real data results

In this section, we provide some experimental results relevant to two different test site areas (Casu, 2009): a wide area in central Nevada (USA) extending for about 600 x 100 km and the highly vegetated region of Gardanne (France).

The Central Nevada SAR data set consists of 264 ERS-1/2 SAR data frames (track 442, frames: from 2781 to 2871), spanning the 1992-2000 time interval, and we generate 148 multilook DInSAR interferograms, with a spatial resolution of about 160 x 160 m. Note that SAR data set and interferogram distribution are the same exploited by (Casu et al., 2008). First of all, we select the set of Seed Points from which the unwrapped phase will be propagated. To do this, we just impose a threshold value to the temporal coherence achieved by applying the EMCF algorithm, in particular we impose a value of 0.8, obtaining more than 850,000 Seed pixels. In Figure 6a it is shown the selected Seed Point mask.

In order to evaluate the phase unwrapping step performances, we compare the achieved RG-EMCF results with those obtained by using a conventional SBAS processing chain. In particular, Figure 6b and Figure 6c show the temporal coherence masks of the reliable pixels (obtained by imposing previous mentioned thresholds for the RG-EMCF algorithm and for the conventional one) relevant to the conventional and new Region-Growing algorithm, respectively. By a first qualitative analysis, it is clear that the RG-EMCF approach permits to obtain a larger number of points at a given coherence threshold. More systematically, we measured an increase of about 88% for the number of "grown" pixels, while the total image increment is of about the 32%.

Moreover, we also compute the temporal coherence histogram, shown in Figure 9, relevant to the common "grown" pixels only. It is clear the coherence improvement achieved by the RG-EMCF results. As mean coherence value we obtain 0.90 for the RG-EMCF approach while 0.85 for the conventional SBAS results. Also in this case, it is clearly visible the improvement of the new proposed algorithm.

As final remark, we present in Figure 7 the mean deformation velocity maps of the two sets of unwrapped phases processed via the SBAS approach. Figure 8 clearly demonstrates the RG-EMCF capability to increment reliable pixels not only in non deforming areas, but also in zone affected by strong and non linear deformation.

In general, Region-Growing is very helpful to unwrap wide areas. Indeed, to manage large set of points can imply PhU infeasibility, due to the complexity of the problem to be solved as well as to the high computational burden. Therefore, a cascade of global and local PhU steps permits to correctly unwrap large areas, by first deliberately reducing the amount of investigated points (Seed Pixels) and, subsequently, reconsidering the removed ones (Candidate Pixels) in a proper Region-Growing step.

However, a Region-Growing approach can be applied also in highly decorrelated region, where the low coherence implies to deal with a reduced amount of Seed Pixels. Therefore, a Region-Growing step can help increasing the unwrapped pixel spatial density.

Accordingly, we applied the proposed RG-EMCF PhU approach to the Gardanne (France) region which is highly vegetated and where an open pit mine is present. Therefore, the zone

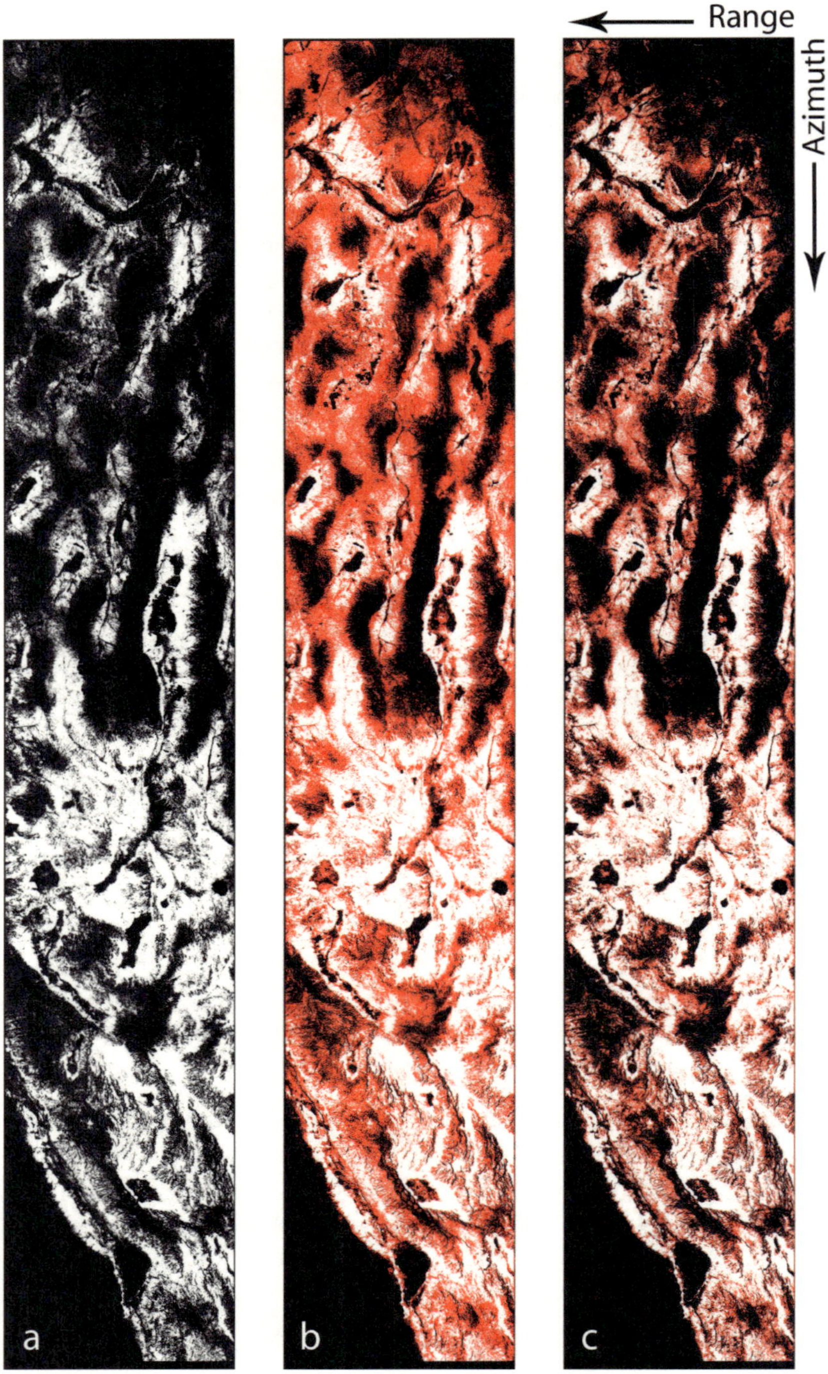

Fig. 6. Temporal coherence masks relevant to Central Nevada area. a) Seed Point mask (~850,000 pixels). b) RG-EMCF results (~1,770,000 pixels). c) Conventional SBAS results (~1,340,000 pixels). Grown points are highlighted in red while in white are represented the Seed Points of Figure 6a. By courtesy of dr. Francesco Casu (Casu, 2009).

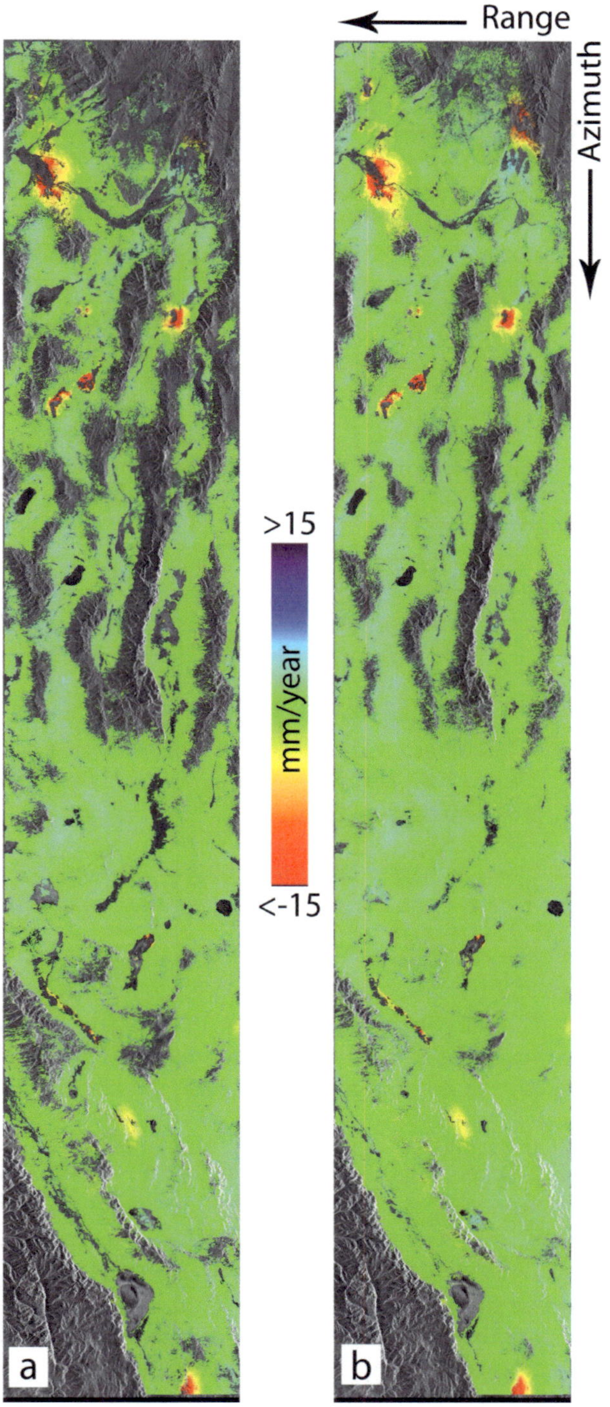

Fig. 7. Mean deformation velocity maps, in SAR coordinates, relevant to Central Nevada area. a) Conventional SBAS results. b) RG-EMCF results. By courtesy of dr. Francesco Casu (Casu, 2009).

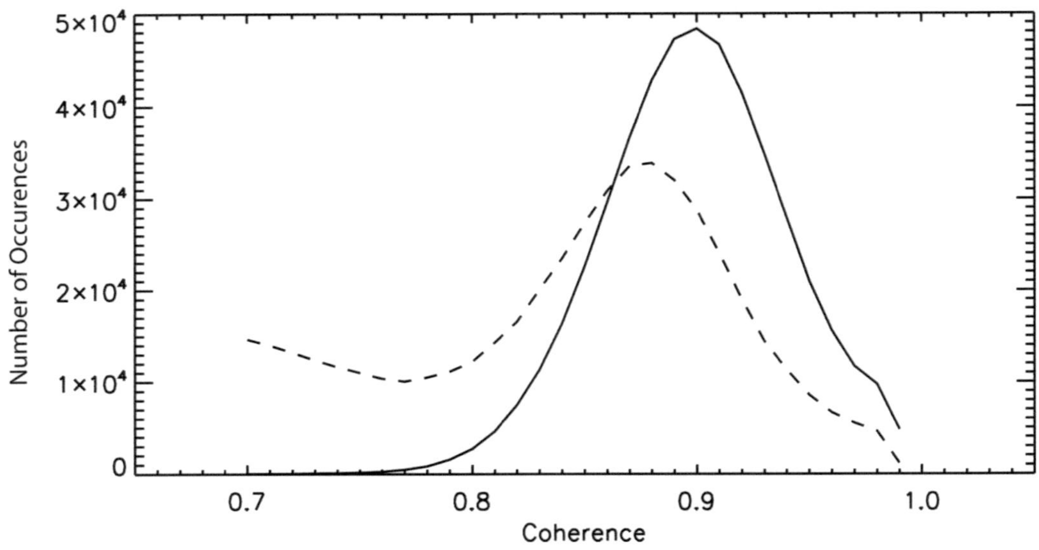

Fig. 8. Temporal coherence histogram relevant to common "grown" pixels obtained by exploiting conventional (dashed line) and new (continuous line) Region-Growing algorithm. By courtesy of dr. Francesco Casu (Casu, 2009).

is characterized by very low temporal coherence and is affected by strongly non linear deformation, implying to be a very critical test area for multi-temporal DInSAR analysis.

In Figure 9, we present the mean deformation velocity maps computed by applying the SBAS approach to a set of 75 ERS-1/2 and 8 ENVISAT images acquired in the 1992-2004 time period and coupled in 243 interferograms, the latter being unwrapped via both the procedure implemented in the SBAS processing chain and the RG-EMCF algorithm. Note that, DInSAR data have been multilook, obtaining a final spatial ground resolution of about 80 by 80 meters.

We just remark that, simultaneous exploitation of ERS and ENVISAT data implies that no cross sensor interferograms can be generated, due to the different signal wavelength of the two sensors. Therefore, the Temporal/Perpendicular baseline network will result decomposed in several subsets (at least two), corresponding to the different sensor interferograms. Also in this case, looking at Figure 9, it is clearly visible the strong increase of the correctly unwrapped pixels. Indeed, "grown" pixel number pass from about 2,500 for the conventional case to more than 87,000 for the RG-EMCF one, with a global improvement (for the whole area, i. e., accounting also for the Seed Points) of about the 113%. It is worthy to remark that the RG approach available in the SBAS processing essentially does not produced any results (only 2,500 "grown" pixels starting from 73,000 Seeds): this is because the strong decorrelation affecting the area.

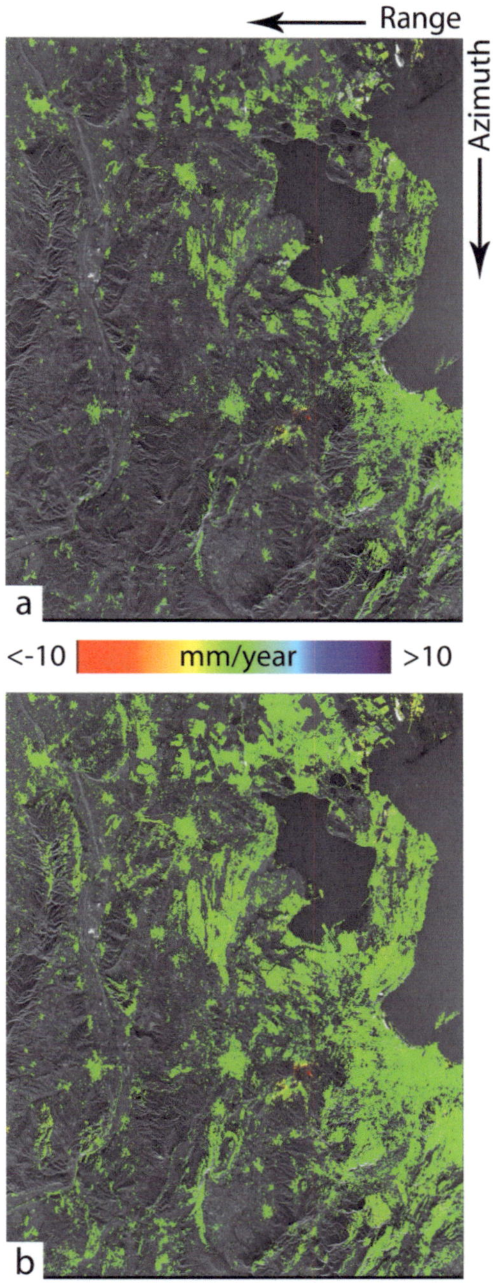

Fig. 9. Mean deformation velocity maps relevant to the Gardanne area (France). a) Conventional SBAS results. b) RG-EMCF results.

4. On unwrapping large interferograms

In this section, we propose to exploit the highly efficient EMCF algorithm as the core procedure to unwrap sequences of large single-look DInSAR interferograms suitable for the generation of deformation time-series through the SBAS approach. This PhU approach allows us to directly apply the SBAS inversion to the unwrapped full resolution DInSAR phase sequences, with no need to pass through the analysis of the corresponding sequences of multi-look DInSAR interferograms (Lanari et. al., 2004). To properly solve this PhU problem, we suggest to apply an effective divide-and-conquer approach to the space-time phase unwrapping problem. The key idea is to divide the complex minimum cost flow network problems, implementing the whole PhU step, into that of simpler sub-networks, which are solved by applying the EMCF approach. More precisely, we start by identifying, and solving, a primary network that involves a selected set of very coherent pixels in our interferograms. The results of this primary network minimization, representing the backbone structure of the overall network, are subsequently used to constrain the solution of the remaining sub-networks, including the entire set of coherent pixels. To achieve this task, the second EMCF PhU step relies on the generation of a Constrained Delaunay Triangulation (CDT) (Chew, 1989), whose constrained edges are relevant to the set of successfully unwrapped pixels analyzed during the first PhU operation. We remark that our approach has some similarities with [6] where a two-scale strategy was also suggested to unwrap large interferograms; however our strategy is similar but inverted because in our scheme the primary PhU step is used to figure out a (global) PhU solution, and the secondary one to locally "propagate" the PhU solution in low coherent areas. The key idea of the proposed approach is to split our complex MCF network problem into that of simpler sub-networks. This solution can be efficiently implemented, as shown in the following, through two subsequent processing steps that are both carried out by using the EMCF technique.

Basically, the first PhU step is carried out on a set of very coherent pixels, used to compute a Delaunay triangulation in the Azimuth/Range plane, and the achieved PhU results are eventually exploited to successfully unwrap the remaining pixels. To achieve this task we solve a "constrained optimization problem" based on the computation of a Constrained Delaunay Triangulation (CDT) in the plane from the grid of the overall coherent pixels. To clarify this issue, let us provide some basic information about the CDT, which is a triangulation of a given set of vertices with the following properties: 1) a pre-specified set of non-crossing edges (referred to as constraints, or constrained edges) is included in the triangulation, and (2) the triangulation is as close as possible to a Delaunay one. As an example, in Figure 10, it is shown a simple CDT relevant to a set of 96 points, see Fig. 10(a); in this case the selected constraints are represented by the 14 edges of the Delaunay triangulation generated from an eight-points subset of the originally 96 ones, see Fig. 10(b); the computed CDT is shown in Fig. 10(c), whereas the corresponding Delaunay triangulation is presented in Fig. 10(d).

Similarly to the case of Figure 10, we compute a CDT for the spatial grid of the coherent pixels , whose constrained edges are the arcs of the previously identified primary network. Since the EMCF Phase Unwrapping algorithm can work with generic triangular irregular grids, not necessarily Delaunay triangulations, it can be also applied to the irregular spatial grid obtained via our CDT. However, in this case we must solve a constrained optimization problem because we want to preserve, for each interferogram, the unwrapped phase values already obtained by solving the primary network. Accordingly, we perform the second unwrapping step again through the EMCF approach but applying the temporal PhU step only to the unconstrained arcs of the generated CDT. Moreover, the spatial PhU step is

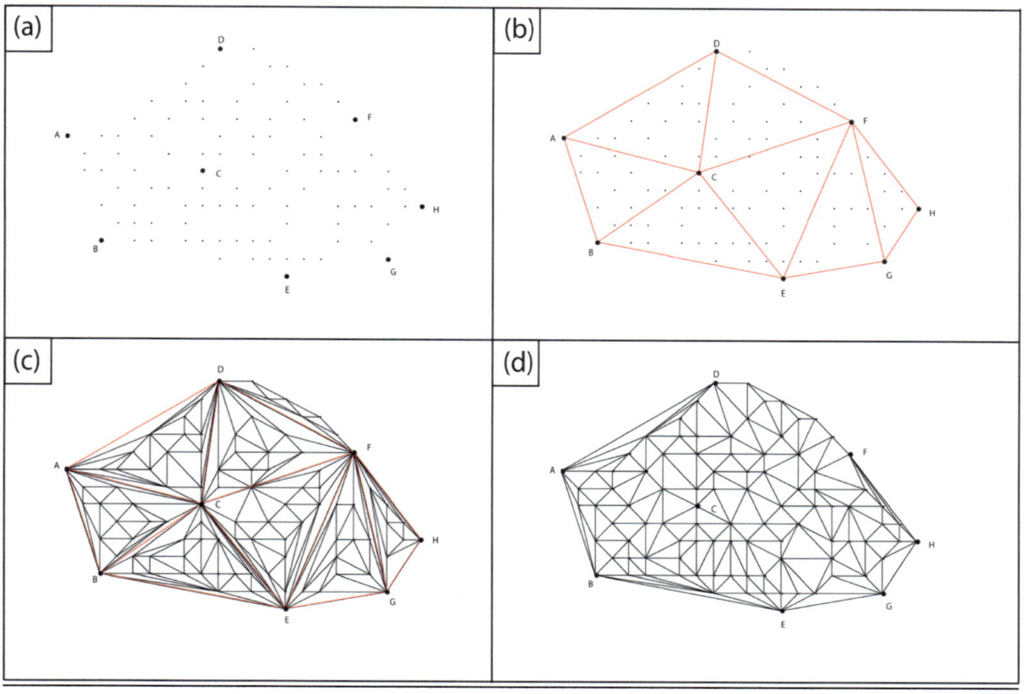

Fig. 10. Examples of Triangulations. (a) Dataset of 96 points with highlighted 8 of them, labeled to as A, B, C, D, E, F, G, H, respectively. (b) The 8-points Delaunay triangulation drawn with red lines (c) Constrained Delaunay Triangulation generated from the set of points of (a), and by using as constraints the triangulation of (b). (d) Delaunay triangulation computed from the 96 points in (a).

carried out on each single interferogram via the basic MCF approach but, in order to preserve the unwrapped phases relevant to the primary network, the weights used for the spatial MCF minimization must be properly set. If we refer to the generic PQ arc, this is easily achieved by imposing:

$$w_{PQ} = \begin{cases} L & PQ \in \{G_{Constrained}\} \\ 100 & \{PQ \notin \{G_{Constrained}\}\} \cap \{Cst_{\min} < \rho\} \\ 1 & \{PQ \notin \{G_{Constrained}\}\} \cap \{Cst_{\min} > \rho\} \end{cases} \tag{21}$$

where $\{G_{Constrained}\}$ is the set of constrained edges, and $Cst_{\min}$ is the temporal minimum network cost relevant to the given spatial arc. Moreover, L is a very large integer number, and ρ is a threshold value that is typically set not greater than 5% of the total number of interferograms. Based on (21), the flow into the constrained MCF network is automatically forced not to cross the constrained arcs and, as a consequence, the estimates of the primary network unwrapped phases are fully preserved. In other words, the presented approach allows us to effectively "propagate" the unwrapped solution from the primary network to the connected sub-networks, largely improving the phase unwrapping performances, and drastically decreasing the overall computational burden.

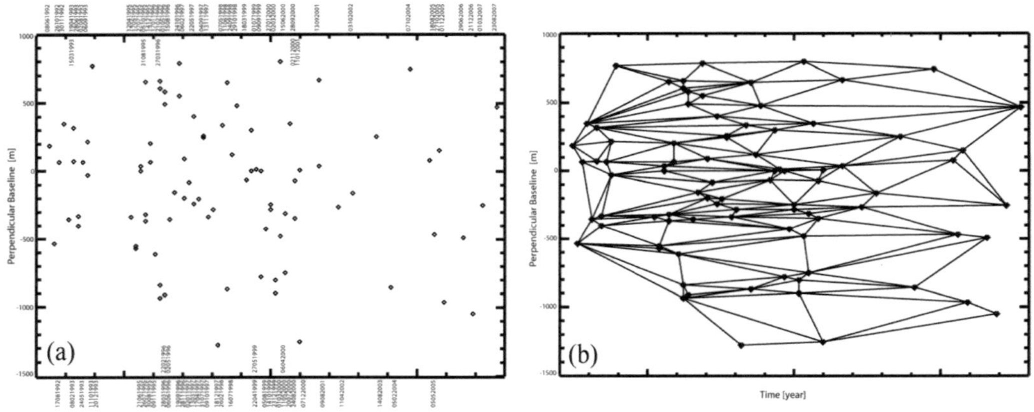

Fig. 11. SAR data representation in the temporal/perpendicular baseline plane for the ERS-1/2 SAR data analyzed in the following experiments, relevant to the Napoli (Italy) bay area. (a) SAR image distribution. (b) Delaunay Triangulation after the removal of triangles with sides characterized by spatial and temporal baseline values, as well as doppler centroid differences, exceeding the selected thresholds (corresponding in our experiments to 400 m, 1500 days and 1000 Hz, respectively).

The proposed approach was validated by analyzing a dataset of 86 ERS-1/2 SAR images acquired on descending orbits (Track 36, Frame 2781) between June 8, 1992 and August 23, 2007, over the Napoli (Italy) Bay area (see Figure 11). From these data, we identified a set of 234 data pairs characterized by perpendicular baseline values smaller than 400 m, and a maximum time interval of 1500 days. Precise satellite orbital information and a 3 arcsec SRTM DEM of the area were used to generate a sequence of single-look DInSAR interferograms. The computed interferograms were then unwrapped by applying the EMCF approach. To achieve this task, we first identified the spatial grid of all the coherent pixels to be unwrapped, composed of about 530 000 pixels, where 50 000 of them are very coherent. From the latter pixels, we generated, in the spatial plane, a Delaunay triangulation. The arcs of this triangulation represent the constrained edges of the implemented CDT: this structure connects the overall set of coherent pixels , and is essential for the second EMCF PhU step, leading to the final estimate of the unwrapped interferograms on the chosen spatial grid.

Figure 12 shows a false color map of the detected mean deformation velocity, where only points with high data quality are included, superimposed on a multi-look SAR amplitude image of the area. Moreover, in order to further investigate the achieved accuracy of the proposed approach, we focused on the Napoli city and surroundings, including Campi Flegrei caldera [see Fig. 3(a)] where independent geodetic information (leveling and GPS measurements) was available. In particular, for our analysis, we considered pixels located in correspondence to continuous GPS stations and leveling benchmarks and, for each of these points, we compared the retrieved DInSAR time-series with those obtained from the geodetic measurements, projected on the radar LOS. Note that, although we did not perform any filtering of the atmospheric phase artifacts affecting the DInSAR time-series, there is a good agreement between the SAR and the geodetic measurements. These results f confirm the effectiveness of the proposed phase unwrapping approach.

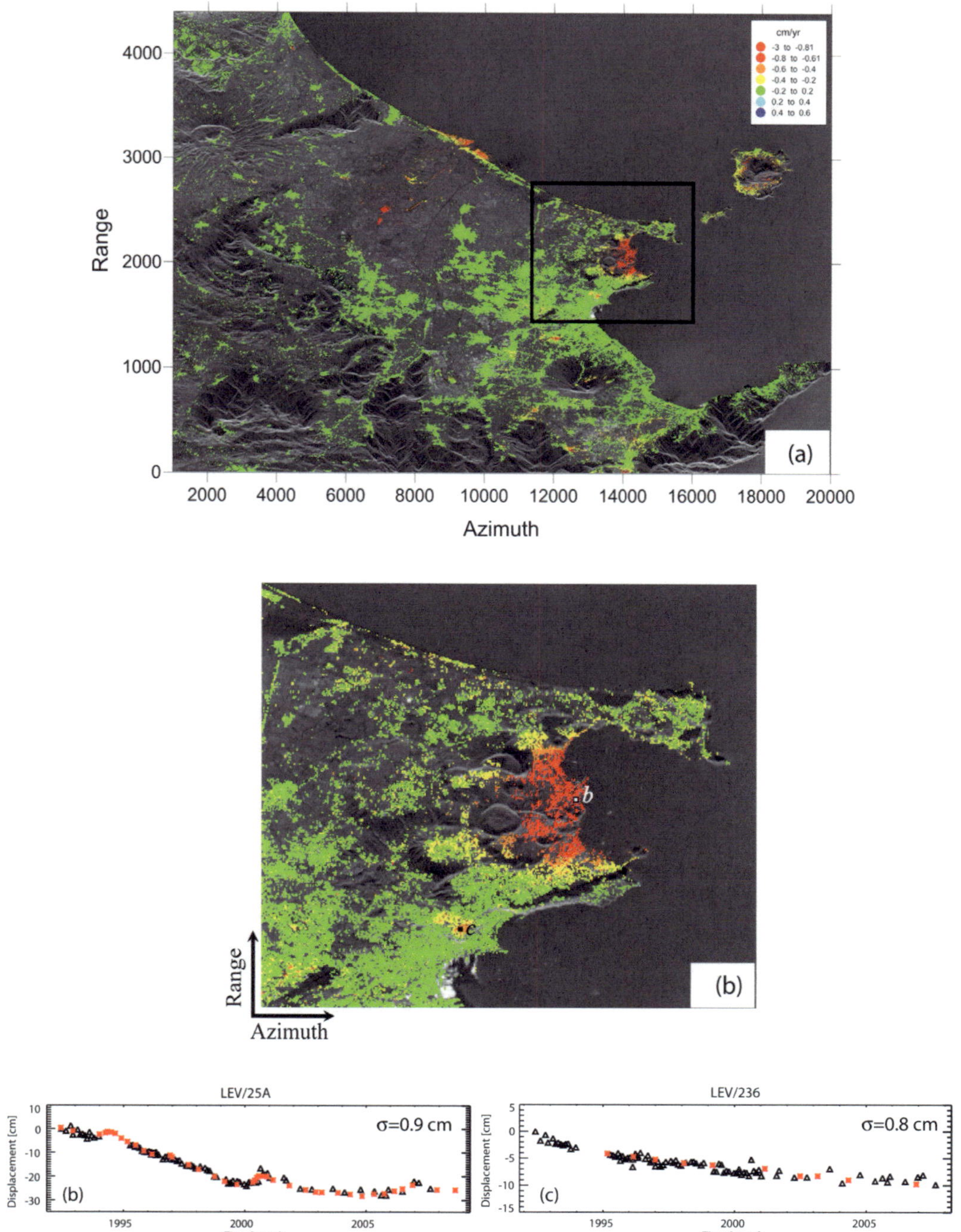

Fig. 12. ERS-1/2 DInSAR results. (a) Geocoded map of the mean deformation velocity of the investigated area (b) Zoomed view of the area of Napoli city and Campi Flegrei caldera (Italy) highlighted by the black box in (a). The plots show the DInSAR/leveling comparison of the deformation time-series corresponding to the pixels in (a) labeled as b (25A leveling benchmark) and c (236 leveling benchmark), respectively.

5. Conclusion

This chapter has presented a short review of some advanced multi-temporal phase unwrapping approaches for the generation of surface deformation time series through the application of the Small Baseline Subset (SBAS) DInSAR technique. Following a description of the basic rationale of the PhU problem, we have described in details the space-time PhU algorithm known as Extended Minimum Cost-Flow (EMCF). Finally, we focused on the recent improvements of this algorithm to analyze large interferograms affected by significant decorrelation effects and/or with severe non-linear deformation signals. The presented results clearly demonstrate the validity of these approaches and their valuable applicability in real cases.

6. Acknowledgment

The ERS-1/2 SAR data were provided by the European Space Agency, and the DEM of the investigated zone has been acquired through the SRTM archive. The exploited precise orbital information was supplied by the Technical University of Delft (The Netherlands). I would like to thank all my colleagues at IREA; a special thank goes in particular to dr. Lanari, dr. Manunta, and dr. Casu who have actively contributed to the development of some of the presented algorithms.

7. References

D. Massonnet, and K. L. Feigl, "Radar Interferometry and its application to changes in the earth's surface," Rev. of Geophys., vol. 36, no. 4, pp. 441–500, doi:10.1029/97RG03139, 1998.

R. Bürgmann, P. A. Rosen, and E. J. Fielding, "Synthetic aperture radar interferometry to measure Earth's surface topography and its deformation," Annual Review Earth Planet Science, vol. 28, pp. 169–209, May 2000.

A. K. Gabriel, R. M. Goldstein, and H. A. Zebker, "Mapping small elevation changes over large areas: Differential interferometry," J. of Geophys. Res., vol. 94, no. B7, pp. 9183–9191, March 1989.

D. Massonnet, M. Rossi, C. Carmona, F. Adragna, G. Peltzer, K. Feigl, and T. Rabaute, "The displacement field of the Landers earthquake mapped by radar interferometry," Nature, vol. 364, no. 6433, pp. 138–142, Jul. 1993.

R. M. Goldstein, H. Engelhardt, B. Kamb, and R. M. Frolich, "Satellite radar interferometry for monitoring ice sheet motion: Application to an antarctic ice stream," Science, vol. 262, no. 5139, pp. 1525–1530, Dec. 1993.

D. Massonnet, P. Briole, and A. Arnaud, "Deflation of Mount Etna monitored by spaceborne radar interferometry," Nature, vol. 375, no. 6532, pp. 567–570, Jun. 1995.

G. Peltzer, and P. A. Rosen, "Surface displacement of the 17 May 1993 Eureka Valley earhtquake observed by SAR interferometry," Science, vol. 268, no. 5215, pp. 1333-1336, Jun. 1995.

A. Ferretti, C. Prati, and F. Rocca, "Permanent scatterers in SAR interferometry," IEEE Trans. Geosci. Remote Sens., vol. 39, no.1, pp. 8-20, Jan. 2001.

P. Berardino, G. Fornaro, R. Lanari, and E. Sansosti, "A new algorithm for surface deformation monitoring based on small baseline differential SAR interferograms," IEEE Trans. Geosci. Remote Sens., vol. 40, no. 11, pp. 2375–2383, Nov. 2002.

A. Hooper, H. Zebker, P. Segall, and B. Kampes, "A new method for measuring deformation on volcanoes and other natural terrains using InSAR persistent scatterers," Geophys. Res. Lett., vol. 31, L23611, doi:10.1029/2004GL021737, Dec. 2004.

O. Mora, J. J. Mallorquí, and A. Broquetas, "Linear and nonlinear terrain deformation maps from a reduced set of interferometric SAR images," IEEE Trans. Geosci. Remote Sens., vol. 41, pp. 2243–2253, Oct. 2003.

H. A. Zebker and J. Villasenor, "Decorrelation in interferometric radar echoes," IEEE Trans. Geosci. Remote Sens., vol. 30, no. 5, pp. 950–959, Sep. 1992.

R. Lanari, O. Mora, M. Manunta, J. J. Mallorquì, P. Berardino, and E. Sansosti, "A Small Baseline Approach for Investigating Deformation on Full resolution Differential SAR Interferograms," IEEE Trans. Geosci Remote Sens., vol. 42, no. 7, Jul. 2004.

H. A. Zebker and Y. Lu "Phase Unwrapping Algorithms for radar interferometry: residue-cut, least-squares, and synthesis algorithm J. Opt Soc. Am A. 15, 586-598, 1998;

R. M. Goldstein, H. A. Zebker and C. L. Werner: "Satellite radar interferometry: two-dimensional phase unwrapping", Radio Sci., 23, 1988;

D. C. Ghiglia and L. A. Romero: "Robust two-dimensional weighted and unweighted phase unwrapping that uses fast transform and iterative methods", J. Opt. Soc. Am. A, 11, 1994;

G. Fornaro, G. Franceschetti, R. Lanari, E. Sansosti: "Robust phaseunwrapping techniques: a comparison", J. Opt. Soc. Am A., 13, 2355,1996;

M. Costantini: "A novel phase unwrapping method based on network programming", IEEE vol 36, 3 May 1998, pp 813-821;

M. Costantini, F. Malvarosa, F. Minati, L. Pietranera and G. Milillo, "A Three-dimensional Phase Unwrapping Algorithm for Processing of Multitemporal SAR Interferometric Measurements," in Proc. Geosc, and Rem. Sensing Symposium, Toronto (Canada), 2002, vol. 3, pp. 1741-1743.

N. Miranda, B. Rosich, C. Santella, and M. Grion: "Review of the impact of ERS-2 piloting modes on the SAR Doppler stability", in Proc. Fringe,Frascati, Italy, Dec. 2003, CD-ROM;

A. Pepe, and R. Lanari R., "On the extension of the minimum cost flow algorithm for phase unwrapping of multitemporal differential SAR interferograms," IEEE Trans. Geosci. Remote Sens., vol. 44, no. 9, pp. 2374-2383, Sept. 2006.

W. Ku, and I. Cumming, "A Region-Growing Algorithm for InSAR Phase Unwrapping, " IEEE Trans. Geosci. Remote Sens., vol. 37, pp. 124–134, Jan. 1999

F. Casu, "The Small BAseline Subset technique: performance assessment and new developments for surface deformation analysis of very extended areas", PhD thesis, 2009

F. Casu, M. Manzo, A. Pepe, and R. Lanari, "SBAS-DInSAR Analysis of Very Extended Areas: First Results on a 60000-km2 Test Site," IEEE Geosci. and Remote Sensing Lett., vol. 5, no. 3, pp. 438-442, Jul. 2008.

L. P. Chew, "Constrained Delaunay triangulations," in Algorithmica, vol. 4, Springer. Verlag. New York 1989, pp. 97–108.

4

Simulation of 3-D Coastal Spit Geomorphology Using Differential Synthetic Aperture Interferometry (DInSAR)

Maged Marghany
Institute for Science and Technology Geospatial (INSTEG)
Universiti Teknologi Malaysia, Skudai, Johor Bahru,
Malaysia

1. Introduction

Interferometric synthetic aperture radar (InSAR or IfSAR), is a geodetic technique uses two or more single look complex synthetic aperture radar (SAR) images to produce maps of surface deformation or digital elevation (Massonnet, and Feigl 1998; Burgmann et al., 2000; Hanssen 2001). It has applications as well, for monitoring of geophysical natural hazards, for instance earthquakes, volcanoes and landslides, also in engineering, in particular recording of subsidence and structural stability. Over time-spans of days to years, InSAR can detect the centimetre-scale of deformation changes (Zebker et al.,1997). Further, the precision DEMs with of a couple of ten meters can produce from InSAR technique compared to conventional remote sensing methods. Nevertheless, the availability of the precision DEMs may a cause of two-pass InSAR; regularly 90 m SRTM data may be accessible for numerous territories (Askne et al.,2003). InSAR, consequently, provides DEMs with 1-10 cm accuracy, which can be improved to millimetre level by DInSAR. Even so, alternative datasets must acquire at high latitudes or in areas of rundown coverage (Nizalapur et al., 2011). However, the baseline decorrelation and temporal decorrelation make InSAR measurements unfeasible (Lee 2001; Luo et al., 2006; Yang et al., 2007; Rao and Jassar 2010). In this regard, Gens (2000) reported the length of the baseline designates the sensitivity to height changes and sum of baseline decorrelation. Further, Gens (2000) stated the time difference for two data acquisitions is a second source of decorrelation. Indeed, the time differences while comparing data sets with a similar baseline length acquired one and 35 days a part suggests only the temporal component of the decorrelation. Therefore, the loss of coherence in the same repeat cycle in data acquisition are most likely because of baseline decorrelation. According to Roa et al. (2006), uncertainties could arise in DEM because of limitation InSAR repeat passes. In addition, the interaction of the radar signal with troposphere can also induce decorrelation. This is explained in several studies of Hanssen (2001); Marghany and Hashim (2009); and Rao and Jassar (2010).

Generally, the propagation of the waves through the atmosphere can be a source of error exist in most interferogram productions. When the SAR signal propagated through a vacuum it should theoretically be subjected to some decent accuracy of timing and cause phase delay (Hanssen 2001). A constant phase difference between the two images caused by the horizontally homogeneous atmosphere was over the length scale of an interferogram and vertically over that of the topography. The atmosphere, however, is laterally heterogeneous on length scales both larger and smaller than typical deformation signals (Lee 2001). In other cases the atmospheric phase delay, however, is caused by vertical inhomogeneity at low altitudes and this may result in fringes appearing to correspond with the topography. Under this circumstance, this spurious signal can appear entirely isolated from the surface features of the image, since the phase difference is measured other points in the interferogram, would not contribute to the signal (Hanssen 2001). This can reduce seriously the low signal-to-noise ratio (SNR) which restricted to perform phase unwrapping. Accordingly, the phases of weak signals are not reliable. According to Yang et al., (2007), the correlation map can be used to measure the intensity of the noise in some sense. It may be overrated because of an inadequate number of samples allied with a small window (Lee 2001). Weights are initiated to the correlation coefficients according to the amplitudes of the complex signals to estimate accurate reliability (Yang et al., 2007).

1.1 Hypothesis and objective of study

Concerning with above prospective, we address the question of decorrelation uncertainties impact on modelling Digital Elevation Model (DEM) for 3-D coastal spit visualization from DInSAR technique. This is demonstrated with RADARSAT-1 fine mode data (F1) using fuzzy B-spline algorithm. Taking advantage of the fact that fuzzy B-spline can use for solving uncertainty problem because of decorrelation and the low signal-to-noise ratio (SNR) in data sets. This work hypothesises that integration of fuzzy B-spline algorithm with phase unwrapping can produce accurately digital elevation of object deformation (Marghany et al., 2010a and Marghany 2012). The aim of this paper is to explore the precision of the digital elevation models (DEM) derived from RADARSAT-1 fine mode data (F1) and, thus, the potential of the sensor for mapping coastal geomorphologic feature changes. Depending on the results, a wider application of F1 mode data for the study of Kuala Terengganu mouth river landscapes is envisaged.

2. Study area

The study area is selected along the mouth river of Kuala Terengganu, Malaysia. According to Marghany et al., (2010a) the coastline appears to be linear and oriented at about 45° along the east coast of Malaysia (Marghany et al. 2010b). In addition, spit was located across the largest hydrological communications between the estuary and the South China Sea the i.e. mouth river of Kuala Terengganu (Fig. 1) which lies on the equatorial region, and is affected by monsoon winds (Marghany and Mazlan 2010a,b). Indeed, during the northeast monsoon period, the strong storm and wave height of 4 m can cause erosion (Marghany et al. 2010b). The 20 km stretches of coastal along the Kuala Terengganu shoreline composed of sandy beach, the somewhat most frequently eroded region. The significant source of sediment is from the Terengganu River which loses to the continental shelf due to the complex movements of waves approached from the north direction (Marghany et al. 2010b).

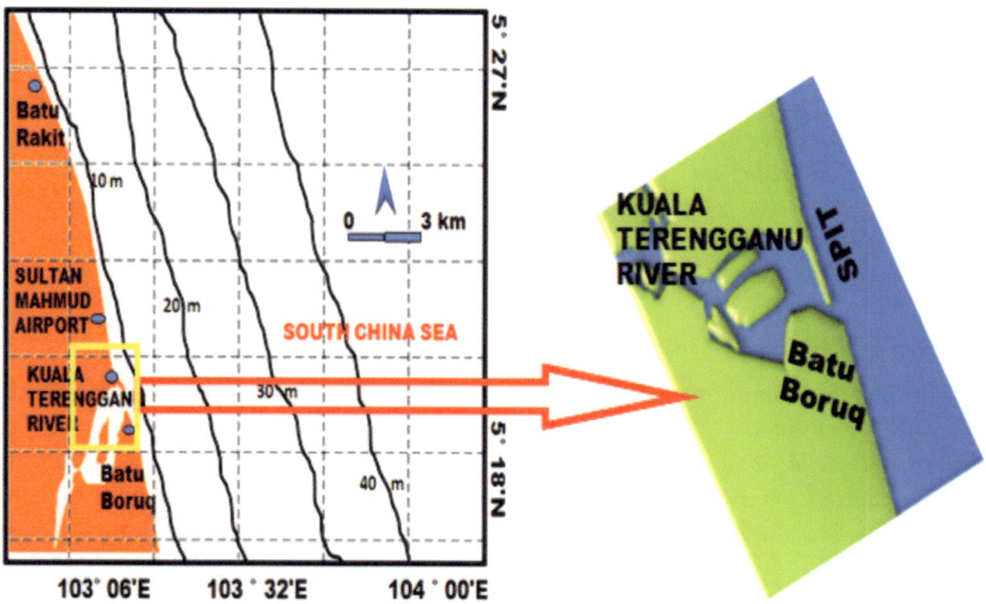

Fig. 1. Location of spit along Kuala Terengganu river mouth.

3. Data sets

3.1 Satellite data

In the present study, RADARSAT-1 SAR data sets of 23 November 1999 (SLC-1), 23 December 2003 (SLC-2) and March 26, 2005, (SLC-3) of Fine mode data (F1) are implemented. These data are C-band and had the lower signal-to-noise ratio owing to their HH polarization with wavelength of 5.6 cm and frequency of 5.3 GHz. The Fine beam mode is intended for applications which require the best spatial resolution available from the RADARSAT-1 SAR system. The azimuth resolution is 8.4 m, and range resolution ranges between 9.1 m to 7.8 m. Originally, five Fine beam positions, F1 to F5, are available to cover the far range of the swath with incidence angle ranges from 37° to 47°. By modifying timing parameters, 10 new positions have been added with offset ground coverage. Each original Fine beam position can either be shifted closer to or farther away from Nadir. The resulting positions are denoted by either an N (Near) or F (Far). For example, F1 is now complemented by F1N and F1F (RADARSAT 2011). Finally, RADARSAT-1 requires 24 days to return to its original orbit path. This means that for most geographic regions, it will take 24 days to acquire exactly the same image (the same beam mode, position, and geographic coverage). However, RADARSAT's imaging flexibility allows images to be acquired on a more frequent basis (RADARSAT 2011).

3.2 Ground data

Following Marghany et al., (2010b), the GPS survey used to: (i) to record exact geographical position of shoreline; (ii) to determine the cross-sections of shore slopes; (iii) to corroborate the reliability of DInSAR data co-registered; and finally, (iv) to create a reference network for future surveys. The geometric location of the GPS survey was obtained by using the new satellite geodetic network, IGM95. After a careful analysis of the places and to identify the reference vertexes, we thickened the network around such vertexes to perform the measurements for the cross sections (transact perpendicular to the coastline). The GPS data collected within 20 sample points scattered along 400 m coastline. The interval distance of 20 m between sample location is considered. In every sample location, Rec-Alta (Recording Electronic Tacheometer) was used to acquire the spit elevation profile. The ground truth data were acquired on 23 December 2003 March 26, 2005, during satellite passes. Then ground data used to validate and find out the level of accuracy for DInSAR and fuzzy B-spline algorithm.

4. DInSAR data processing

The DInSAR technique measures the block displacement of land surface caused by subsidence, earthquake, glacier movement, and volcano inflation to cm or even mm accuracy (Luo et al., 2006). According to Lee (2001), the surface displacement can estimate using the acquisition times of two SAR images S_1 and S_2. The component of surface displacement thus, in the radar-look direction, contributes to further interferometric phase (φ) as

$$\phi = \frac{4\pi}{\lambda}(\Delta R + \zeta) \tag{1}$$

where ΔR is the slant range difference from satellite to target respectively at different time, λ is the RADARSAT-1 SAR fine mode wavelength which is about 5.6 cm for CHH- band. According to Lee (2001), for the surface displacement measurement, the zero-baseline InSAR configuration is the ideal as $\Delta R = 0$, so that

$$\phi = \phi_d = \frac{4\pi}{\lambda}\zeta \tag{2}$$

In practice, zero-baseline, repeat-pass InSAR configuration is hardly achievable for either spaceborne or airborne SAR. Therefore, a method to remove the topographic phase as well as the system geometric phase in a non-zero baseline interferogram is needed. If the interferometric phase from the InSAR geometry and topography can strip of from the interferogram, the remnant phase would be the phase from block surface movement, providing the surface maintains high coherence (Luo et al., 2006).

Zebker et al. (1994) and Luo et al., (2006) used the three-pass method to remove topographic phase from the interferogram. This method requires a reference interferogram, which is promised to contain the topographic phase only. The three-pass approach has the advantage in that all data is kept within the SAR data geometry while DEM method can produce errors by misregistration between SAR data and cartographic DEM. The three-pass approach is restricted by the data availability. The three-passes DInSAR technique uses another InSAR pair as a reference interferogram that does not contain any surface movement event as

$$\phi' = \frac{4\pi}{\lambda}\Delta R' \; . \tag{3}$$

Incorporating equations 2 and 3 gives the phase difference, only from the surface displacement as

$$\phi_d = \phi - \frac{\Delta R}{\Delta R'}\phi' = \frac{4\pi}{\lambda}\zeta \,. \tag{4}$$

For an exceptional case where $\frac{\Delta R}{\Delta R'}$ in equation 4 there is a positive integer number, phase unwrapping may not be necessary (Massonnet et al., 1998). However, this is not practical and it is difficult to achieve from the system design for a repeat-pass interferometer. From equation 4 the displacement sensitivity of DInSAR is given as

$$\frac{\partial \phi_d}{\partial \zeta} = \frac{4\pi}{\lambda} \,. \tag{5}$$

Marghany and Mazlan (2009) introduces a method to construct 3-D object visualisation from unwrapping phase as follow,

$$S(p,q) = \frac{\sum_{i=0}^{M}\sum_{j=0}^{O} \phi_d C_{ij} \beta_{i,4}(p)\beta_{j,4}(q) w_{i,j}}{\sum_{m=0}^{M}\sum_{l=0}^{O} \beta_{m,4}(p)\beta_{l,4}(q) w_{ml}} = \sum_{i=0}^{M}\sum_{j=0}^{O} \phi_d C_{ij} S_{ij}(p,q) \tag{6}$$

where $\beta_{i,4}(p)$ and $\beta_{j,4}(q)$ are two bases of B-spline functions, $\{C_{ij}\}$ is the bidirectionally control net and $\{w_{ij}\}$ is the weighted correlation coefficient which was estimated based on Marghany (2011) as

$$w_{ij} = \frac{\left| \frac{\sum S_M(i,j) S_s(i,j)}{\sqrt{\sum |S_M(i,j)|^2 \sum |S_s(i,j)|^2}} \right| x \min(|S_M(i,j)|, |S_s(i,j)|) - t_1}{t_2 - t_1} \tag{7}$$

where $|S_M(i,j)|$ and $S_s(i,j)$ are master and slave complex data while t_1 and t_2 are thresholds. The curve points *S(p,q)* are affected by $\{w_{ij}\}$ in case of $p \in [r_i, r_{i+P+1}]$ and $q \in [r_j, r_{j+P'+1}]$, where *P* and *P′* are the degree of the two B-spline basis functions constituted the B-spline surface. Two sets of knot vectors are knot *p*=[0,0,0,0,1,2,3,……,*O,O,O,O*], and *knot q*=[0,0,0,0,1,2,3,….,*M,M,M,M*]. Fourth the order B-spline basis are used $\beta_{j,4}(.)$ to ensure the continuity of the tangents and curvatures on the whole surface topology including at the patches' boundaries (Marghany 2011).

5. Three-dimensional SPIT visualization using DInSAR technique

The new fuzzy B-spline formula for 3D coastal features' reconstruction from DInSAR retrieved unwrapping phase was trained on three RADARSAT-1 SAR fine mode data (Fig.2). The master data was acquired on 23 November 1999, the slave 1 data was acquired on 23 December 2003, while slave 2 data acquisition was on 26 March 2005, respectively. The master data was ascending while both slave data were descending. Figure 2 shows the variation of the backscatter intensity for the F1 mode data along Terengganu's estuary. The urban areas have the highest backscatter of-10 dB as compared to water body and the vegetation area (Fig. 2).

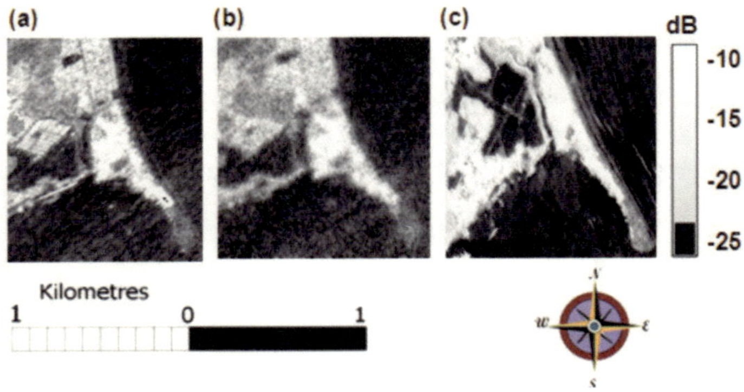

Fig. 2. RADARSAT-1 SAR fine mode data acquisition (a) master data, (b) data slave 1 data and (c) slave 2 data.

It is interesting to find the coherence image coincided with backscatter variation along the coastal zone. Fig. 3(a) shows that urban zone dominated with higher coherence of 0.8 than vegetation and sand areas. The coastal spit has lower backscatter and coherence of 0.3 dB and 0.25, respectively. Since three F1 mode data acquired in wet north-east monsoon period, there is an impact of wet sand on radar signal penetration which causing weak penetration of radar signal because of dielectric. Figure 3b shows the ratio coherence image, clearly the total topographic decorrelation effects along the radar-facing slopes are dominant and highlighted as bright features of 3 over a grey background. This is caused by the micro-scale movement of the sand particles driven by the coastal hydrodynamic, and wind continuously changes the distribution of scatterers resulting in rapid temporal decorrelation which has contributed to decorrelation in the spit zone.

Clearly, the random changes in the surface scatterer locations among data acquisitions with a wavelength of 5.6 cm for C-band are sufficient to decorrelate the interferometric signal. Under this circumstance, it will be visible in the coherence data (Fig.3). Since vegetation and wet sand changes may also reduce the coherence because the estuary area has tides and water lines that are so highly variable, this can be defined in fuzzy or probabilistic terms. The geomorphology feature of spit is rendered meaningless or unreliable in the long term because of their high variability. This confirms the studies of of Hanssen (2001); Marghany and Hashim (2009); and Rao and Jassar (2010); Marghany (2011).

Further, the estimated baseline is varied between master data and both slave data. The estimated baseline between master data and second slave data is 400 m which is larger than slave 1 data (Table 1). In this context, Gens (2000) reported the length of the baseline designates the sensitivity to height changes and sum of baseline decorrelation. Further, Nizalapur et al., (2011) stated the time difference for two data acquisitions is a second source of decorrelation. Indeed, the time difference while comparing data sets with a similar baseline length acquired one and 35 days a part suggest only the temporal component of

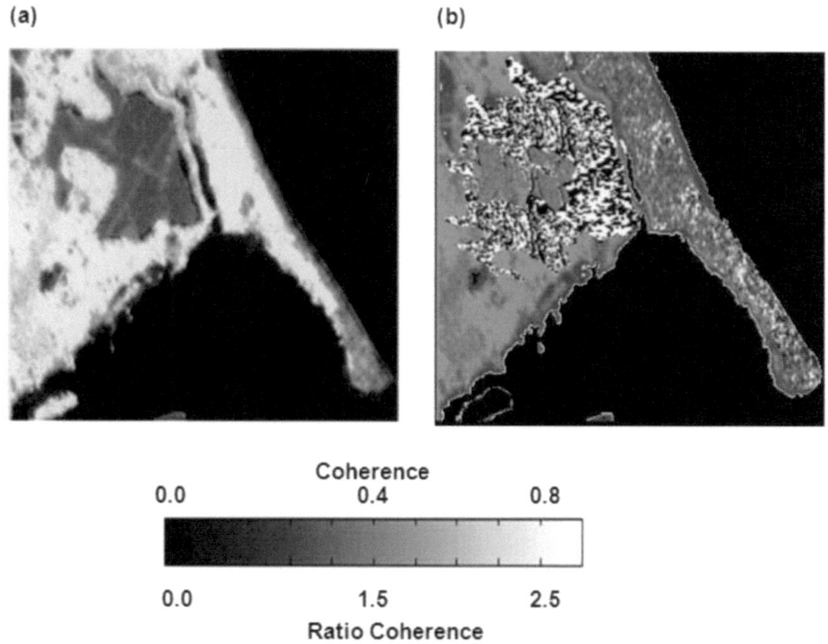

Fig. 3. Variation of (a) coherence and (b) ratio coherence in F1 mode data.

the decorrelation. Therefore, the loss of coherence in the same repeat cycle in data acquisition are most likely because of baseline decorrelation.

Acquisition Data	Baseline	Wind Speed (m/s)	Tidal (m)
23 November 1999	345	7.3	1.2
23 December 2003	266	9	1.5
26 March 2005	400	11	1.8

Table 1. Baseline estimations with wind and tidal conditions during acquisition time.

Evidently, wind speed of 11 m/s affects the scattering from certain vegetation classes and sandy regions and consequently produce poor coherence (Table 1). The overall scene is highly incoherent, not only because of the meteorological conditions and the vegetation cover at the time but also because of ocean surface turbulent changes. This decorelation caused poor detection of spit which induce large ambiguities because of poor coherence and scattering phenomenology. The ground ambiguity and ideal assumption that volume-only coherence can be acquired in at least one polarization. This assumption may fail when vegetation is thick, dense, or the penetration of electromagnetic wave is weak. This is agreed with study of Lee (2001).

Fig. 4 shows the interferogram created from F1 data. For three data sets, only small portion of the scene processed because of temporal decorrelation. According to Luo et al.,(2007), the SAR interferogram is considered to be difficult to unwrap because of its large areas of low coherence, which caused by temporal decorrelation. These areas of low coherence segment the interferogram into many pieces, which creates difficulties for the unwrapping algorithms (Fig.4). In this context, Lee (2001) reported that when creating an interferogram of surface deformation by using InSAR, it is not always true that an interference pattern (fringes) of an initial interferogram directly shows surface deformation. Indeed, the difference in phase between two observations is influenced by things outside surface deformation.

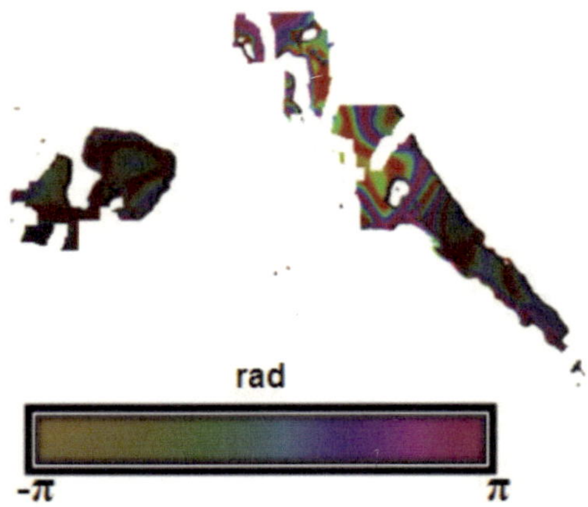

Fig. 4. Interfeorgram generated from F1 mode data.

Figure 5 shows the interferogram created using fuzzy B-spline algorithm. The full color cycle represents a phase cycle, covering range between -π to π. In this context, the phase difference given module 2 π; is color encoded in the fringes. Seemingly, the color bands change in the reverse order, indicating that the center has a great deformation along the spit. This shift corresponds to 0.4 centimetres (cm) of coastal deformation over the distance of 500 m. The urban area dominated by deformation of 2.8 cm.

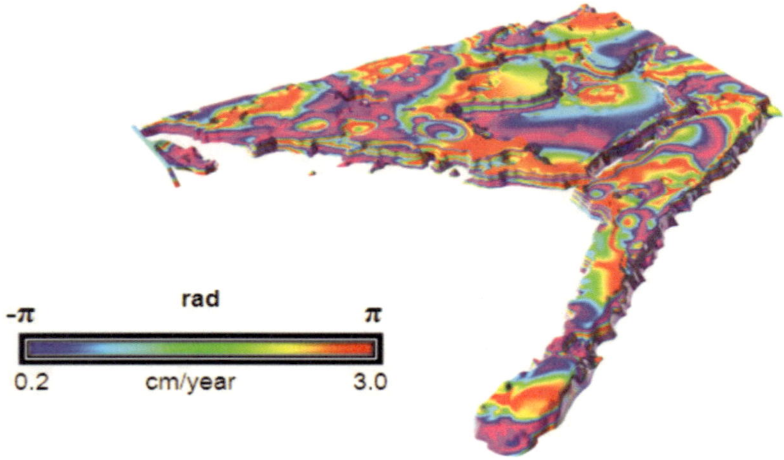

Fig. 5. Fringe Interferometry generated by fuzzy B-spline.

Fig. 6 represents 3-D spit reconstruction using fuzzy B-spline with the maximum spit 's elevation is 3 m with gentle slope of 0.86 m. The rate change of spit is 3 m/year with maximum elevation height of 2.4 m (Figs 5 and 6). Clearly, Terengganu 's spit was generated due to the deposition of sediment due to hydrodynamic changes between estuary and ocean. According to Marghany (2012) Terengganu 's mouth river is the largest hydrologic communication between an estuary and the South China Sea. This spit occurred when longshore drift reaches a section of Terengganu's River where the turn is greater than 30 degrees. It continued out therefore, into the sea until water pressure from a Terengganu 's River becomes too much to allow the sand to deposit. The spit be then grown upon and become stable and often fertile. As spit grows, the water behind them is sheltered from wind and waves. This could be due the high sediment transport through the water outflow from the river mouth, or northerly net sediment transport due to northeast monsoon wave effects (Marghany et al., 2010b). Longshore drift (also called littoral drift) occurs due to waves meeting the beach at an oblique angle, and back washing perpendicular to the shore, moving sediment down the beach in a zigzag pattern. Longshore drifting is complemented by longshore currents, which transport sediment through the water alongside the beach. These currents are set in motion by the same oblique angle of entering waves that cause littoral drift and transport sediment in a similar process. The hydrodynamic interaction between the longshore current and water inflow from the Terengganu Mouth River is causing the changes in spit's geomorphology characteristics. This finding confirms the study of Marghany et al., (2010a) ; Marghany et al., (2010b) and Marghany (2011). The increasing growth of spit across the estuary is due to impact of sedimentation due to littoral drift. According to Marghany and Mazlan (2010a) net littoral drift along Kuala Terengganu coastal water is towards the southward which could induce the growth of spit length.

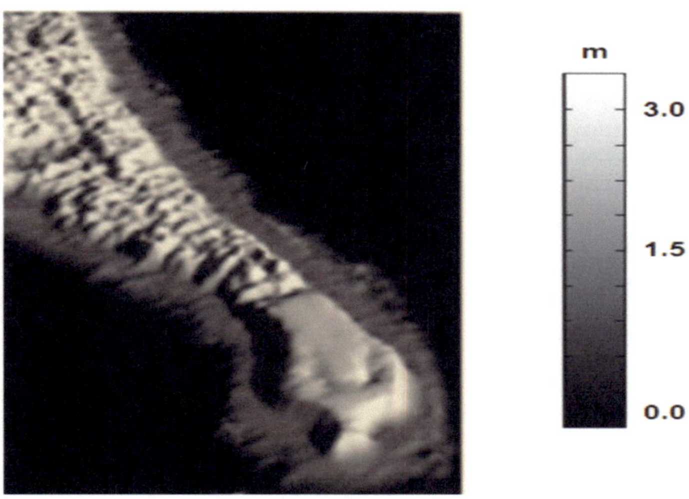

Fig. 6. DEM of coastal spit.

Finally, a difference statistical comparison confirms the results of Figs 4,5 and 6. Table 2 shows the statistical comparison between the simulated DEM from the DInSAR, real ground measurements and with using fuzzy B-spline. This table represents the bias (averages mean the standard error, 90 and 95% confidence intervals, respectively. Evidently, the DInSAR using fuzzy B-spline performance has bias of-0.05 m, lower than ground measurements and the DInSAR method. Therefore, fuzzy B-spline has a standard error of mean of ± 0.034 m, lower than ground measurements and the DInSAR method. Overall performances of DInSAR method using fuzzy B-spline is better than DInSAR technique which is validated by a lower range of error (0.02±0.21 m) with 90% confidence intervals.

Statistical Parameters	DInSAR techniques			
	DInSAR		DInSAR-Fuzzy B-spline	
Bias	2.5		-0.05	
Standard error of the mean	1.5		0.034	
	Lower	**Upper**	**Lower**	**Upper**
90 % (90 % confidence interval)	1.2	2.6	0.02	0.16
95 % (95 % confidence interval)	0.98	2.35	0.03	0.21

Table 2. Statistical Comparison between DInSAR and DInSAR-Fuzzy B-spline Techniques.

Fuzzy B-spline produced perfect pattern of fringe interfeormetry compared with one produced by DInSAR technique (Fig. 5). It shows there are many deformations of over several centimetres. In these deformations, it is known the deformation in spit because of coastline sedimentation. The other deformations, however, are caused not by the movement of the coastal sediment but the spatial fluctuation of water vapour in the atmosphere. In addition, the growths of urban area induces also land cover changes. Further, it can be noticed that fuzzy B-spline preserves detailed edges with discernible fringes (Russo 1998 and Rövid et al. 2004). Indeed, Fig 5. shows smooth interferogram, in terms of spatial resolution maintenance, and noise reduction, compared to conventional methods (Zebker et al.,1997; Massonnet, and Feigl 1998; Burgmann et al., 2000; Hanssen 2001; Yang et al., 2007; Rao and Jassar (2010).

This has been contributed since each operation on a fuzzy number becomes a sequence of corresponding operations on the respective μ -levels , and the multiple occurrences of the same fuzzy parameters evaluated because of the function on fuzzy variables (Anile et L., 1995 and Anile et al., 2000). It is easy to distinguish between small and long fringes. Typically, in computer graphics, two objective quality definitions for fuzzy B-spline were used: triangle-based criteria and edge-based criteria. Triangle-based criteria follow the rule of maximization or minimization, respectively, of the angles of each triangle (Fuchs et al. 1997). The so-called max-min angle criterion prefers short triangles with obtuse angles. This result agrees confirms the studies of Anile et al. (2000); Marghany et al.,(2010a); and Marghany (2011).

6. Conclusions

Synthetic Aperture Radar interferometry (InSAR) is a relatively new technique for 3-D topography mapping. This study presents a new approach for 3-D object simulation using Differential synthetic aperture interferometry (DInSAR). This work has demonstrated the 3-D spit reconstruction from DInSAR using three C-band SAR images acquired by RADARSAT-1 SAR F 1 mode data. The conventional method of DInSAR used to create 3-D coastal geomorphology reconstruction. Nevertheless, it was difficult to generate phase and interfeorgram using conventional DInSAR because of temporal decorrelation. The result shows that spit and vegetation zone have poor coherence of 0.25 as compared to the urban area. In addition, only small portion of the F1 mode scene was processed because of temporal decorrelation. Finally, the fuzzy B-spline algorithm used to reconstruct fringe pattern, and 3-D from decorrelate unwrap phase. The fringe pattern shows the deformation of 0.4 cm along spit and 1.4 cm in urban area. Further, the maximum 3-D spit elevation is 3 m with the standard error of mean of ± 0.034 m. In conclusion, the integration between the conventional DInSAR method and the FBSs could be an excellent tool for 3-D coastal geomorphology reconstruction from SAR data the under circumstance of temporal decorrelation.

7. References

Anile, A.M., B. Falcidieno, G. Gallo, M. Spagnuolo, S. Spinello, (2000). "Modeling uncertain data with fuzzy B-splines", Fuzzy Sets and Syst. 113, 397-410.

Anile, AM, Deodato, S, Privitera, G, (1995) *Implementing fuzzy arithmetic*, Fuzzy Sets and Systems, 72,123-156.

Askne, J., M. Santoro, G. Smith, and J. E. S. Fransson (2003). "Multitemporal repeat-pass SAR interferometry of boreal forests," *IEEE Trans. Geosci. Remote Sens.* 41, 1540–1550.

Burgmann, R., P.A. Rosen, and E.J. Fielding (2000). "Synthetic aperture radar interferometry to measure Earth's surface topography and its deformation", *Ann. Rev.of Earth and Plan. Sci.* 28: 169–209.

Fuchs, H. Z.M. Kedem, and Uselton, S.P., (1977). Optimal Surface Reconstruction from Planar Contours. *Communications of the ACM*, 20, 693-702.

Gens,R., (2000)."The influence of input parameters on SAR interferometric processing and its implication on the calibration of SAR interferometric data", *Int. J. Remote Sens.* 2,11767–1771.

Hanssen R.F., (2001). *Radar Interferometry: Data Interpretation and Error Analysis*, Kluwer Academic, Dordrecht, Boston.

Lee H., (2001). "Interferometric Synthetic Aperture Radar Coherence Imagery for Land Surface Change Detection" Ph.D theses, University of London.

Luo, X., F.Huang, and G. Liu, (2006). "Extraction co-seismic Deformation of Bam earthquake with Differential SAR Interferometry". *J. New Zea. Inst. of Surv.* 296:20-23.

Massonnet, D. and K. L. Feigl (1998).,"Radar interferometry and its application to changes in the earth's surface,"*Rev. Geophys.* 36, 441–500 .

Marghany M (2012). 3-D Coastal Bathymetry Simulation from Airborne TOPSAR Polarized Data. "*Bathymetry and Its Applications*". In Ed., Blondel P., InTech - Open Access Publisher, University Campus STeP Ri, Croatia, 57-76.

Marghany M (2011).Three-dimensional visualisation of coastal geomorphology using fuzzy B-spline of dinsar technique. Int. J. of the Phys. Sci. 6(30):6967 - 6971.

Marghany,M., M. Hashim and A. P. Cracknell, (2010a)."3-D visualizations of coastal bathymetry by utilization of airborne TOPSAR polarized data". *Int. J. of Dig. Earth*, 3,187 - 206.

Marghany, M., and M. Hashim (2009)."Differential Synthetic Aperture radar Interferometry (DInSAR) for 3D Coastal Geomorphology Reconstruction". IJCSNS Int. J. of Comp. Sci. and Network Secu., 9,59-63.

Marghany, M., and M. Hashim (2010a)."Different polarised topographic synthetic aperture radar (TOPSAR) bands for shoreline change mapping. *Int. J. Phys. Sci. 5,* 1883-1889.

Marghany, M.,Z. Sabu and M. Hashim,(2010b) "Mapping coastal geomorphology changes using synthetic aperture radar data". *Int. J. Phys. Sci. 5,*1890-1896.

Marghany M., and M. Hashim (2010b). "Velocity bunching and Canny algorithms for modelling shoreline change rate from synthetic aperture radar (SAR). *Int. J. Phys. Sci. 5,*1908-1914.

Nizalapur, V., R. Madugundu, and C. Shekhar Jha (2011). "Coherence-based land cover classification in forested areas of Chattisgarh, Central India, using environmental satellite—advanced synthetic aperture radar data", *J. Appl.Remote Sens.* 5, 059501-1-059501-6.

RADARSAT International, (2011)"RADARSAT application [online] Available from http:\www.rsi.ca [Accessed 8 June 2011].

Rao, K.S., H. K. Al Jassar, S. Phalke, Y. S. Rao, J. P. Muller, and. Z. Li, (2006). "A study on the applicability of repeat pass SAR interferometry for generating DEMs over several Indian test sites," *Int. J. Remote Sens.* 27, 595–616.

Rao, K.S.,and H. K. Al Jassar (2010). "Error analysis in the digital elevation model of Kuwait desert derived from repeat pass synthetic aperture radar interferometry", *J. Appl.Remote Sens.*4,1-24.

Russo, F., (1998).Recent advances in fuzzy techniques for image enhancement. IEEE Transactions on Instrumentation and measurement. 47, pp: 1428-1434.

Rövid, A., Várkonyi, A.R. andVárlaki, P., (2004). 3D Model estimation from multiple images," IEEE International Conference on Fuzzy Systems, FUZZ-IEEE'2004, July 25-29, 2004, Budapest, Hungary, pp. 1661-1666.

Yang, J., T.Xiong, and Y. Peng, (2007). "A fuzzy Approach to Filtering Interferometric SAR Data". *Int. J. of Remote Sens.*, 28, 1375-1382.

Zebker, H.A., C.L.,Werner, P.A. Rosen, and S. Hensley, (1994). "Accuracy of Topographic Maps Derived from ERS-1 Interferometric Radar", *IEEE Geosci. Remote Sens.*,2,823-836.

Zebker, H.A., P.A. Rosen, and S. Hensley (1997)."Atmospheric effects in inteferometric synthetic aperture radar surface deformation and topographic maps", *J. Geophys. Res.*102, 7547–7563.

5

Local, Fine Co-Registration of SAR Interferometry Using the Number of Singular Points for the Evaluation

Ryo Natsuaki and Akira Hirose
Department of Electrical Engineering and Information Systems, The University of Tokyo
Japan

1. Introduction

Synthetic aperture radar (SAR) has great advantage of being able to observe large area accurately in any weather. It can measure various properties of the earth Boerner (2003), e.g., the topography, the vegetation Hajinsek et al. (2009), and the landscape changes caused by earthquakes or volcanoes Gabriel et al. (1989). One of the common usages of SAR is Interferometric synthetic aperture radar (InSAR), which can measure the landscape with an interferogram of SAR images. The SAR interferogram is made from two complex-valued maps obtained by observing identical place, named "master" and "slave." The phase information of the interferogram corresponds to the ground topography. To generate digital elevation model (DEM) from interferogram, phase unwrapping process is required. However, an unwrapping process is disturbed seriously by singular points (SPs), the rotational points existing in the phase map.

Most of the SPs hinder us from creating accurate DEM Ghiglia & Pritt (1998). Many researchers have tried to solve this problem by proposing novel methods concerning branch-cut techniques Costantini (1998), least squares Pritt & Shipman (1994), and singularity spreading technique Yamaki & Hirose (2007). These major methods have numerous efficient improvements e.g., Fornaro et al. (1996) Reigber & Moreia (1997) Suksmono & Hirose (2006). Various SP elimination filters have also developed and applied to the interferogram. For example, Lee filter Lee et al. (1998), Goldstein filter Goldstein & Werner (1998), Bayesian filter Ferraiuolo & Poggi (2004), and Markov random field modeled filter Suksmono & Hirose (2002) Yamaki & Hirose (2009) are the popular.

There are three origins in the SP generation. One is the low SNR caused by low scattering reflectance. Another one is the sharp cliff and layover. The landscape can be so rough that the aliasing occurs. The last one is the local distortion in the co-registration of the master and the slave. That is, the reflection, scattering and fore-shortening can be different in the two observations with slightly different sight angle, resulting in local phase distortion in the interferogram. Filtering and unwrapping methods can solve first two origins. On the other hand, the last origin, the local distortion, has been generally ignored.

Without the distortion, no SP is expected through an appropriate co-registration of non-aliasing master and slave maps. Usually the co-registration is realized by maximizing the amplitude cross-correlation of the maps in macro scale, while by maximizing the complex-amplitude correlation in micro scale. The correlations require an averaging process over a certain area for sufficient reduction of included noise. However, a wide-area averaging degrades the locality in the matching required to eliminate the distortion. This trade-off brings a limitation in the co-registration performance. In short, the difference in the reflection, scattering and fore-shortening yields local distortion, and SPs are generated inevitably by the cross-correlation process.

In this chapter, we firstly introduce the basis of InSAR and its SP problem. Secondly, we introduce a local and fine co-registration method which employs the number of SPs as evaluation criterion (SPEC method Natsuaki & Hirose (2011)). Finally, we demonstrate the effectiveness of the improvement by comparing the DEMs generated from interferograms which co-registered with and without the SPEC method. For experiment, we use the data of Mt. Fuji observed by JERS-1 which was launched by JAXA (Japan Aerospace Exploration Agency). Mt. Fuji has an ordinary single volcanic cone shape.

2. InSAR and SP problem

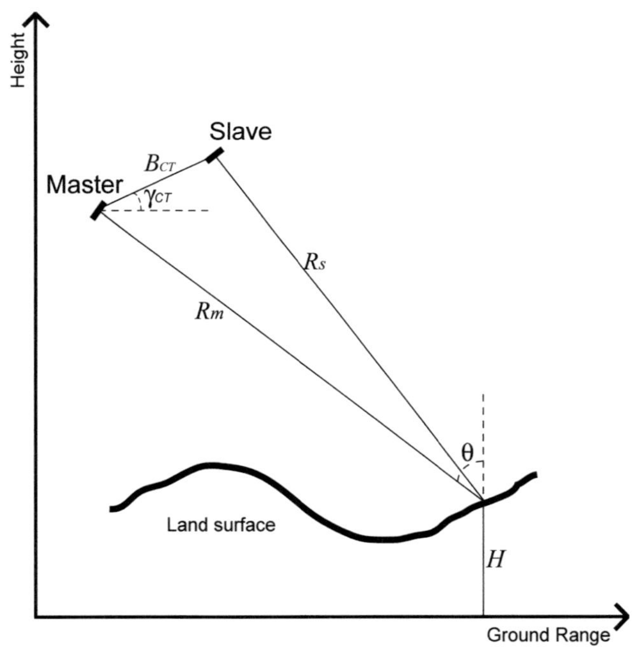

Fig. 1. Schematic diagram of InSAR

Figure 1 shows the observation system of InSAR. We define wave length as λ, elevation angle as θ, distances from ground to the master and the slave satellite as R_m and R_s, distance between the master and the slave as B_{CT}, relative angle of the master and the slave as γ_{CT}.

The phase value Φ of the interferogram corresponds to R_m and R_s as

$$\frac{\lambda\Phi}{2\pi} = 2\left(R_m - R_s\right) \tag{1}$$

Geometrically, $R_m - R_s$ corresponds to B_{CT} and γ_{CT} as

$$\left(R_m - R_s\right) = B_{CT}\sin\left(\theta - \gamma_{CT}\right) \tag{2}$$

From (1) and (2), the relationship between Φ and θ is expressed as

$$\frac{\lambda\Phi}{2\pi} = 2B_{CT}\sin\left(\theta - \gamma_{CT}\right) \tag{3}$$

If there is a height increment between the neighbor pixels, the increment of Φ is

$$\frac{\Delta\Phi}{\Delta\theta} = \frac{4\pi B_{CT}\cos(\theta - \gamma_{CT})}{\lambda} \tag{4}$$

and the increment of H is

$$\frac{\Delta H}{\Delta\theta} = R_m\sin(\theta) \tag{5}$$

From (4) and (5), the relationship between the interferogram phase increment $\Delta\Phi$ and the height increment of the observation point ΔH can be expressed as

$$\Delta H = \frac{\lambda R_m\sin(\theta)}{4\pi B_{CT}\cos(\theta - \gamma_{CT})}\Delta\Phi \tag{6}$$

which indicates that the phase gradient of 2π corresponds to the height gradient of $\frac{\lambda R_m\sin(\theta)}{2B_{CT}\cos(\theta-\gamma_{CT})}$.

In order to analyze the ground topography, we have to unwrap, line integrate, the phase information. As the ground topography is the conservative field, its contour integral should be zero.

$$\oint_c \Delta\Phi ds' = 0 \tag{7}$$

However, there are many non-zero rotational points, namely singular points (SPs), in the interferogram. As shown in Fig.2, if there is a rotational point in the interferogram, phase unwrapping will fail. We assume that there are three origins of the SP emergence.

1. Low SNR caused by low scattering reflectance
2. Sharp cliff and layover
3. Local distortion in the co-registration of the master and the slave

Origin (i) is generally thought as the main reason of SPs which should be erased. SPs generated by origin (ii) should remain. The third reason (iii) has been conventionally ignored. A pixel in the master representing a small local area should completely correspond to a pixel in the slave that represents the same area. However, we assume that the slight difference of the radar incidence direction between the master and the slave distorts this correspondence

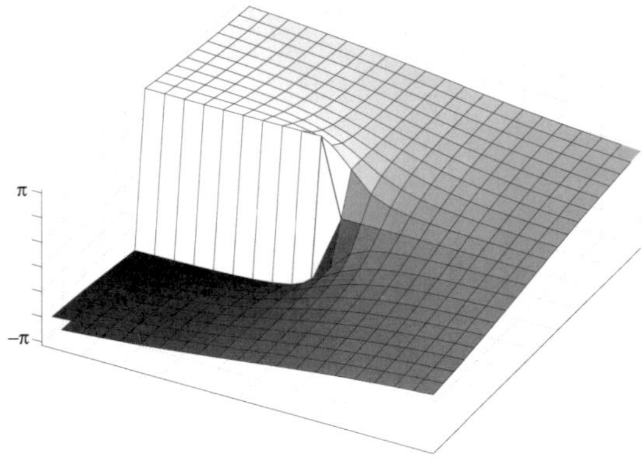

Fig. 2. Failure of unwrapping due to the existence of the SP.

in sub-pixel order, and that interferograms show these local distortions as massive SPs. In the next section, we introduce the local-o-registration method to solve the local distortion.

3. Singular points as evaluation criterion

3.1 Typical co-registration technique and the changes of SP distributions with subpixel shifts

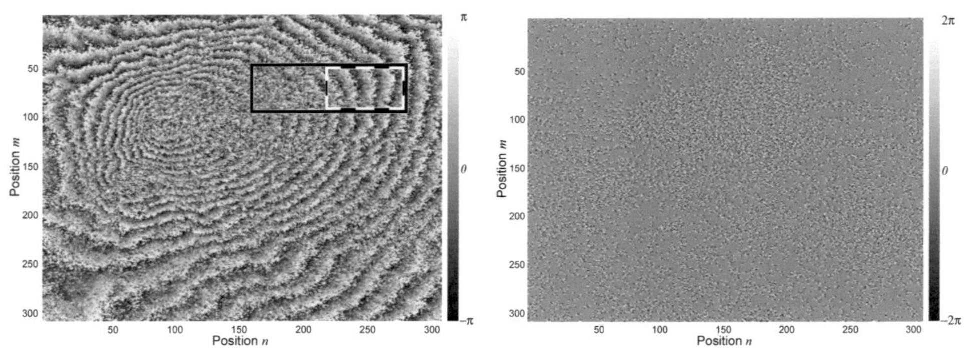

Fig. 3. (a)Interferogram of Mt. Fuji (black square corresponds to Fig.8.) and (b)its SP plot made by the typical method (# SPs = 11,518).

To create an interferogram with master and slave, we have to co-register them in advance as they observe the same place from slightly different angle. The typical co-registration process is explained as follows Tobita et al. (1999). First, we affine-transform the slave map adaptively to maximize the cross correlation between the master and slave amplitude maps in a macro scale, e.g., 64×64 pixels. Next, we maximize the complex-valued cross correlation locally in 1/32 subpixels with interpolation, e.g., 8×8 pixels. Figure 3(a) shows the interferogram of Mt. Fuji created by this method, and Fig.3(b) gives its SP plot. This interferogram has 304×304 pixels and contains 11,518 SPs. In Fig.3(a), the phase value is shown in gray scale, in which a white dot stands for π as the phase value and a black dot means $-\pi$. In all figures in this article, the up-down direction is the azimuth and the left-right direction is the range. In Fig.3(b), a white dot indicates a clockwise SP, while a black dot shows a counterclockwise one.

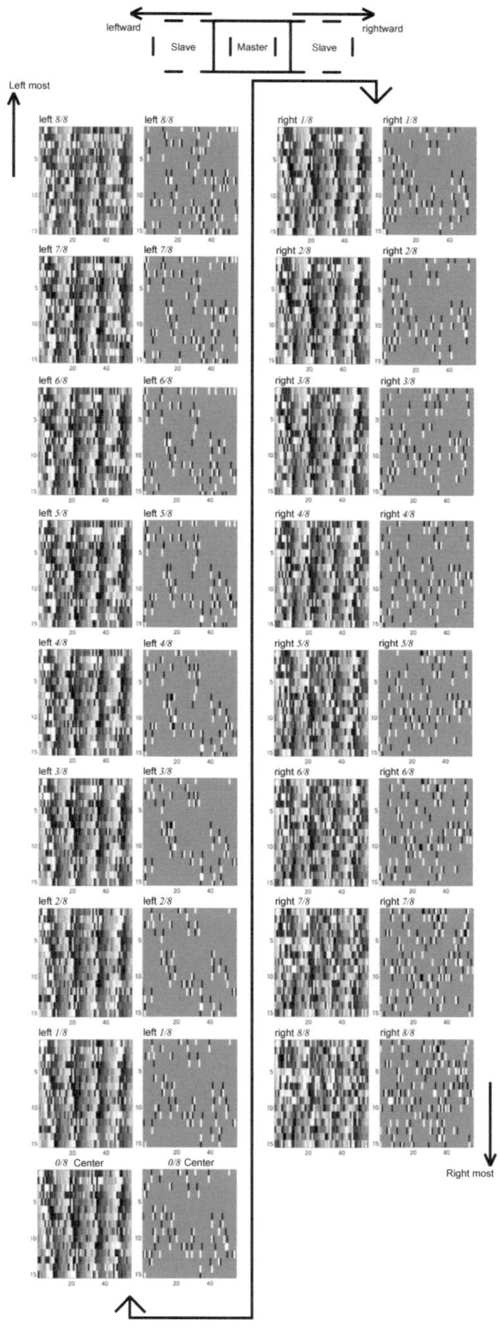

Fig. 4. Interferograms and SP plots obtained by moving the slave map from the position of maximum cross correlation leftward or rightward by integral multiple of 1/8 subpixels for the area of the right half of the black square in Fig.3(a).

Figure 4 represents how the interferogram and its SP plot change when the slave shifts leftward or rightward by integral multiple of 1/8 pixel from the maximum cross-correlation position for the right half area in the black square in Fig.3(a). It is obvious that many SPs move, emerge, or disappear, but the fringes in the interferogram show only small changes.

The emergence and the disappearance occur rather locally than the scale of correlation calculation. This fact suggests that we need a more local, and consequently nonlinear, co-registration process in addition to the conventional one. The changes in the fringes is not so large, which means that the rough landscape is not changed by this additional process, but the local precise co-registration is expected to improve the local landscape since, basically, no SPs are expected in non-distorted interferogram.

The aim of our proposal is to improve the accuracy of DEM by removing the local distortion. In this removal process, we pay attention to the number of SPs as explained in the following section. Based on this idea which came from the result of the above preliminary experiment, we introduce our new method below.

3.2 Proposal of the SPEC method

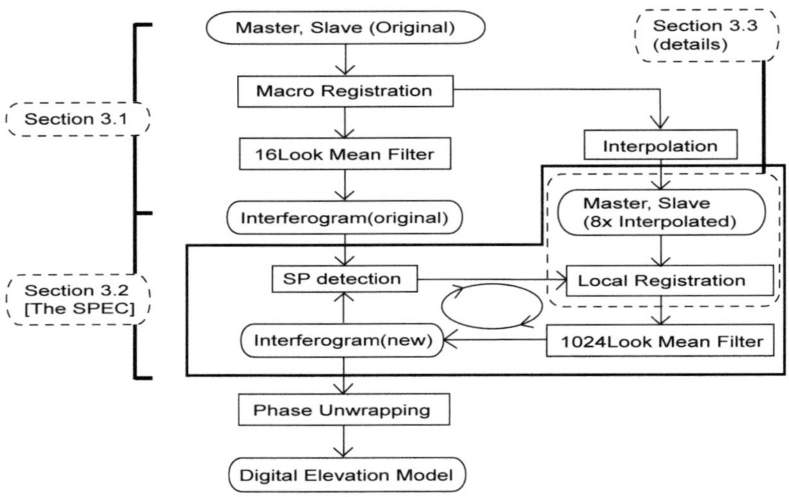

Fig. 5. Flowchart of the whole co-registration process including the SPEC method.

Figure 5 is the flowchart of the whole co-registration process including our SPEC method. First, we co-register master M and slave S by the conventional method as explained in Section 3.1. Then we make an interferogram I from master M and slave S with 16-look mean-filtering as

$$I(m,n) = \frac{1}{16} \sum_{q=1}^{8} \sum_{p=1}^{2} M(p,q;m,n) S^*(p,q;m,n). \tag{8}$$

Fig. 6(a) explains the process. We regard the 8×2 pixels in the master and the slave as single blocks, $\mathbf{M}(m,n)$ and $\mathbf{S}(m,n)$, respectively, each of which pair yields one pixel in 16-look interferogram shown in Fig. 6(a). $M(p,q;m,n)$ represents the p-th top and q-th left pixel in the block $\mathbf{M}(m,n)$. The 16-look mean-filtering works to decrease the noise by averaging

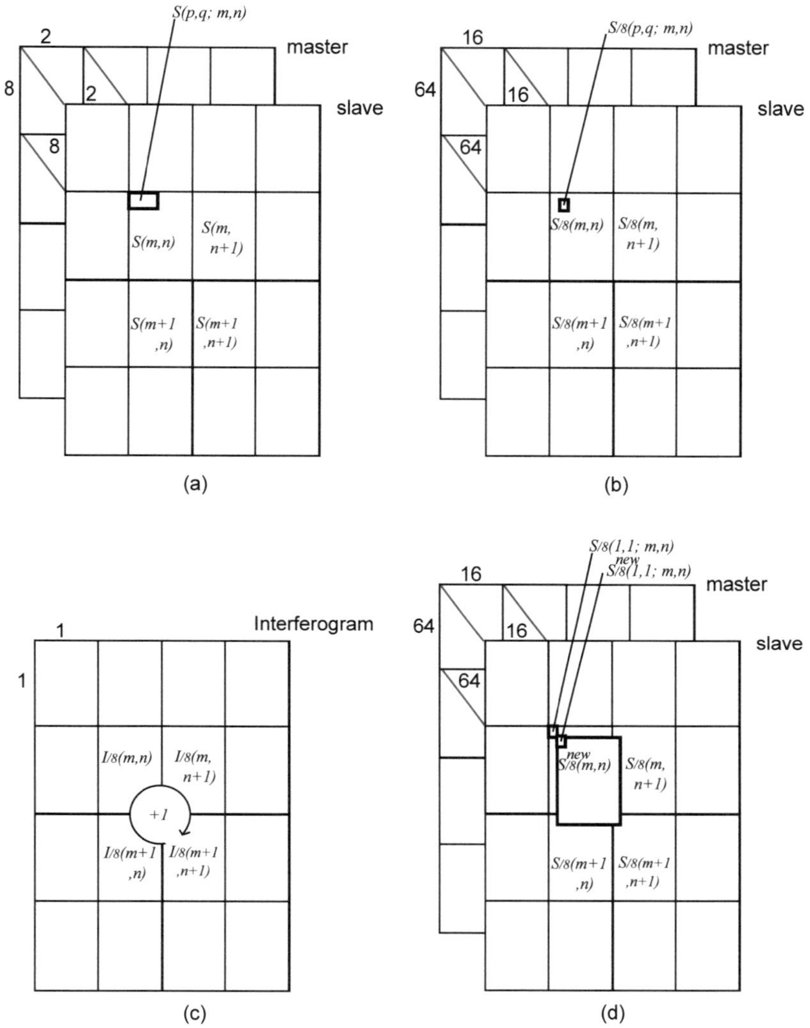

Fig. 6. Relationship between the blocks and the interferogram pixels: (a)Blocks for making regular 16-look mean-filtered interferogram in 1 pixel coordinate system, (b)blocks equal to (a) in 1/8-pixel coordinate system, (c)a SP in the interferogram made in 1/8-pixel coordinate system, and (d)local movement of the interpolated slave to delete the SP.

16 (=8×2) interferogram pixels. That is, to make an interferogram, in this paper, 8 times azimuth compression and 2 times range compression are required (i.e., to make a 304×304 pixels interferogram, 2432×608 pixels master and slave are required). Next, we find SPs in it and co-register interpolated master with interpolated slave locally and nonlinearly as follows.

3.3 Details of the local and nonlinear co-registration based on the number of SPs

Figures 6 and 7 are schematic diagrams of our local and nonlinear co-registration based on the number of SPs. We call the 8-times interpolated master and slave maps as "1/8-pixel

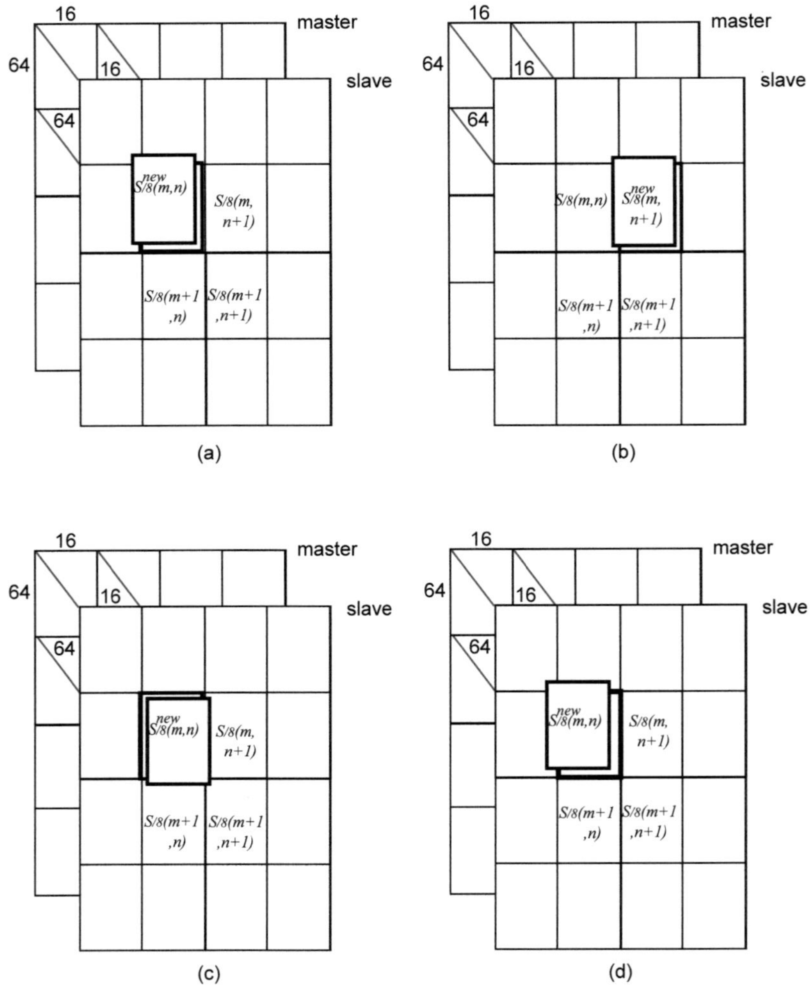

Fig. 7. The process of moving the slave: (a)Move $\mathbf{S}_{/8}(m,n)$ by 1/8 pixel as $S^{\text{new}}_{/8}(p,q;m,n) \leftarrow S_{/8}(p-1,q-1;m,n)$ in order to erase the SP made by the 4 blocks, (b)replace back $S^{\text{new}}_{/8}(p,q;m,n)$ and do the same process to $\mathbf{S}_{/8}(m,n+1)$, (c)move $S^{\text{new}}_{/8}(p,q;m,n)$ by 1/8 pixel to a different direction, and (d)move $S^{\text{new}}_{/8}(p,q;m,n)$ by 2/8 pixels.

coordinate-system" maps, and the original ones as "1-pixel coordinate-system" maps. Table 1 lists the definitions used here, in which $I_{/8}(m,n)$ denotes the pixel value of interferogram in the 1/8-pixel coordinate system, and $I(m,n)$ denotes the value in the 1-pixel coordinate system simply. Coordinate (m,n) stands for global position in the (16-look) mean-filtered interferogram, while (p,q) represents the local position in the 1/8-pixel coordinate system.

To get $I_{/8}(m,n)$, on the other hand, we need $(8 \times 8) \times (2 \times 8) = 1024$ subpixels. Hence we regard these 1024 pixels as single blocks, $\mathbf{M}_{/8}(m,n)$ and $\mathbf{S}_{/8}(m,n)$, respectively.

Map	Status	Notation	Unit
interferogram	16-look mean-filtered	$I(m,n)$	mean-filtered pixel
interferogram	1024-look mean-filtered in 8-times interpolation	$I_{/8}(m,n)$	mean-filtered pixel
original master, slave	single look	$M(p,q;m,n)$ $S(p,q;m,n)$	pixel
8×2 pixel block in master, slave	single look	$\mathbf{M}(m,n)$ $\mathbf{S}(m,n)$	block
8-times interpolated master, slave	8-times interpolated	$M_{/8}(p,q;m,n)$ $S_{/8}(p,q;m,n)$	subpixel
8-times interpolated block	8-times interpolated	$\mathbf{M}_{/8}(m,n)$ $\mathbf{S}_{/8}(m,n)$	block

Table 1. Map status and notation

We first make an 1/8-pixel coordinate-system interferogram locally. The pixel value $I_{/8}(m,n)$, which corresponds to a pixel of a 16-look mean-filtered interferogram $I(m,n)$, is calculated as

$$I_{/8}(m,n) = \frac{1}{1024} \sum_{q=1}^{64} \sum_{p=1}^{16} M_{/8}(p,q;m,n) S^*_{/8}(p,q;m,n) \tag{9}$$

where $M_{/8}(p,q;m,n)$ represents the p-th top and q-th left pixel in the block $\mathbf{M}_{/8}(m,n)$, while $S_{/8}(p,q;m,n)$ represents the p-th and q-th pixel in $\mathbf{S}_{/8}(m,n)$ in the same manner. For example, if there is a SP at $I_{/8}(m,n), I_{/8}(m+1,n), I_{/8}(m,n+1), I_{/8}(m+1,n+1)$ in the interferogram, as shown in Fig.6(c), we find non-zero rotation as

$$\begin{aligned} &\big|(\theta_{/8}(m+1,n) - \theta_{/8}(m,n)) + (\theta_{/8}(m,n) - \theta_{/8}(m,n+1)) + \\ &(\theta_{/8}(m+1,n+1) - \theta_{/8}(m+1,n)) + (\theta_{/8}(m+1,n+1) - \theta_{/8}(m,n+1))\big| \geq 2\pi \end{aligned} \tag{10}$$

where

$$\theta_{/8}(m,n) \equiv \arg(I_{/8}(m,n)). \tag{11}$$

Simultaneously, in this square $I_{/8}(m,n)$—$I_{/8}(m+1,n+1)$, there is local distortion in the master and / or the slave. Then we therefore move one or some of the blocks among the four $\mathbf{S}_{/8}(m,n)$—$\mathbf{S}_{/8}(m+1,n+1)$ blocks to modify the interferogram. For example, we move $\mathbf{S}_{/8}(m,n)$ locally in such a manner that $S^{\text{new}}_{/8}(p,q;m,n) \leftarrow S_{/8}(p+1,q+1;m,n)$ to try to erase the SP as shown in Fig.6(d). That is, in this particular case, we shift the pixels as

$$\begin{pmatrix} S^{\text{new}}_{/8}(1,1;m,n) & S^{\text{new}}_{/8}(1,2;m,n) & \ldots & S^{\text{new}}_{/8}(1,16;m,n) \\ S^{\text{new}}_{/8}(2,1;m,n) & S^{\text{new}}_{/8}(2,2;m,n) & \ldots & S^{\text{new}}_{/8}(2,16;m,n) \\ \vdots & \vdots & \ddots & \vdots \\ S^{\text{new}}_{/8}(64,1;m,n) & S^{\text{new}}_{/8}(64,2;m,n) & \ldots & S^{\text{new}}_{/8}(64,16;m,n) \end{pmatrix}$$

$$\leftarrow \begin{pmatrix} S_{/8}(1,1;m,n) & S_{/8}(1,1;m,n) & \dots & S_{/8}(1,16;m,n) \\ S_{/8}(2,1;m,n) & S_{/8}(1,1;m,n) & \dots & S_{/8}(1,15;m,n) \\ \vdots & \vdots & \ddots & \vdots \\ S_{/8}(64,1;m,n) & S_{/8}(63,1;m,n) & \dots & S_{/8}(63,15;m,n) \end{pmatrix} \tag{12}$$

There are nesting stages in our method. First, we move the slave by 1/8 pixel and replace $\mathbf{S}_{/8}(m,n)$ as $S^{\text{new}}_{/8}(p,q;m,n) \leftarrow S_{/8}(p-1,q-1;m,n)$ as shown in Fig.7(a). Then we check whether the SP disappears with this operation or not. If it does not, we move $\mathbf{S}_{/8}(m,n+1), \mathbf{S}_{/8}(m+1,n)$, and $\mathbf{S}_{/8}(m+1,n+1)$ in the slave in turn, in the same direction (Fig.7(b)). If it is impossible to erase the SP with the above up-leftward movements, we employ other seven directions (Fig.7(c)). Then for the remaining SPs, we try 2/8 shifts in the same way (Fig.7(d)). If the SP cannot be removed with up to 8/8(=1) shifts, we abandon the elimination of the SP there, and try to erase the next one in the interferogram.

We apply the above process to all of the SPs in the interferogram iteratively. If we find that there is no erasable SP any more, we apply a similar shifting process for 4 (=2 × 2) big blocks. For example, we shift $\mathbf{S}_{/8}(m,n)$, $\mathbf{S}_{/8}(m,n+1)$, $\mathbf{S}_{/8}(m+1,n)$, and $\mathbf{S}_{/8}(m+1,n+1)$ simultaneously as a single large block.

$$\begin{pmatrix} S^{\text{new}}_{/8}(1,1;m,n) & S^{\text{new}}_{/8}(1,2;m,n) & \dots & S^{\text{new}}_{/8}(1,16;m,n+1) \\ S^{\text{new}}_{/8}(2,1;m,n) & S^{\text{new}}_{/8}(2,2;m,n) & \dots & S^{\text{new}}_{/8}(2,16;m,n+1) \\ \vdots & \vdots & \ddots & \vdots \\ S^{\text{new}}_{/8}(64,1;m+1,n) & S^{\text{new}}_{/8}(64,2;m+1,n) & \dots & S^{\text{new}}_{/8}(64,16;m+1,n+1) \end{pmatrix}$$

$$\leftarrow \begin{pmatrix} S_{/8}(1,1;m,n) & S_{/8}(1,1;m,n) & \dots & S_{/8}(1,16;m,n+1) \\ S_{/8}(2,1;m,n) & S_{/8}(1,1;m,n) & \dots & S_{/8}(1,15;m,n+1) \\ \vdots & \vdots & \ddots & \vdots \\ S_{/8}(64,1;m+1,n) & S_{/8}(63,1;m+1,n) & \dots & S_{/8}(63,15;m+1,n+1) \end{pmatrix} \tag{13}$$

We can also apply a 9 (=3 × 3) bigger block movement afterward, if needed.

4. Experimental results

Figure 8 presents the changes of the phase map and corresponding SP distributions when we apply the SPEC method. The shown area is the black-squared part in Fig.3(a). As shown in Figs.8(a) and (b), there were 1,014 SPs in the original interferogram. The first iteration in our proposed method erased more than 60 percent of them, resulting in 396 SPs (Figs.8(c) and (d)). In the second iteration, 324 points left (Figs.8(e) and (f)). For the present data, with 1-block movement, no SP was erased anymore. With the additional 4-block move, our proposed method decreased the SP number to 184 (Figs.8(g) and (h)). With the 9-block move, our method decreased the SP number to 171 (Figs.8(i) and (j)), where our method finally erased about 83% of the SPs in the original interferogram. Figures 9(a) and 9(b) show the resulting interferogram and its SP plot for the data in Fig.3(a). The number of the SPs decreased from 11,518 to 1,865. The decreasing ratio was about 83% again for the whole interferogram.

Figure 10 compares (a) the true height map and the results of the unwrapping by the iterative least-square (LS) techniqueSuksmono & Hirose (2006) (b) with and (c) without the SPEC

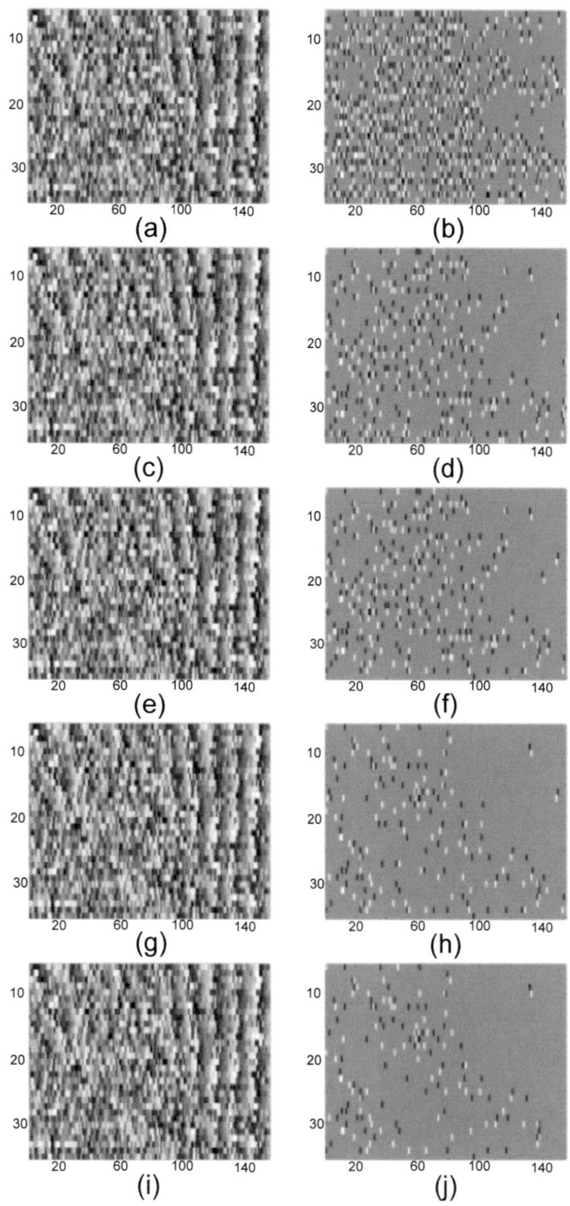

Fig. 8. Interferogram and its SP plot for the black-square area in Fig.3(a): (a)Interferogram before the local co-registration process and (b)its SP plot (1,014 SPs), (c)result of the first iteration with 1-block shifts and (d)its SP plot (396 SPs), (e)result of the second iteration with 1-block shifts and (f)its SP plot (324 SPs), (g)result of 4-block shifts in addition and (h)its SP plot (184 SPs), and (i)result of 9-block shifts in addition and (j)its SP plot (171 SPs).

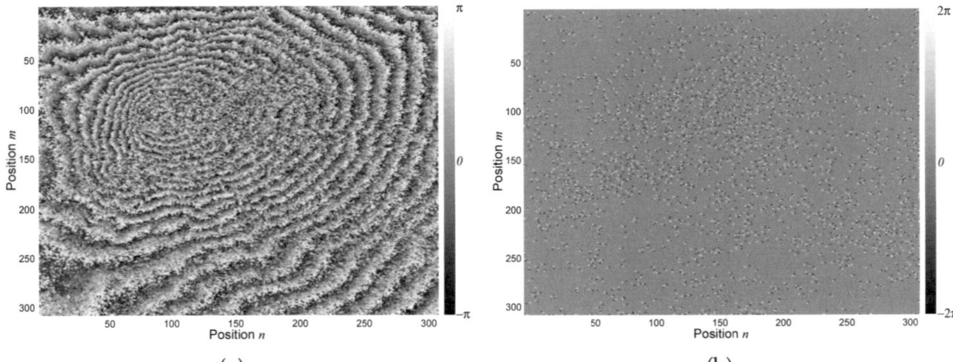

Fig. 9. (a)Interferogram made by the proposed SPEC method for the data shown in Fig.3(a) and (b)its SP plot (# SPs = 1,865).

method. We find improvement in some regions in the SPEC result in (b). For example, the dotted-square region in Fig.10(b) (zoomed in Fig. 11) shows more accurate ridges than those in Fig.10(c). It is obvious that the valleys and edges in Fig.11(b) are more distinct than those in Fig.11(c). We compared the mean signal-to-noise ratio (MSNR) ($\equiv$ squared height range / mean squared error) and the peak signal-to-noise ratio (PSNR) ($\equiv$ squared height range / peak squared error) based on the true data. The DEM with the SPEC method resulted in $MSNR = 29.5$[dB] and $PSNR = 14.5$[dB]. On the other hand, the DEM without our method resulted in $MSNR = 27.2$[dB] and $PSNR = 14.3$[dB]. The SPEC method improved the quality of the DEM both in average and at the peak.

We intend that our SPEC method compensates the local phase distortions in the interferogram. As this ability is similar to filtering (i.e., phase estimation), we calculated whether the SNR of the filtered interferogram changes if we co-register the interferogram with our proposed method. We used the iterative LS technique for unwrapping and the complex-valued Markov random field model (CMRF) filter Yamaki & Hirose (2009) for this evaluation. We compared the results for filtered interferograms of Mt. Fuji with those of non-filtered ones, which are co-registered with and without our SPEC method.

Filtering \ Co-registration		Without SPEC method	With SPEC method
Without filter	MSNR [dB]	27.2	29.5
	PSNR [dB]	14.3	14.5
With the CMRF filter	MSNR [dB]	35.8	34.9
	PSNR [dB]	19.2	19.7

Table 2. MSNR and PSNR of Mt. Fuji's DEMs unwrapped by the iterative LS

Table 2 shows the results of this experiment. The CMRF filter increased the SNRs of the DEMs of Mt. Fuji. Table 2 shows that the use of the SPEC method could not improve the MSNR of Mt. Fuji. The reason of this result is that the shape of the mountains and the phase ambiguity led to this result. Mt. Fuji has an ordinary single volcanic cone shape with clear fringes. Basically, phase estimation works good when the interferogram has clear fringes. As our SPEC method is an additional step of co-registration, it decreases the number of SPs and make fringes clearer

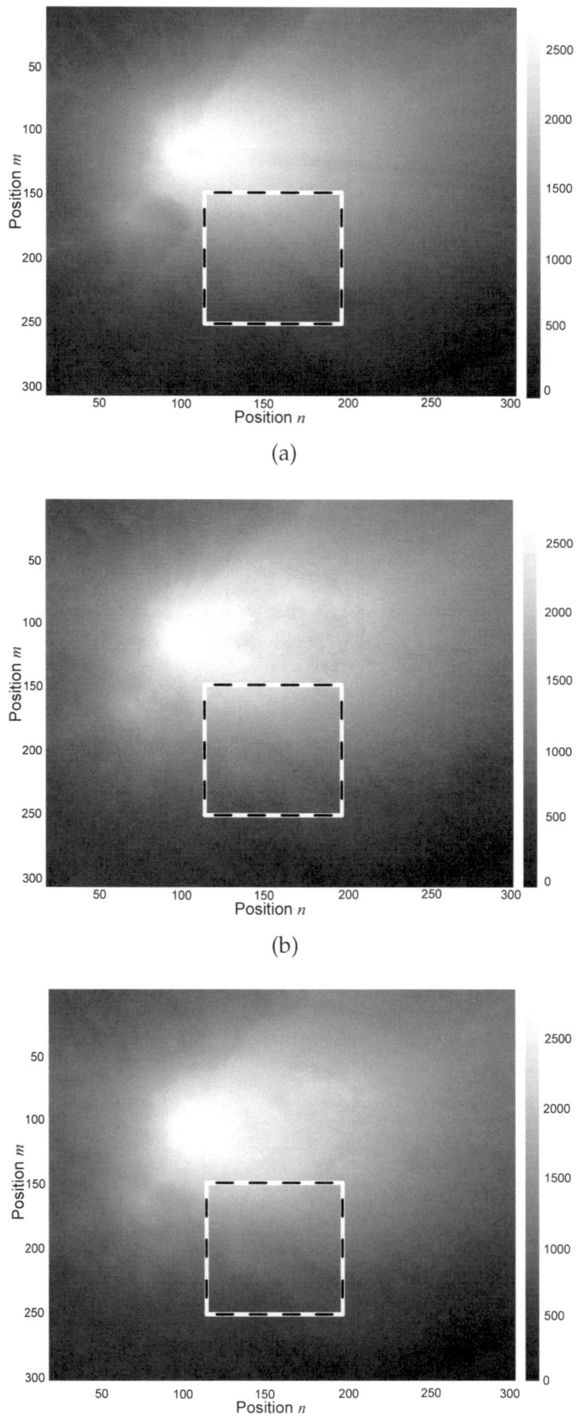

Fig. 10. (a)True height map and DEMs obtained by the iterative least square technique(b)with and (c)without the SPEC method (Mt. Fuji).

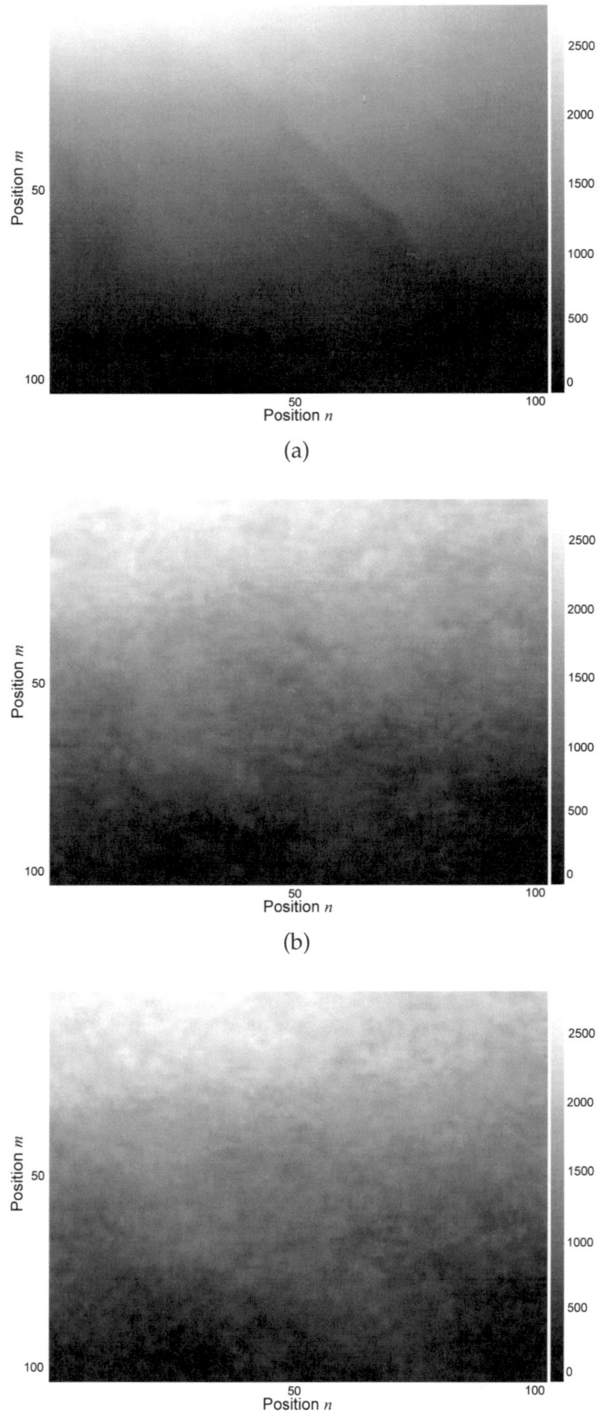

Fig. 11. Dotted square part of Fig.10. (a)True height map and DEMs obtained (b)with and (c)without the SPEC method.

before the CMRF filter works. That is why the CMRF filter could estimate the fringes of the interferogram more precisely.

5. Conclusion

We proposed a new method of co-registration, namely, the SPEC method. This method uses the number of SPs in the temporary interferogram as the evaluation criterion to co-register the master and slave maps locally and nonlinearly. By applying our method to real data, we found that the SPEC method successfully improves the quality of the DEM in many cases.At the same time, we found that the SPEC method make ambiguous fringes clearer in its appearance. Our present method uses only the number of singular points as the evaluation criterion. In the future, we have the possibility to use other information, in addition to the SP number, to improve the performance further.

6. Acknowledgment

The authors would like to thank Dr. M. Shimada of EORC/JAXA, Japan, for supplying the InSAR data and the height data for evaluation.

7. References

Boerner, W. M. (2003). "Recent advances in extra-wide-band polarimetry, interferometry and polarimetric interferometry in synthetic aperture remote sensing and its applications," *IEE Proceedings Radar, Sonar and Navigation*, Vol.150, No.3, pp.113-124, June 2003.

Costantini, M. (1998). "A Novel Phase Unwrapping Method Based on Network Programming," *IEEE Transactions on Geoscience and Remote Sensing*, vol. 36, no. 3, pp. 813-821, May 1998.

Ferraiuolo, G. & Poggi, G. (2004), "A Bayesian Filtering Technique for SAR Interferometric Phase Fields," *IEEE Transactions on Image Processing*, Vol.13, No.10, pp.1368-1378, October 2004.

Fornaro, G.; Franceschetti, G., & Lanari, G. (1996) "Interferometric SAR phase unwrapping using Green's formulation,"*IEEE Transactions on Geoscience and Remote Sensing*, vol. 34, no. 3, pp. 720-727, May 1996.

Gabriel, A. K.; Goldstein, R. M. & Zebker, H. A. (1989). "Mapping small elevation changes over large areas: Differential interferometry," Journal of Geophysical Research, vol. 94, pp. 9183-9191, July 1989.

Ghiglia, D. C. & Pritt, M. D. (1998). "Two-Dimensional Phase Unwrapping : Theory," *Algorithms, and Software. John Wiley and Sons, Inc.*, 1998.

Goldstein, R. and Werner, C. (1998). "Radar interferogram filtering for geophysical applications,"*Geophysical Research Letters*, vol. 25, no. 21, pp. 4035-4038, November 1998.

Hajnsek, I.; Jagdhuber, T.; Schon, H. & Papathanassiou, K. P. (2009). "Potential of Estimating Soil Moisture Under Vegetation Cover by PolSAR," *IEEE Transactions on Geoscience and Remote Sensing*, Vol.47, No.2, pp. 442-454, February 2009.

Lee, J. S.; Papathanassiou, K. P.; Ainsworth, T. L.; Grunes, M. R. & Reigber, A. (1998). "A new technique for noise filtering of sar interferometric phase images,"*IEEE Transactions on Geoscience and Remote Sensing*, vol. 36, pp. 1456-1464, Sept. 1998.

Natsuaki, R. & Hirose A. (2011). "SPEC Method – A fine co-registration method for SAR interferometry," *IEEE Transactions Geoscience and Remote Sensing,* Vol.49, No.1, pp. 28-37, January 2011.

Pritt, M. & Shipman, j. (1994). "Least-squares two-dimensional phase unwrapping using FFT's," *IEEE Transactions on Geoscience and Remote Sensing,* vol. 32, no. 3, pp. 706-708, May 1994.

Reigber, A. & Moreia, J. (1997). "Phase unwrapping by fusion of local and global methods," in *International Geoscience and Remote Sensing Symposium (IGARSS) 1997*, pp. 869-871, August 1997.

Suksmono, A. B. & A. Hirose, A. (2002). "Adaptive noise reduction of InSAR images based on a complex-valued MRF model and its application to phase unwrapping problem," *IEEE Transactions on Geoscience and Remote Sensing,* Vol.40, No.3, pp.699-709, March 2002.

Suksmono, A. B. and Hirose, A. (2006). "Progressive transform-based phase unwrapping utilizing a recursive structure,"*IEICE Transactions on Communications,* vol. E89-B, no. 3, pp. 929-936, Mar. 2006.

Tobita, M.; Fujiwara, S.; Murakami, M.; Nakagawa, H. & Rosen, P. A. (1999). "Accurate offset estimation between two SLC images for SAR interferometry" (in Japanese), *Journal of the Geodetic Society of Japan*, 45, 297-314, 1999.

Yamaki, R. & Hirose, A. (2007). "Singularity-Spreading Phase Unwrapping," *IEEE Transactions on Geoscience and Remote Sensing*, Vol.45, No.10, Oct. 2007.

Yamaki, R. & Hirose, A. (2009). "Singular Unit Restoration in Interferograms Based on Complex-valued Markov Random Field Model for Phase Unwrapping," *IEEE Geoscience and Remote Sensing Letters*, Vol.6, No.1, pp.18-22, January 2009.

6

Robust Interferometric Phase Estimation in InSAR via Joint Subspace Projection

Hai Li and Renbiao Wu
Tianjin Key Lab for Advanced Signal Processing, Civil Aviation University of China, Tianjin, P.R. China

1. Introduction

Synthetic aperture radar interferometry (InSAR) is an important remote sensing technique to retrieve the terrain digital elevation model (DEM)[1][2]. Image coregistration, InSAR interferometric phase estimation (or noise filtering) and interferometric phase unwrapping[3][4][5][6] are three key processing procedures of InSAR. It is well known that the performance of interferometric phase estimation suffers seriously from poor image coregistration.

Image coregistration is an important preprocessing operation that aligns the pixels of one image to the corresponding pixels of another image. A review of recent as well as classic image registration methods can be found in Ref.[7]. Mutual information used for the registration of remote sensing imagery are presented in the literature[8]. In literature[9], the feature-based registration methods are presented. A new direct Fourier-transform-based algorithm for subpixel registration is proposed in Ref.[10]. In Ref.[11], a geometrical approach for image registration of SAR images is proposed, and the algorithm has been tested on several real data. A image registration method based on isolated point scatterers is proposed in Ref.[12].

Almost all the conventional InSAR interferometric phase estimation methods are based on interferogram filtering[13][14][15][16][17][18], such as pivoting mean filtering[13], pivoting median filtering[14], adaptive phase noise filtering[15], and adaptive contoured window filtering[18]. However, when the quality of an interferogram is very poor due to a large coregistration error, it is very difficult for these methods to retrieve the true terrain interferometric phases. In fact, the interferometric phases are random in nature with their variances being inversely proportional to the correlation coefficients between the corresponding pixel pairs of the two coregistered SAR images[2]. Therefore, the terrain interferometric phases should be estimated statistically.

In this chapter, the interferometric phase estimation method based on subspace projection and its modified version were proposed. Theoretical analysis and computer simulation results show that the methods can provide accurate estimation of the terrain interferometric phase (interferogram) even if the coregistration error reaches one pixel. The remainder of this chapter is organized as follows. Section 2 presents the signal model of a single pixel pair and the problem formulation. In Section 3, we discuss the interferometric phase estimation method based on subspace projection. In Section 4, the modified interferometric phase estimation method via subspace projection is presented in details. Finally, numerical and experemental results are presented in Section 5. Section 6 concludes the whole chapter.

2. Data model and problem formulation

Assuming that the SAR images are accurately coregistered and the interferometric phases are flattened with a zero-height reference plane surface. The complex data vector, denoted as $\mathbf{s}(i)$, of a pixel pair i (corresponding to the same ground area) of the coregistered SAR images can be formulated as follows[19],

$$\mathbf{s}(i) = [s_1(i), s_2(i)]^{\mathrm{T}} = \mathbf{a}(\varphi_i) \odot [x_1(i), x_2(i)]^{\mathrm{T}} + \mathbf{n}(i) = \mathbf{a}(\varphi_i) \odot \mathbf{x}(i) + \mathbf{n}(i) \quad (1)$$

where $\mathbf{a}(\varphi_i) = [1, e^{j\varphi_i}]^T$ is the spatial steering vector (i.e., the array steering vector) of the pixel i, superscript T denotes the vector transpose operation, φ_i is the terrain interferometric phase to be estimated, $\odot$ denotes the Hadamard product, $\mathbf{x}(i)$ is the complex magnitude vector (i.e., complex reflectivity vector of scene received by the satellites) of the pixel i, and $\mathbf{n}(i)$ is the additive noise term. The complex data vector $\mathbf{s}(i)$ can be modeled as a joint complex circular Gaussian random vector with zero-mean and the corresponding covariance matrix $\mathbf{C}_s(i)$ is given by

$$\begin{aligned} \mathbf{C}_s(i) &= E\{\mathbf{s}(i)\mathbf{s}^H(i)\} \\ &= \mathbf{a}(\varphi_i)\mathbf{a}^H(\varphi_i) \odot E\{\mathbf{x}(i)\mathbf{x}^H(i)\} + \sigma_n^2\mathbf{I} \\ &= \sigma_s^2(i)\mathbf{a}(\varphi_i)\mathbf{a}^H(\varphi_i) \odot \mathbf{R}_s(i) + \sigma_n^2\mathbf{I} \end{aligned} \quad (2)$$

$$\mathbf{R}_s(i) = \begin{bmatrix} r_{11}(i), & r_{12}(i) \\ r_{21}(i), & r_{22}(i) \end{bmatrix} \quad (3)$$

where $\mathbf{R}_s(i)$ is called the correlation coefficient matrix, $\mathbf{I}$ is a 2×2 identity matrix, $r_{mn}(i)$ ($0 \le r_{mn}(i) \le 1$, $n = 1,2$, and $m = 1,2$) are the correlation coefficients between the satellites m and n, $E\{\}$ denotes the statistical expectation, superscript H denotes vector conjugate-transpose, $\sigma_s^2(i)$ is the echo power of the pixel i and σ_n^2 is the noise power. In order to simplify the mathematical expressions, the denotation i (denoting the pixel) in the right side of the following expressions is omitted. In practice, the statistical covariance matrix of (2) can be adaptively estimated using the sample covariance matrix.

If the SAR images are accurately coregistered and the cross-correlation coefficients (i.e., the nondiagonal elements) of $\mathbf{R}_s(i)$ are large enough, the number of the principal eigenvalues of the covariance matrix $\mathbf{C}_s(i)$ is one; i.e., the dimensions of the signal subspace and the noise subspace are both one (in the absence of the layover). In this case, the eigen-decomposition of the covariance matrix $\mathbf{C}_s(i)$ is as follows:

$$\mathbf{C}_s(i) = (\lambda_{rs} + \sigma_n^2)(\mathbf{a}(\varphi_i) \odot \boldsymbol{\beta}_{rs})(\mathbf{a}(\varphi_i) \odot \boldsymbol{\beta}_{rs})^H + \sigma_n^2\boldsymbol{\beta}_n\boldsymbol{\beta}_n^H \quad (4)$$

where λ_{rs} and $\boldsymbol{\beta}_{rs}$ are the principal eigenvalue and the corresponding eigenvector (i.e., the signal eigenvector) of $\mathbf{R}_s(i)$, respectively, and $\boldsymbol{\beta}_n$ is the noise eigenvector corresponding to the insignificant eigenvalue of $\mathbf{C}_s(i)$. From (4) we can note that $(\mathbf{a}(\varphi_i)\odot\boldsymbol{\beta}_{rs})$ is in the signal subspace, $\boldsymbol{\beta}_n$ is in the noise subspace, and $(\mathbf{a}(\varphi_i)\odot\boldsymbol{\beta}_{rs})$ are orthogonal to $\boldsymbol{\beta}_n$, which is used to estimate the interferometric phase φ_i.

The definition of cost function is,

$$J_1 = (\mathbf{a}(\varphi_i)\odot\boldsymbol{\beta}_{rs})^H\boldsymbol{\beta}_n\boldsymbol{\beta}_n^H(\mathbf{a}(\varphi_i)\odot\boldsymbol{\beta}_{rs}) \tag{5}$$

The minimization of J_1 can provide the optimum estimate of the interferometric phase φ_i.

The eigenvalues of $C_s(i)$ result in a low dispersion due to the increase of the coregistration error. Actually, it is induced by the increase of noise eigenvalue, i.e., the signal component spreads into the noise space. Moreover, the 2-dimensional space will be fully occupied by signal component with a worse coregistration error of one pixel. At this instant, the noise eigenvalue is almost equal to the bigger one. The noise subspace dimension becomes zero. The degree of the signal component spreading to the noise subspace is smaller, the estimation of the InSAR interferometric phase is better(the subspace projection technique is used to estimate the InSAR interferometric phase). On the contrary, the degree of the signal component spreading to the noise subspace is larger, the estimation of the InSAR interferometric phase is worse. The conclusions obtained from the preceding analysis are briefly summarized as follows: the degree of the signal component spreading to the noise subspace becomes larger and larger as the coregistration error increases when the example of formula (1) is used to build the data vector. In other words, the degree of dispersion is heavily impacted by the coregistration error. The simulation result shown in Fig.1 demonstrates the above conclusions.

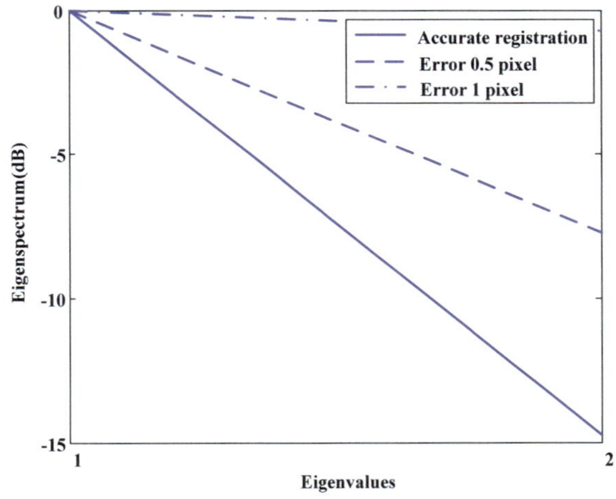

Fig. 1. Eigenspectra of the covariance matrix for accurate coregistration, coregistration errors of 0.5 and 1 pixels, respectively.

The cost function given by (5) can be used to estimate the InSAR interferometric phase when the SAR images are accurately coregistered. However, in the presence of coregistration error, the cross-correlation coefficients are smaller than 1 and the noise eigenvalue becomes large. In the completely misregistered case, the rank of $\mathbf{R}_s(i)$ becomes 2, which means that the noise subspace vanishes in the eigenspace of $\mathbf{C}_s(i)$. So the cost function given by (5) can not be used to estimate the InSAR interferometric phase in the presence of coregistration error.

3. Interferometric phase estimation via subspace projection

Considering the difficulties in accurate coregistration, we use not only the corresponding pixel pair i of the coarsely coregistered SAR image pair (as given in (1)) but also the neighboring pixel pairs centered on the pixel pair i to jointly construct the data vector. An example of the construction method for the data vector is shown in Fig.2, where a circle represents a SAR image pixel and i denotes the centric pixel pair (i.e., the desired pixel pair whose interferometric phase is to be estimated). We call this extended data vector $\mathbf{si}(i)$ the joint data vector. The number of the neighboring pixel pairs to construct the joint data vector is 8 as shown in Fig.2.

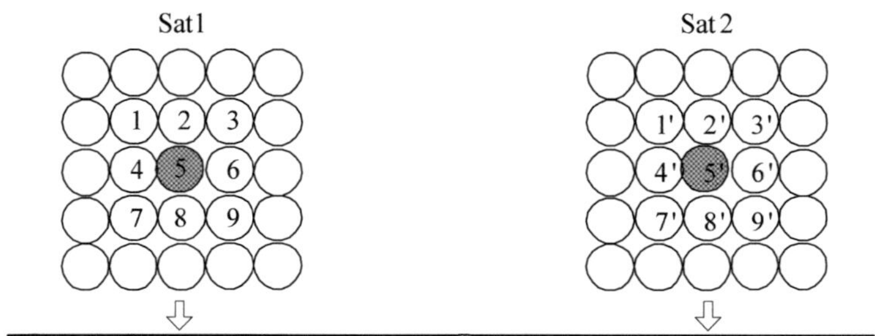

Fig.2. A construction method for the joint data vector.

The joint data vector $\mathbf{si}(i)$ as shown in Fig.2 can be written as:

$$\mathbf{si}(i) = [\mathbf{s}(i-4)^T, \mathbf{s}(i-3)^T, \cdots, \mathbf{s}(i)^T, \cdots, \mathbf{s}(i+4)^T]^T \tag{6}$$

The corresponding joint covariance matrix is given by

$$\begin{aligned}\mathbf{C}_{\mathbf{si}}(i) &= E\{\mathbf{si}(i)\mathbf{si}^H(i)\} \\ &= \boldsymbol{\alpha}(\varphi_{i-4}, \varphi_{i-3}, \cdots, \varphi_{i+4})\boldsymbol{\alpha}^H(\varphi_{i-4}, \varphi_{i-3}, \cdots, \varphi_{i+4}) \odot \mathbf{R}_{\mathbf{si}}(i) + \sigma_n^2 \mathbf{I}\end{aligned} \tag{7}$$

where $\boldsymbol{\alpha}(\varphi_{i-4}, \varphi_{i-3}, \cdots, \varphi_{i+4}) = [\mathbf{a}^T(\varphi_{i-4}), \mathbf{a}^T(\varphi_{i-3}), \cdots, \mathbf{a}^T(\varphi_{i+4})]^T$ and $\mathbf{R}_{\mathbf{si}}(i)$ are referred to as the joint steering vector and the joint correlation function matrix of the pixel pair i, respectively. In fact, most of the nature terrain can be approximated by a local plane. After we estimate the local slopes and correct the neighboring pixels, we can assume that the neighboring pixels have the almost identical terrain height[20]. That is, the spatial steering vectors of the pixel pairs in $\mathbf{si}(i)$ are assumed to be identical, i.e., $\mathbf{a}(\varphi_{i-4}) = \mathbf{a}(\varphi_{i-3}) = \cdots = \mathbf{a}(\varphi_{i+4}) = [1, e^{j\varphi_i}]^T$ and $\boldsymbol{\alpha}(\varphi_{i-4}, \varphi_{i-3}, \cdots, \varphi_{i+4}) = [\mathbf{a}^T(\varphi_i), \mathbf{a}^T(\varphi_i), \cdots, \mathbf{a}^T(\varphi_i)]^T = [1, e^{j\varphi_i} \cdots, 1, e^{j\varphi_i}]^T$. The simplified joint steering vector of the pixel i is denoted by $\mathbf{ai}(\varphi_i)=[\mathbf{a}^T(\varphi_i),\mathbf{a}^T(\varphi_i),\cdots,\mathbf{a}^T(\varphi_i)]^T$ (18×1). Substituting $\mathbf{ai}(\varphi_i)$ for $\boldsymbol{\alpha}(\varphi_{i-4},\varphi_{i-3},\cdots,\varphi_{i+4})$ in (7), we have,

$$\begin{aligned}\mathbf{C_{si}}(i) &= \mathbf{ai}(\varphi_i)\mathbf{ai}^H(\varphi_i)\odot\mathbf{R_{si}}(i)+\sigma_n^2\mathbf{I}\\ &\underline{\underline{EVD}}\sum_{k=1}^{K}\lambda_{csi}^{(k)}\boldsymbol{\beta}_{csi}^{(k)}\boldsymbol{\beta}_{csi}^{(k)H}+\sum_{l=1}^{18-K}\sigma_n^2\boldsymbol{\beta}_{nsi}^{(l)}\boldsymbol{\beta}_{nsi}^{(l)H}\\ &=\sum_{k=1}^{K}(\lambda_{rsi}^{(k)}+\sigma_n^2)(\mathbf{ai}(\varphi_i)\odot\boldsymbol{\beta}_{rsi}^{(k)})(\mathbf{ai}(\varphi_i)\odot\boldsymbol{\beta}_{rsi}^{(k)})^H+\sum_{l=1}^{18-K}\sigma_n^2\boldsymbol{\beta}_{nsi}^{(l)}\boldsymbol{\beta}_{nsi}^{(l)H}\end{aligned} \tag{8}$$

where K is the number of the principal eigenvalues of $\mathbf{C_{si}}(i)$, $\boldsymbol{\beta}_{csi}^{(k)}$ ($k=1,2,\cdots,K$) are the eigenvectors corresponding to the principal eigenvalues $\lambda_{csi}^{(k)}$ of $\mathbf{C_{si}}(i)$, $\boldsymbol{\beta}_{rsi}^{(k)}$ ($k=1,2,\cdots,K$) are the eigenvectors corresponding to the principal eigenvalues $\lambda_{rsi}^{(k)}$ of $\mathbf{R_{si}}(i)$. σ_n^2 and $\boldsymbol{\beta}_{nsi}^{(l)}$ ($l=1,2,\cdots,18-K$) are the noise eigenvalue and the corresponding eigenvectors of $\mathbf{C_{si}}(i)$,respectively. From (8) we can note that $\mathbf{ai}(\varphi_i)\odot\boldsymbol{\beta}_{rsi}^{(k)}$ ($k=1,2,\cdots,K$) are in the signal subspace of $\mathbf{C_{si}}(i)$, $\boldsymbol{\beta}_{nsi}^{(l)}$ ($l=1,2,\cdots,18-K$) are in the noise subspace of $\mathbf{C_{si}}(i)$, and $\mathbf{ai}(\varphi_i)\odot\boldsymbol{\beta}_{rsi}^{(k)}$ are orthogonal to $\boldsymbol{\beta}_{nsi}^{(l)}$.

If the SAR images are accurately coregistered, the structure form of the joint correlation function matrix $\mathbf{R_{si}}(i)$ are given by

$$\mathbf{R_{si}}(i)=\begin{bmatrix}\sigma_s^2(i-4)\mathbf{R_s}(i-4) & 0 & \cdots & 0\\ 0 & \sigma_s^2(i-3)\mathbf{R_s}(i-3) & & \\ \vdots & & \ddots & \\ 0 & & & \sigma_s^2(i+4)\mathbf{R_s}(i+4)\end{bmatrix} \tag{9}$$

where $\mathbf{R_s}(m)$ and $\sigma_s^2(m)$ ($m=i-4,i-3,\cdots,i+4$) in the diagonal of the $\mathbf{R_{si}}(i)$ are the coherence coefficient matrix (given by (3)) and the echo power of the pixel pair m , respectively. We can notice from (9) that when the SAR images are accurately coregistered, only the elements in $\mathbf{R_s}(m)$ are nonzero, while all the other elements of $\mathbf{R_{si}}(i)$ (or $\mathbf{C_{si}}(i)$) are zero (assuming the complex reflectivity is independent from pixel to pixel and neglecting the noise). In other words, $\mathbf{R_{si}}(i)$ is a block diagonal matrix. However, if the SAR images are not accurately coregistered, the nonzero elements in the submatrices $\mathbf{R_s}(m)$ are diffused to other non-diagonal element positions of $\mathbf{R_{si}}(i)$, as shown in Fig.3(b) and Fig.3(c). The dimensions of the joint correlation function matrix $\mathbf{R_{si}}(i)$ are 18×18. Fig.3(a) is the structure of $\mathbf{R_{si}}(i)$ for accuracy coregistration; Fig.3(b) is the structure of $\mathbf{R_{si}}(i)$ for the coregistration error of 0.5 pixel; Fig.3(c) is the structure of $\mathbf{R_{si}}(i)$ for the coregistration error of 1 pixel. From Fig.3 we can see that when the SAR images are not accurately coregistered, the same imaged ground area can be coregistered to different pixel positions; thus its correlation information appears in other non-diagonal element position of $\mathbf{R_{si}}(i)$. The gray level denotes the magnitude of each element with the white strongest and black zero.

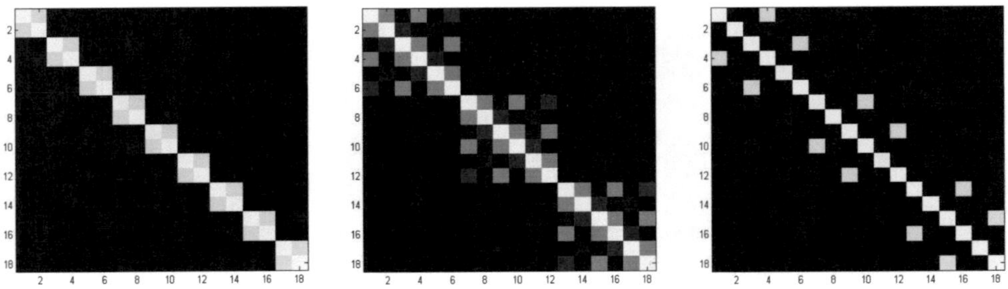

(a) (b) (c)

Fig. 3. Structures of joint correlation function matrix for different coregistration errors: (a) accurate coregistration; (b) coregistration error of 0.5 pixels; (c) coregistration error of 1 pixel.

Comparing Fig.3 with (4), the conventional estimation approaches based on the single pixel pair will be not feasible if the coregistration error is very large, for example, larger than 0.5 pixel, while our approach can still achieve the optimum estimation due to the use of multiple neighboring pixel pairs.

From the literature[21], we can know that: for the accurate coregistration, the dimensions of the signal subspace and the noise subspace of the joint covariance matrix $\mathbf{C_{si}}(i)$ (18×18) are both 9. For the coregistration error of 1 pixel, the dimensions of the signal subspace and the noise subspace of the joint covariance matrix $\mathbf{C_{si}}(i)$ are changed to 12 and 6, respectively. The method can also provide accurate estimation of the terrain interferometric phase even if the coregistration error reaches one pixel only by changing the dimension of the noise subspace from 9 to 6.

As mentioned above, the signal subspace $\mathbf{ai}(\varphi_i) \odot \boldsymbol{\beta}_{rsi}^{(k)}$ $(k = 1, 2, \cdots, K)$ is orthogonal to the noise subspace $\mathbf{N_c} = span\{\boldsymbol{\beta}_{nsi}^{(1)}, \boldsymbol{\beta}_{nsi}^{(2)}, \cdots \boldsymbol{\beta}_{nsi}^{(M-K)}\}$, which is used to estimate the interferometric phase φ_i.

The definition of cost function is,

$$J_2 = \sum_{k=1}^{K} \sum_{l=1}^{M-K} (\mathbf{ai}(\varphi_i) \odot \boldsymbol{\beta}_{rsi}^{(k)})^H \boldsymbol{\beta}_{nsi}^{(l)} \boldsymbol{\beta}_{nsi}^{(l)H} (\mathbf{ai}(\varphi_i) \odot \boldsymbol{\beta}_{rsi}^{(k)}) \tag{10}$$

The minimization of J_2 can provide the optimum estimate of the interferometric phase φ_i.

Fig.s 4 shows the simulation results for various coregistration errors by the subspace projection method. Comparing these figures, we can observe that the large coregistration error has almost no effect on the interferograms obtained by the proposed method. We can see that the subspace projection method in this chapter is robust to large coregistration errors (up to one pixel).

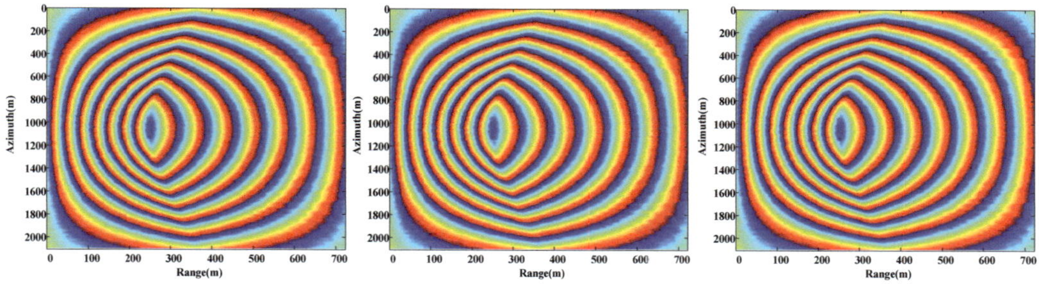

(a) (b) (c)

Fig. 4. The interferograms obtained by the subspace projection method for the accurate coregistration (a), the coregistration error of 0.5 pixels (b) and the coregistration error of one pixel (c).

4. Modified interferometric phase estimation via subspace projection

The joint subspace projection method mentioned above employs the projection of the joint signal subspace onto the corresponding joint noise subspace which is obtained from the eigendecomposition of the joint covariance matrix to estimate the terrain interferometric phase, and takes advantage of the coherence information of the neighboring pixel pairs to auto-coregister the SAR images, where the phase noise is reduced simultaneously. However, the noise subspace dimension of the covariance matrix changes with the coregistration error. For accurate estimating the InSAR interferometric phase, the noise subspace dimension of the covariance matrix must be known, and the performance of the method (i.e. subspace projection method) degrades when the noise subspace dimension is not estimated correctly. In this chapter, an modified joint subspace projection method for InSAR interferometric phase estimation is proposed. In this method, the benefit from the new formulation of joint data vector is that the noise subspace dimension of the covariance matrix is not affected by the coregistration error (i.e., the noise subspace dimension of the corresponding covariance matrix with the coregistration error μ $(0 < \mu \leq 1)$ pixel is the same as that of the covariance matrix with accurate coregistration). So the method does not need to calculate the noise subspace dimension before estimating the InSAR interferometric phase.

4.1 Data modeling of the modified method

For avoiding the trouble of calculating the noise subspace dimension before estimating the InSAR interferometric phase, the new formulation of joint data vector is used in the proposed method. An example to construct the new joint data vector $\mathbf{si}(i)$ is shown in Fig.5.

The joint data vector $\mathbf{si}(i)$ shown in Fig.5 can be written as

$$\begin{aligned}\mathbf{si}(i) = [&s_1(i-1), s_2(i-6), s_2(i-5), s_2(i-2), s_2(i-1), s_1(i), s_2(i-4), s_2(i-3), s_2(i), s_2(i+1),\\ &s_1(i+3), s_2(i+2), s_2(i+3), s_2(i+6), s_2(i+7), s_1(i+4), s_2(i+4), s_2(i+5), s_2(i+8), s_2(i+9)]^T\end{aligned} \tag{11}$$

The corresponding covariance matrix $\mathbf{C}_{\mathbf{si}}(i)$ is given by

$$\mathbf{C_{si}}(i) = E\{\mathbf{si}(i)\mathbf{si}^H(i)\} \\ = \mathbf{ai}(\varphi_i)\mathbf{ai}^H(\varphi_i) \odot \mathbf{R_{si}}(i) + \sigma_n^2\mathbf{I} \tag{12}$$

where

$\mathbf{ai}(\varphi_i) = \left[1, e^{j\varphi_i}, e^{j\varphi_i}, e^{j\varphi_i}, e^{j\varphi_i}, 1, e^{j\varphi_i}, e^{j\varphi_i}, e^{j\varphi_i}, e^{j\varphi_i}, 1, e^{j\varphi_i}, e^{j\varphi_i}, e^{j\varphi_i}, e^{j\varphi_i}, 1, e^{j\varphi_i}, e^{j\varphi_i}, e^{j\varphi_i}, e^{j\varphi_i}\right]^T$ and $\mathbf{R_{si}}(i)$ are referred to as the joint generalized steering vector and the joint correlation function matrix of the pixel pair *i*, respectively. The deduction of equation (12) is presented in appendix A.

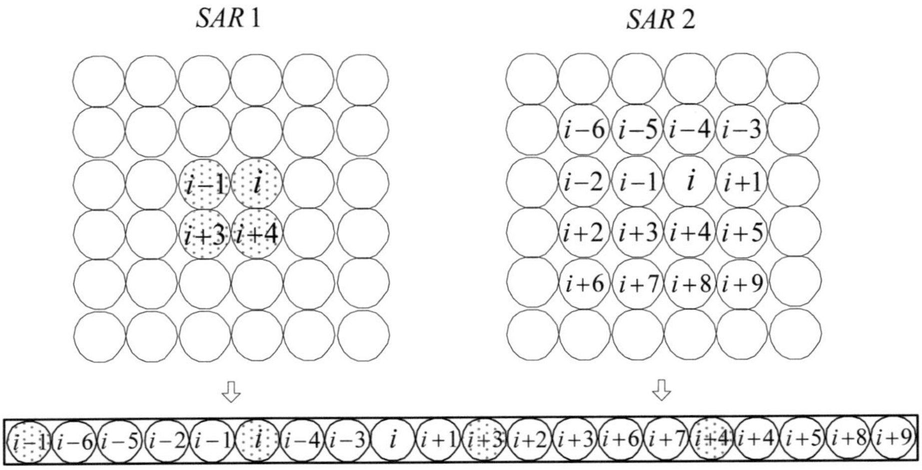

Fig. 5. Formulation of the joint data vector.

In the following the noise subspace dimension of the joint covariance matrix $\mathbf{C_{si}}(i)$ for different coregistration errors are discussed.

a. Accurate coregistration.

For the accurate coregistration, the joint data vector, $\mathbf{si}(i)$, that is shown in Fig.6, where circles represent SAR image pixels pair whose interferometric phase has to be estimated.

We can rearrange the elements (pixels) of $\mathbf{si}(i)$ to obtain $\mathbf{si}'(i)$ as shown in Fig.6, which does not change the eigenvalues of the corresponding covariance matrix[21]. The joint data vector $\mathbf{si}'(i)$, shown in Fig.6, can be written as

$$\mathbf{si}'(i) = [s_1(i-1), s_2(i-1), s_1(i), s_2(i), s_1(i+3), s_2(i+3), s_1(i+4), s_2(i+4), s_2(i-6), s_2(i-5), \\ s_2(i-2), s_2(i-4), s_2(i-3), s_2(i+1), s_2(i+2), s_2(i+6), s_2(i+7), s_2(i+5), s_2(i+8), s_2(i+9)]^T \tag{13}$$

The corresponding covariance matrix $\mathbf{C_{si'}}(i)$ (20×20) is given by

$$\mathbf{C_{si'}}(i) = E\{\mathbf{si}'(i)\mathbf{si}'(i)^H\} \\ = \mathbf{ai}(\varphi_i)\mathbf{ai}^H(\varphi_i) \odot \mathbf{R_{si'}}(i) + \sigma_n^2\mathbf{I} \tag{14}$$

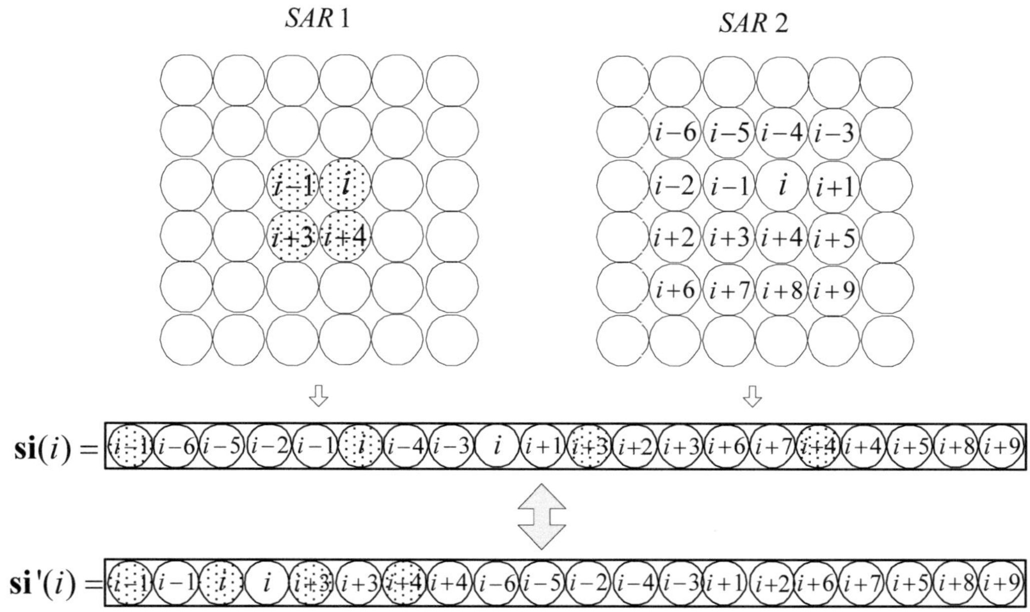

Fig. 6. Joint data vector for the accurate coregistration.

$$
\mathbf{R}_{\mathbf{si'}}(i)=\begin{bmatrix}
\sigma_s^2(i-1)\mathbf{R_s}(i-1) & 0 & \cdots & 0 & \cdots & 0 \\
0 & \sigma_s^2(i)\mathbf{R_s}(i) & & & & \vdots \\
\vdots & & \sigma_s^2(i+3)\mathbf{R_s}(i+3) & & & \\
 & & & \sigma_s^2(i+4)\mathbf{R_s}(i+4) & & \\
0 & & & \sigma_s^2(i-6) & \cdots\ 0\ \cdots & 0 \\
 & & & & \sigma_s^2(i-5) & \\
 & & & & \sigma_s^2(i-2) & \\
\vdots & & & & \sigma_s^2(i-4) & \vdots \\
 & & & & \sigma_s^2(i-3) & \\
 & & & & \sigma_s^2(i+1) & \\
 & & & & \sigma_s^2(i+2) & \\
0 & \cdots & 0 & & \sigma_s^2(i+6) & 0 \\
 & & & & \sigma_s^2(i+7) & \\
 & & & & \sigma_s^2(i+5) & \\
 & & & & \sigma_s^2(i+8) & \\
0 & \cdots & 0 & \cdots & 0 & \sigma_s^2(i+9)
\end{bmatrix} \tag{15}
$$

From (14) we can see that the number of the principal eigenvalues of $\mathbf{C}_{\mathbf{si'}}(i)$ is 16 Ref.[21]. The noise subspace dimension of the covariance matrix $\mathbf{C}_{\mathbf{si'}}(i)$ (20×20) is 4, thus, the noise subspace dimension of the covariance matrix $\mathbf{C}_{\mathbf{si}}(i)$ estimated from the joint data vector $\mathbf{si}(i)$ is 4 Ref.[21].

b. Coregistration error of one pixel.

When the azimuth coregistration error is one pixel and its direction is upwards (i.e., the pixel of the image from the second satellite is shifted upwards compared to the pixel in the first satellite image), the joint data vector, $\mathbf{si}(i)$, is shown in Fig.7.

We can rearrange the elements (pixels) of $\mathbf{si}(i)$ to obtain $\mathbf{si}^{'}(i)$ as shown in Fig.7, which does not change the eigenvalues of the corresponding covariance matrix[21]. The joint data vector $\mathbf{si}^{'}(i)$, shown in Fig.7, can be written as

$$\begin{aligned}\mathbf{si}^{'}(i) = [&s_1(i-1), s_2(i-1), s_1(i), s_2(i), s_1(i+3), s_2(i+3), s_1(i+4), s_2(i+4), s_2(i-2), s_2(i+2),\\ &s_2(i+1), s_2(i+5), s_2(i+6), s_2(i+7), s_2(i+10), s_2(i+11), s_2(i+8), s_2(i+9), s_2(i+12), s_2(i+13)]^T\end{aligned} \quad (16)$$

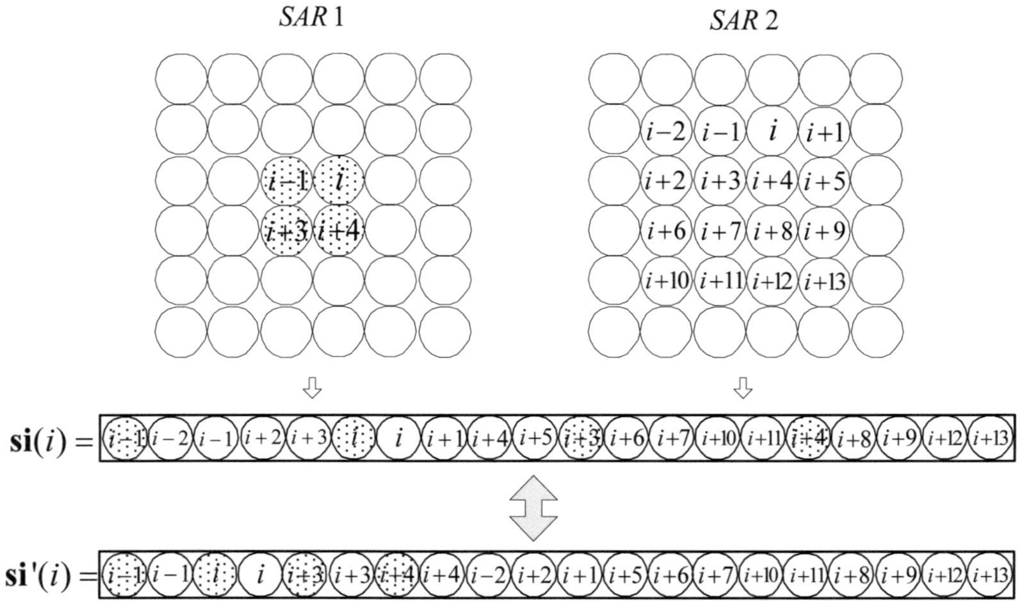

Fig. 7. Joint data vector for the coregistration error of one pixel.

The corresponding covariance matrix $\mathbf{C}_{\mathbf{si'}}(i)$ is given by

$$\begin{aligned}\mathbf{C}_{\mathbf{si'}}(i) &= E\{\mathbf{si}'(i)\mathbf{si}'(i)^H\}\\ &= \mathbf{ai}(\varphi_i)\mathbf{ai}^H(\varphi_i) \odot \mathbf{R}_{\mathbf{si'}}(i) + \sigma_n^2\mathbf{I}\end{aligned} \quad (17)$$

$$\mathbf{R}_{\mathbf{si'}}(i)=\begin{bmatrix}
\sigma_s^2(i-1)\mathbf{R}_\mathbf{s}(i-1) & 0 & \cdots & 0 & \cdots & 0\\
0 \quad \sigma_s^2(i)\mathbf{R}_\mathbf{s}(i) & & & & & \vdots\\
\vdots \quad \sigma_s^2(i+3)\mathbf{R}_\mathbf{s}(i+3) & & & & & \\
\sigma_s^2(i+4)\mathbf{R}_\mathbf{s}(i+4) & & & & & \\
0 & \sigma_s^2(i-2) & \cdots & 0 & \cdots & 0\\
 & \sigma_s^2(i+2) & & & & \\
 & \sigma_s^2(i+1) & & & & \\
\vdots & & \sigma_s^2(i+5) & & & \vdots\\
 & & \sigma_s^2(i+6) & & & \\
 & & \sigma_s^2(i+7) & & & \\
 & & & \sigma_s^2(i+10) & & \\
0 & \cdots & 0 & \sigma_s^2(i+11) & & 0\\
 & & & & \sigma_s^2(i+8) & \\
 & & & & \sigma_s^2(i+9) & \\
 & & & & & \sigma_s^2(i+12)\\
0 & \cdots & 0 & \cdots & 0 & \sigma_s^2(i+13)
\end{bmatrix} \tag{18}$$

From the above discussion, we know the noise subspace dimension of the covariance matrix $\mathbf{C}_{\mathbf{si}}(i)$ estimated from the joint data vector $\mathbf{si}(i)$ is 4.

c. Coregistration error of μ $(0<\mu<1)$ pixel.

When the azimuth coregistration error is μ $(0<\mu<1)$ pixel and its direction is upwards (i.e., the pixel of the image from the second satellite is shifted upwards compared to the pixel in the first satellite image), the joint data vector, $\mathbf{si}(i)$,is shown in Fig.8(a).

From the literature[21], we can know that the eigenspectrum (i.e., the distribution of eigenvalues) of a covariance matrix is invariant no matter how the elements of the corresponding data vector are permuted. So we can rearrange the elements of $\mathbf{si}(i)$ to obtain $\mathbf{js}(i)$ as shown in Fig.8(b) [21], which does not change the eigenvalues of the corresponding covariance matrix. The joint data vector $\mathbf{js}(i)$, shown in Fig.8(b), can be written as

$$\begin{aligned}\mathbf{js}(i)=[&s_1(i-1),s_2(i-1),s_1(i),s_2(i),s_1(i+3),s_2(i+3),s_1(i+4),s_2(i+4),s_2(i-2),s_2(i+2),\\ &s_2(i+1),s_2(i+5),s_2(i+6),s_2(i+7),s_2(A),s_2(B),s_2(i+8),s_2(i+9),s_2(C),s_2(D)]^T\end{aligned} \tag{19}$$

The corresponding covariance matrix $\mathbf{C}_{\mathbf{js}}(i)$ is given by

$$\begin{aligned}\mathbf{C}_{\mathbf{js}}(i)&=E\{\mathbf{js}(i)\mathbf{js}^H(i)\}\\ &=\mathbf{ai}(\varphi_i)\mathbf{ai}^H(\varphi_i)\odot\mathbf{R}_{\mathbf{js}}(i)+\sigma_n^2\mathbf{I}\end{aligned} \tag{20}$$

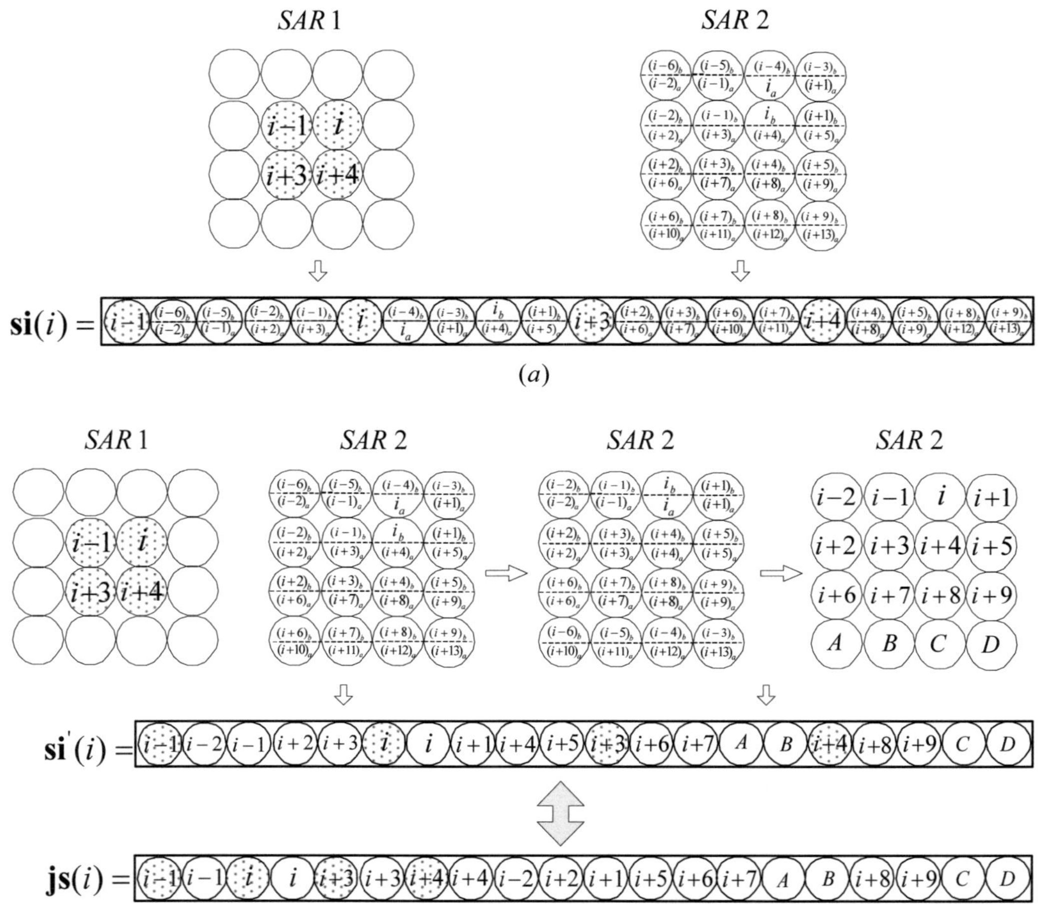

Fig. 8. Joint data vector for the coregistration error of μ pixels.

$$\mathbf{R}_{\mathbf{js}}(i)=\begin{bmatrix}
\sigma_s^2(i-1)\mathbf{R}_{\mathbf{s}}(i-1) & 0 & & & \cdots & & & 0 & & & \cdots & & & & & 0 \\
0 & \sigma_s^2(i)\mathbf{R}_{\mathbf{s}}(i) & & & & & & & & & & & & & & \vdots \\
\vdots & & \sigma_s^2(i+3)\mathbf{R}_{\mathbf{s}}(i+3) & & & & & & & & & & & & & \\
 & & & \sigma_s^2(i+4)\mathbf{R}_{\mathbf{s}}(i+4) & & & & & & & & & & & & \\
0 & & & & \sigma_s^2(i-2) & & & \cdots & & 0 & & \cdots & & & & 0 \\
 & & & & & \sigma_s^2(i+2) & & & & & & & & & & \\
 & & & & & & \sigma_s^2(i+1) & & & & & & & & & \\
\vdots & & & & & & & \sigma_s^2(i+5) & & & & & & & & \vdots \\
 & & & & & & & & \sigma_s^2(i+6) & & & & & & & \\
 & & & & & & & & & \sigma_s^2(i+7) & & & & & & \\
 & & & & & & & & & & \sigma_s^2(A) & & & & & \\
0 & & \cdots & & 0 & & & & & & & \sigma_s^2(B) & & & & 0 \\
 & & & & & & & & & & & & \sigma_s^2(i+8) & & & \\
 & & & & & & & & & & & & & \sigma_s^2(i+9) & & \\
 & & & & & & & & & & & & & & \sigma_s^2(C) & \\
0 & & \cdots & & 0 & & \cdots & & & 0 & & & & & & \sigma_s^2(D)
\end{bmatrix} \tag{21}$$

We know the noise subspace dimension of the covariance matrix $\mathbf{C}_{\mathbf{js}}(i)$ (20×20) is 4, thus, the noise subspace dimension of the covariance matrix $\mathbf{C}_{\mathbf{si}}(i)$ estimated from the joint data vector $\mathbf{si}(i)$ is also 4[21].

From the results derived above, we can see that: the new formulation of joint data vector proposed in this chapter has the advantage that the noise subspace dimension of the corresponding covariance matrix is independent of the coregistration error. That is to say, the noise subspace dimension of the corresponding covariance matrix with the coregistration error μ $(0<\mu\leq 1)$ pixel is the same as that of the accurate covariance matrix. Therefore, it is not required to calculate the noise subspace dimension, thus avoiding the trouble of calculating the noise subspace dimension before estimating the InSAR interferometric phase.

4.2 Summary of the modified method

In this section, we give the detailed steps for the modified interferometric phase estimation method based on subspace projection.

Step 1. Coregister SAR images. The SAR images are coarsely coregistered using the crosscorrelation information of the SAR image intensity or other strategies[1][2] after SAR imaging of the echoes acquired by each satellite. However, the allowable coregistration error of the proposed method can be very large (such as one pixel), which is useful in practice. The low coregistration accuracy requirement can greatly mitigate the complexity in image coregistration processing.

Step 2. Estimate the covariance matrix. The covariance matrix $\mathbf{C_{si}}(i)$ can be estimated by using joint data vector $\mathbf{si}(i)$ shown in Fig.5. Under the assumption that the neighboring pixels have the identical terrain height and the complex reflectivity is independent from pixel to pixel[20][21], the covariance matrix $\mathbf{C_{si}}(i)$ can be estimated by its sample covariance matrix $\hat{\mathbf{C}}_{\mathbf{si}}(i)$, i.e.,

$$\hat{\mathbf{C}}_{\mathbf{si}}(i) = \frac{1}{2K+1} \sum_{k=-L}^{L} \mathbf{si}(i+k)\mathbf{si}^H(i+k) \tag{22}$$

where $2L+1$ is the number of i.i.d. samples from the neighboring pixel pairs.

Remark 1: It is easy to obtain enough i.i.d. samples for locally flat terrains. However, an imaging terrain in practice can not be relied upon to be so flat that the adjacent pixels have the identical terrain height. If the local terrain slope is available in advance or can be estimated[20], the steering vector (i.e., the interferometric phase) variation due to the different terrain height from pixel to pixel can be compensated, which greatly enlarges the size of the sample window.

Step 3. Subspace estimation by Eigendecomposing. The estimated covariance matrix $\hat{\mathbf{C}}_{\mathbf{si}}(i)$ of the dimensions 20×20 can be eigendecomposed into

$$\hat{\mathbf{C}}_{\mathbf{si}}(i) = \sum_{m=1}^{K} \hat{\lambda}_{csi}^{(m)} \hat{\boldsymbol{\beta}}_{csi}^{(m)} \hat{\boldsymbol{\beta}}_{csi}^{(m)H} + \sum_{l=1}^{20-K} \hat{\lambda}_{csi}^{(l+K)} \hat{\boldsymbol{\beta}}_{nsi}^{(l)} \hat{\boldsymbol{\beta}}_{nsi}^{(l)H} \tag{23}$$

where K is the number of the principal eigenvalues of $\hat{\mathbf{C}}_{\mathbf{si}}(i)$, $\hat{\lambda}_{csi}^{(1)} > \hat{\lambda}_{csi}^{(2)} > \cdots > \hat{\lambda}_{csi}^{(K)} >> \hat{\lambda}_{csi}^{(K+1)} > \cdots > \hat{\lambda}_{csi}^{(20)}$, eigenvectors $\hat{\boldsymbol{\beta}}_{nsi}^{(l)}$ ($l = 1,2,\cdots,20-K$) corresponding to the smaller eigenvalues $\hat{\lambda}_{csi}^{(l+K)}$ ($l = 1,2,\cdots,20-K$) span the noise subspace, i.e,

$$\mathbf{N_c} = span\left\{\hat{\boldsymbol{\beta}}_{nsi}^{(1)}, \hat{\boldsymbol{\beta}}_{nsi}^{(2)}, \cdots, \hat{\boldsymbol{\beta}}_{nsi}^{(20-K)}\right\} \tag{24}$$

whereas the larger eigenvectors $\hat{\boldsymbol{\beta}}_{csi}^{(m)}$ ($m = 1,2,\cdots,K$) corresponding to the principal eigenvalues $\hat{\lambda}_{csi}^{(m)}$ ($m = 1,2,\cdots,K$) span the signal subspace, i.e.,

$$\mathbf{S_c} = span\left\{\hat{\boldsymbol{\beta}}_{csi}^{(1)}, \hat{\boldsymbol{\beta}}_{csi}^{(2)}, \cdots, \hat{\boldsymbol{\beta}}_{csi}^{(K)}\right\} \tag{25}$$

The noise power is often estimated by

$$\hat{\sigma}_n^2 = \frac{1}{20-K} \sum_{l=1}^{20-K} \hat{\lambda}_{csi}^{(l+K)} \tag{26}$$

The joint correlation function matrix $\hat{\mathbf{R}}_{\mathbf{si}}(i)$ can be approximated as the amplitude (i.e., the absolute value) of the estimated covariance matrix $\hat{\mathbf{C}}_{\mathbf{si}}(i)$ [20], i.e.,

$$\hat{\mathbf{R}}_{\mathbf{si}}(i) = \left|\hat{\mathbf{C}}_{\mathbf{si}}(i) - \hat{\sigma}_n^2 \mathbf{I}\right| \tag{27}$$

By eigen-decomposing $\hat{\mathbf{R}}_{\mathbf{si}}(i)$, we obtain K principal eigenvectors $\hat{\boldsymbol{\beta}}_{rsi}^{(m)}$ ($m=1,2,\cdots,K$). As shown by (8), the same signal subspace spanned by the principal eigenvectors $\hat{\boldsymbol{\beta}}_{csi}^{(m)}$ ($m=1,2,\cdots,K$) of $\hat{\mathbf{C}}_{\mathbf{si}}(i)$ can be spanned by the Hadamard product vectors $\mathbf{ai}(\varphi_i)\odot\hat{\boldsymbol{\beta}}_{rsi}^{(m)}$ ($m=1,2,\cdots,K$) , i.e.,

$$\mathbf{S_c} = span\left\{\mathbf{ai}(\varphi_i)\odot\hat{\boldsymbol{\beta}}_{rsi}^{(1)}, \mathbf{ai}(\varphi_i)\odot\hat{\boldsymbol{\beta}}_{rsi}^{(2)}, \cdots, \mathbf{ai}(\varphi_i)\odot\hat{\boldsymbol{\beta}}_{rsi}^{(K)}\right\} \tag{28}$$

Step 4. Projection of signal subspace onto noise subspace. The projection of the signal subspace onto the corresponding noise subspace is performed as follows:

$$J_3 = \sum_{m=1}^{K}\sum_{l=1}^{20-K}(\mathbf{ai}(\phi_i)\odot\hat{\boldsymbol{\beta}}_{rsi}^{(m)})^H\hat{\boldsymbol{\beta}}_{nsi}^{(l)}\hat{\boldsymbol{\beta}}_{nsi}^{(l)H}(\mathbf{ai}(\phi_i)\odot\hat{\boldsymbol{\beta}}_{rsi}^{(m)}) \tag{29}$$

where

$$\mathbf{ai}(\phi_i) = \left[1, e^{j\phi_i}, e^{j\phi_i}, e^{j\phi_i}, e^{j\phi_i}, 1, e^{j\phi_i}, e^{j\phi_i}, e^{j\phi_i}, e^{j\phi_i}, 1, e^{j\phi_i}, e^{j\phi_i}, e^{j\phi_i}, e^{j\phi_i}, 1, e^{j\phi_i}, e^{j\phi_i}, e^{j\phi_i}, e^{j\phi_i}\right]^T \tag{30}$$

The cost function given by (29) is used to estimate the terrain interferometric phase φ_i . And the minimization of J_3 can provide the optimum estimate of the interferometric phase φ_i , i.e., $\hat{\phi}_i = \varphi_i$.

Remark 2: The computational burden will be high if the minimization of J_3 is obtained via search of ϕ_i in the principal phase interval $[-\pi,+\pi]$. To reduce the computational burden, a fast algorithm to compute the minimization of J_3 is developed in Appendix B, where the closed-form solution to the estimate of ϕ_i is directly obtained by using the fast algorithm.

Using the above four steps, the terrain interferogram can be recovered after the pixel pairs of the SAR images are processed separately.

5. Numerical and experemental results

In this section we demonstrate the robustness of the modified method via subspace projection to coregistration errors by using simulated data and real data.

The simulated data are described as follows. We assume there are two formation-flying satellites in the cartwheel formation, and we select one orbit position for simulation, with an effective cross-track baseline of 281.46 m, an orbit height of 750 kilometers and an incidence angle of 45°. We use a two-dimensional window to simulate the terrain and use the statistical model to generate the complex SAR image pairs[23]. The signal-to-noise ratio (SNR) of the SAR images is 18 dB.

Here, the number of the samples to estimate the covariance matrix is 7 (in range) $\times$7 (in azimuth)=49.

Fig.s 9-12 compare the simulation results for various techniques and coregistration errors. Comparing Fig. 9,10 ,11and 12, we can observe that the large coregistration error heavily affects the interferograms obtained by pivoting median filtering, pivoting mean filtering and adaptive contoured window filtering. On the contrary, the large coregistration error has almost no effect on the interferograms obtained by the modified method. We can see that the modified method in this chapter is robust to large coregistration errors (up to one pixel).

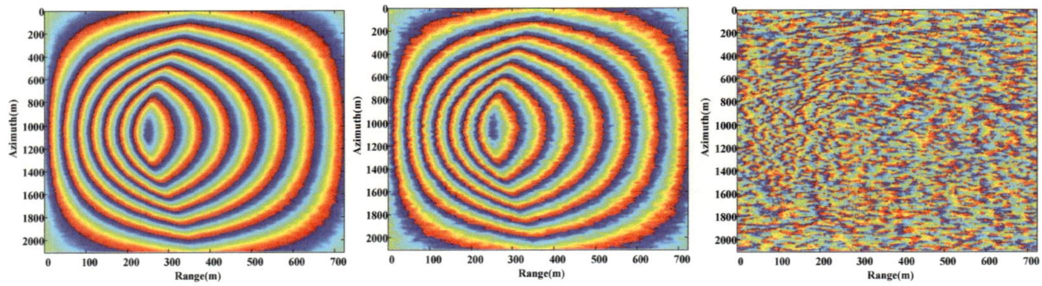

(a) (b) (c)

Fig. 9. The interferograms obtained by the pivoting median filtering for the accurate coregistration (a), the coregistration error of 0.5 pixels (b) and the coregistration error of one pixel (c).

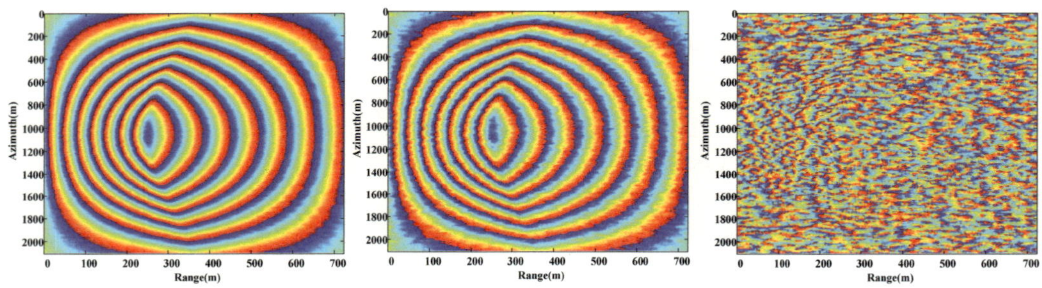

(a) (b) (c)

Fig. 10. The interferograms obtained by the pivoting mean filtering for the accurate coregistration (a), the coregistration error of 0.5 pixels (b) and the coregistration error of one pixel (c).

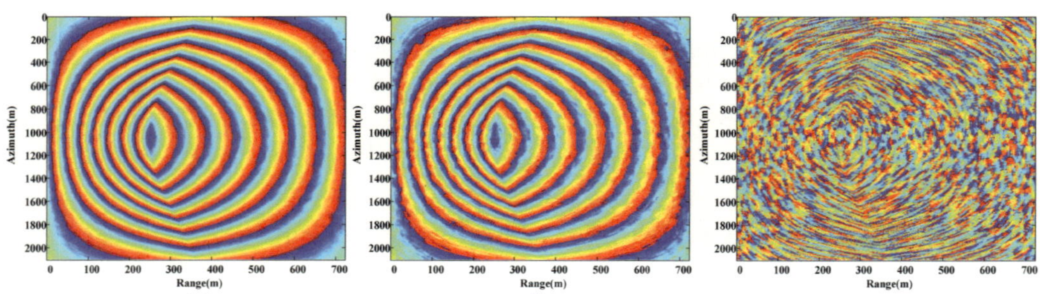

(a) (b) (c)

Fig. 11. The interferograms obtained by the adaptive contoured window filtering for the accurate coregistration (a), the coregistration error of 0.5 pixels (b) and the coregistration error of one pixel (c).

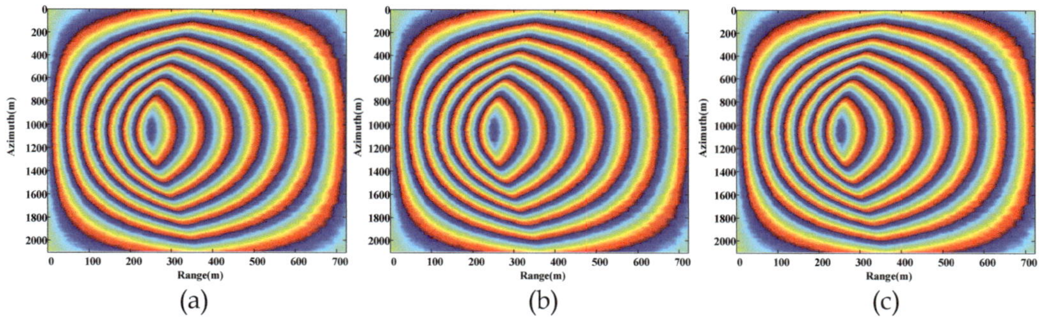

Fig. 12. The interferograms obtained by the modified method for the accurate coregistration (a), the coregistration error of 0.5 pixels (b) and the coregistration error of one pixel (c).

In the following, we will verify the validity of the modified method with the ERS1/ERS2 (European Remote Sensing 1 and 2 tandem satellites) real data.

Fig. 13 shows the interferograms generated from the ERS1/ERS2 real data. Fig.13(a) is the interferogram obtained by the conventional processing, and Fig.13(b) is that obtained by the modified method proposed in this chapter.

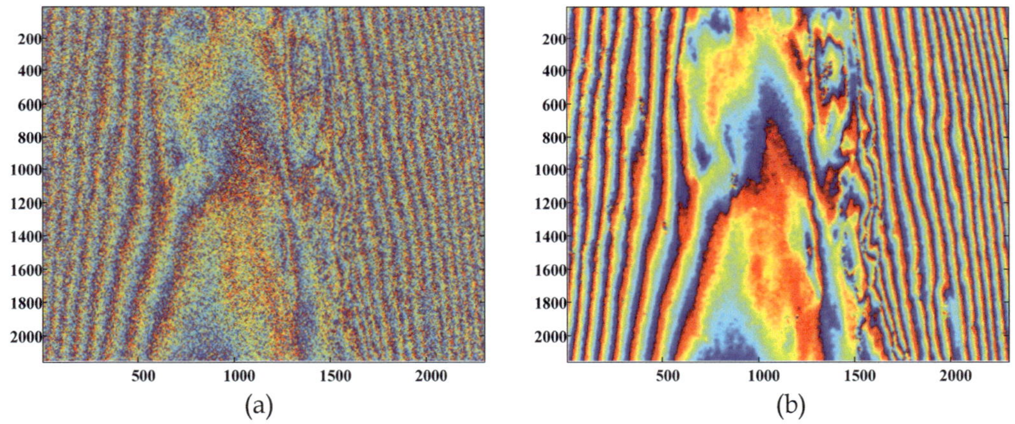

Fig. 13. The interferograms obtained by (a) the conventional processing, and (b) the modified method for the ERS1/ ERS2 real data.

6. Conclusions

In this chapter, the interferometric phase estimation method based on subspace projection and its modified method were presented. The interferometric phase estimation methods based on joint subspace projection can provide accurate estimate of the terrain interferometric phase (interferogram) even if the coregistration error reaches one pixel. Benefiting from the new formulation of joint data vector, the modified method does not need to calculate the noise subspace dimension, thus avoiding the trouble of calculating the noise subspace dimension before estimating the InSAR interferometric phase. A fast algorithm is developed to implement the modified method, which can significantly reduce the computational burden. Theoretical analysis and simulations demonstrate the efficiency of the proposed new algorithms.

7. Acknowledgement

This work is supported in part by the National Natural Science Foundations of China under grant 60736009, 61071194 and 60979002, by the Fund of Civil Aviation University of China under grant ZXH2009D018 and 2011kyE06.

8. Appendix A

8.1 Proof of equation (12)

For easy discussion, we assume that the structure of joint data vector $\mathbf{ss}(i)$ is shown in Fig.A.1, where circles represent SAR image pixels and i denotes the desired pixel pair whose interferometric phase is to be estimated.

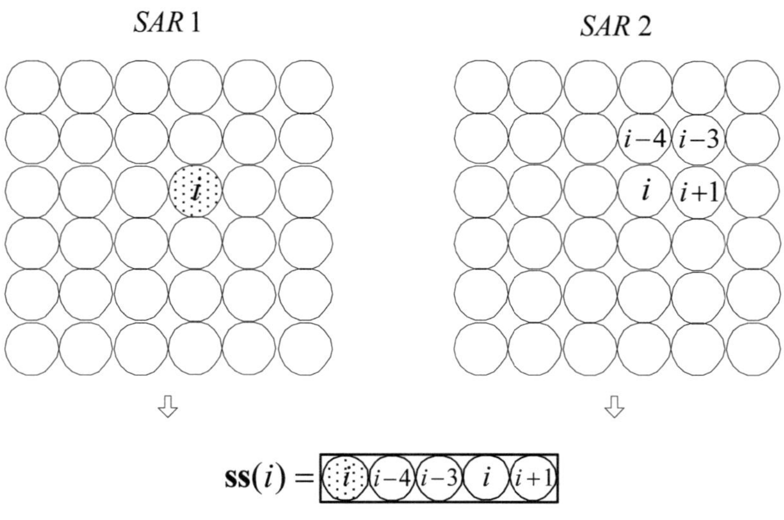

Fig. A.1. Formulation of the joint data vector $\mathbf{ss}(i)$.

The joint data vector $\mathbf{ss}(i)$, shown in Fig.A.1, can be written as

$$\mathbf{ss}(i) = [s_1(i), s_2(i-4), s_2(i-3), s_2(i), s_2(i+1)]^T \tag{A.1}$$

The corresponding covariance matrix $\mathbf{C}_{\mathbf{ss}}(i)$ is given by

$$\begin{aligned}
&\mathbf{C}_{\mathbf{ss}}(i) = E\{\mathbf{ss}(i)\mathbf{ss}^H(i)\}\\
&= E\{[s_1(i), s_2(i-4), s_2(i-3), s_2(i), s_2(i+1)]^T[s_1(i), s_2(i-4), s_2(i-3), s_2(i), s_2(i+1)]^*\}\\
&= \begin{bmatrix}
E\{s_1(i)s_1^*(i)\}, & E\{s_1(i)s_2^*(i-4)\}, & E\{s_1(i)s_2^*(i-3)\}, & E\{s_1(i)s_2^*(i)\}, & E\{s_1(i)s_2^*(i+1)\}\\
E\{s_2(i-4)s_1^*(i)\}, & E\{s_2(i-4)s_2^*(i-4)\}, & E\{s_2(i-4)s_2^*(i-3)\}, & E\{s_2(i-4)s_2^*(i)\}, & E\{s_2(i-4)s_2^*(i+1)\}\\
E\{s_2(i-3)s_1^*(i)\}, & E\{s_2(i-3)s_2^*(i-4)\}, & E\{s_2(i-3)s_2^*(i-3)\}, & E\{s_2(i-3)s_2^*(i)\}, & E\{s_2(i-3)s_2^*(i+1)\}\\
E\{s_2(i)s_1^*(i)\}, & E\{s_2(i)s_2^*(i-4)\}, & E\{s_2(i)s_2^*(i-3)\}, & E\{s_2(i)s_2^*(i)\}, & E\{s_2(i)s_2^*(i+1)\}\\
E\{s_2(i+1)s_1^*(i)\}, & E\{s_2(i+1)s_2^*(i-4)\}, & E\{s_2(i+1)s_2^*(i-3)\}, & E\{s_2(i+1)s_2^*(i)\}, & E\{s_2(i+1)s_2^*(i+1)\}
\end{bmatrix}\\
&= \sigma_s^2(i)\begin{bmatrix}
1, & r_{12}(i,i-4)e^{-j\varphi_i}, & r_{12}(i,i-3)e^{-j\varphi_i}, & r_{12}(i,i)e^{-j\varphi_i}, & r_{12}(i,i+1)e^{-j\varphi_i}\\
r_{21}(i-4,i)e^{j\varphi_i}, & 1, & r_{22}(i-4,i-3), & r_{22}(i-4,i), & r_{22}(i-4,i+1)\\
r_{21}(i-3,i)e^{j\varphi_i}, & r_{22}(i-3,i-4), & 1, & r_{22}(i-3,i), & r_{22}(i-3,i+1)\\
r_{21}(i,i)e^{j\varphi_i}, & r_{22}(i,i-4), & r_{22}(i,i-3), & 1, & r_{22}(i,i+1)\\
r_{21}(i+1,i)e^{j\varphi_i}, & r_{22}(i+1,i-4), & r_{22}(i+1,i-3), & r_{22}(i+1,i), & 1
\end{bmatrix} + \sigma_n^2\mathbf{I}\\
&= \begin{bmatrix}
1, & e^{-j\varphi_i}, & e^{-j\varphi_i}, & e^{-j\varphi_i}, & e^{-j\varphi_i}\\
e^{j\varphi_i}, & 1, & 1, & 1, & 1\\
e^{j\varphi_i}, & 1, & 1, & 1, & 1\\
e^{j\varphi_i}, & 1, & 1, & 1, & 1\\
e^{j\varphi_i}, & 1, & 1, & 1, & 1
\end{bmatrix} \odot \left\{\sigma_s^2(i)\begin{bmatrix}
1, & r_{12}(i,i-4), & r_{12}(i,i-3), & r_{12}(i,i), & r_{12}(i,i+1)\\
r_{21}(i-4,i), & 1, & r_{22}(i-4,i-3), & r_{22}(i-4,i), & r_{22}(i-4,i+1)\\
r_{21}(i-3,i), & r_{22}(i-3,i-4), & 1, & r_{22}(i-3,i), & r_{22}(i-3,i+1)\\
r_{21}(i,i), & r_{22}(i,i-4), & r_{22}(i,i-3), & 1, & r_{22}(i,i+1)\\
r_{21}(i+1,i), & r_{22}(i+1,i-4), & r_{22}(i+1,i-3), & r_{22}(i+1,i), & 1
\end{bmatrix}\right\} + \sigma_n^2\mathbf{I}\\
&= \boldsymbol{\rho}(\varphi_i)\boldsymbol{\rho}^H(\varphi_i) \odot \mathbf{R}_{\mathbf{ss}}(i) + \sigma_n^2\mathbf{I}
\end{aligned} \tag{A.2}$$

where

$$\boldsymbol{\rho}(\varphi_i) = \left[1, e^{j\varphi_i}, e^{j\varphi_i}, e^{j\varphi_i}, e^{j\varphi_i}\right]^T \tag{A.3}$$

$$\mathbf{R}_{\mathbf{ss}}(i) = \sigma_s^2(i)\begin{bmatrix}
1, & r_{12}(i,i-4), & r_{12}(i,i-3), & r_{12}(i,i), & r_{12}(i,i+1)\\
r_{21}(i-4,i), & 1, & r_{22}(i-4,i-3), & r_{22}(i-4,i), & r_{22}(i-4,i+1)\\
r_{21}(i-3,i), & r_{22}(i-3,i-4), & 1, & r_{22}(i-3,i), & r_{22}(i-3,i+1)\\
r_{21}(i,i), & r_{22}(i,i-4), & r_{22}(i,i-3), & 1, & r_{22}(i,i+1)\\
r_{21}(i+1,i), & r_{22}(i+1,i-4), & r_{22}(i+1,i-3), & r_{22}(i+1,i), & 1
\end{bmatrix} \tag{A.4}$$

So the covariance matrix $\mathbf{C}_{\mathbf{si}}(i)$ of the joint data vector $\mathbf{si}(i)$, shown in Fig.5, can be given by

$$\begin{aligned}
&\mathbf{C}_{\mathbf{si}}(i) = E\{\mathbf{si}(i)\mathbf{si}^H(i)\}\\
&= \mathbf{ai}(\varphi_i)\mathbf{ai}^H(\varphi_i) \odot \mathbf{R}_{\mathbf{si}}(i) + \sigma_n^2\mathbf{I}
\end{aligned} \tag{A.5}$$

9. Appendix B

9.1 Fast algorithm for optimal interferometric phase estimation

If U, V and W are arbitrary complex column vectors, then[20]

$$(U \odot V)^H W \cdot W^H (U \odot V) = U^H \left[(W \cdot W^H) \odot (V^* \cdot (V^*)^H) \right] U \tag{B.1}$$

Using the equation (B.1), we can rewrite the cost function of (29) as

$$\begin{aligned}
J_3 &= \sum_{m=1}^{K} \sum_{l=1}^{20-K} (\mathbf{ai}(\phi_i) \odot \hat{\boldsymbol{\beta}}_{rsi}^{(m)})^H \hat{\boldsymbol{\beta}}_{nsi}^{(l)} \hat{\boldsymbol{\beta}}_{nsi}^{(l)H} (\mathbf{ai}(\phi_i) \odot \hat{\boldsymbol{\beta}}_{rsi}^{(m)}) \\
&= \sum_{m=1}^{K} \sum_{l=1}^{20-K} \left\{ \mathbf{ai}^H(\phi_i) [(\hat{\boldsymbol{\beta}}_{nsi}^{(l)} \hat{\boldsymbol{\beta}}_{nsi}^{(l)H}) \odot (\hat{\boldsymbol{\beta}}_{rsi}^{(m)*} (\hat{\boldsymbol{\beta}}_{rsi}^{(m)*})^H)] \mathbf{ai}(\phi_i) \right\} \\
&= \mathbf{ai}^H(\phi_i) \left\{ \sum_{m=1}^{K} \sum_{l=1}^{20-K} [(\hat{\boldsymbol{\beta}}_{nsi}^{(l)} \hat{\boldsymbol{\beta}}_{nsi}^{(l)H}) \odot (\hat{\boldsymbol{\beta}}_{rsi}^{(m)*} (\hat{\boldsymbol{\beta}}_{rsi}^{(m)*})^H)] \right\} \mathbf{ai}(\phi_i)
\end{aligned} \tag{B.2}$$

Let $\mathbf{A} = \sum_{m=1}^{K} \sum_{l=1}^{20-K} [(\hat{\boldsymbol{\beta}}_{nsi}^{(l)} \hat{\boldsymbol{\beta}}_{nsi}^{(l)H}) \odot (\hat{\boldsymbol{\beta}}_{rsi}^{(m)*} (\hat{\boldsymbol{\beta}}_{rsi}^{(m)*})^H)]$. It can be easily proved that $\mathbf{A}$ (20×20) is a Hermitian matrix. Then (B.2) can be rewritten as

$$\begin{aligned}
J_3 &= \mathbf{ai}^H(\phi_i) \left\{ \sum_{m=1}^{K} \sum_{l=1}^{20-K} [(\hat{\boldsymbol{\beta}}_{nsi}^{(l)} \hat{\boldsymbol{\beta}}_{nsi}^{(l)H}) \odot (\hat{\boldsymbol{\beta}}_{rsi}^{(m)*} (\hat{\boldsymbol{\beta}}_{rsi}^{(m)*})^H)] \right\} \mathbf{ai}(\phi_i) \\
&= \mathbf{ai}^H(\phi_i) \mathbf{A} \mathbf{ai}(\phi_i) \\
&= \left[\boldsymbol{\beta}^H(\phi_i), \boldsymbol{\beta}^H(\phi_i), \boldsymbol{\beta}^H(\phi_i), \boldsymbol{\beta}^H(\phi_i) \right] \begin{bmatrix} \mathbf{A}_1 & \mathbf{A}_2 \\ \mathbf{A}_3 & \mathbf{A}_4 \end{bmatrix} \begin{bmatrix} \boldsymbol{\beta}(\phi_i) \\ \boldsymbol{\beta}(\phi_i) \\ \boldsymbol{\beta}(\phi_i) \\ \boldsymbol{\beta}(\phi_i) \end{bmatrix} \\
&= \sum_{n=1}^{4} \boldsymbol{\beta}^H(\phi_i) \mathbf{A}_n \boldsymbol{\beta}(\phi_i) \\
&= \boldsymbol{\beta}^H(\phi_i) \left\{ \sum_{n=1}^{4} \mathbf{A}_n \right\} \boldsymbol{\beta}(\phi_i)
\end{aligned} \tag{B.3}$$

where

$$\boldsymbol{\beta}(\phi_i) = \left[1, e^{j\phi_i}, e^{j\phi_i}, e^{j\phi_i}, e^{j\phi_i} \right]^T \tag{B.4}$$

Let

$$\mathbf{B} = \sum_{n=1}^{4} \mathbf{A}_n = \begin{bmatrix} b_{11} & b_{12} & b_{13} & b_{14} & b_{15} \\ b_{21} & b_{22} & b_{23} & b_{24} & b_{25} \\ b_{31} & b_{32} & b_{33} & b_{34} & b_{35} \\ b_{41} & b_{42} & b_{43} & b_{44} & b_{45} \\ b_{51} & b_{52} & b_{53} & b_{54} & b_{55} \end{bmatrix} \tag{B.5}$$

It can be easily proved that $\mathbf{B}$ is a Hermitian matrix, i.e.,

$$b_{1n}^* = b_{n1} \; (n = 2,3,4,5) \tag{B.6}$$

so we can get $(\sum_{n=2}^{5} b_{1n})^* = \sum_{n=2}^{5} b_{n1}$, Let $\sum_{n=2}^{5} b_{1n} = \left|\sum_{n=2}^{5} b_{1n}\right| e^{j\mu}$ where $\mu = angle(\sum_{n=2}^{5} b_{1n})$ and $-\pi \le \mu < \pi$.

Using (B.5) and (B.6), the cost function of (B.3) can be rewritten as

$$
\begin{aligned}
J_3 &= \boldsymbol{\beta}^H(\phi_i)\left\{\sum_{n=1}^{4} \mathbf{A}_n\right\}\boldsymbol{\beta}(\phi_i) \\
&= \boldsymbol{\beta}^H(\phi_i)\mathbf{B}\boldsymbol{\beta}(\phi_i) \\
&= \left[1, e^{-j\phi_i}, e^{-j\phi_i}, e^{-j\phi_i}, e^{-j\phi_i}\right] \begin{bmatrix} b_{11} & b_{12} & b_{13} & b_{14} & b_{15} \\ b_{21} & b_{22} & b_{23} & b_{24} & b_{25} \\ b_{31} & b_{32} & b_{33} & b_{34} & b_{35} \\ b_{41} & b_{42} & b_{43} & b_{44} & b_{45} \\ b_{51} & b_{52} & b_{53} & b_{54} & b_{55} \end{bmatrix} \begin{bmatrix} 1 \\ e^{j\phi_i} \\ e^{j\phi_i} \\ e^{j\phi_i} \\ e^{j\phi_i} \end{bmatrix} \\
&= b_{11} + \sum_{m=2}^{5}\sum_{n=2}^{5} b_{mn} + (\sum_{n=2}^{5} b_{n1})e^{-j\phi_i} + (\sum_{n=2}^{5} b_{1n})e^{j\phi_i} \\
&= b_{11} + \sum_{m=2}^{5}\sum_{n=2}^{5} b_{mn} + ((\sum_{n=2}^{5} b_{1n})e^{j\phi_i})^* + (\sum_{n=2}^{5} b_{1n})e^{j\phi_i} \\
&= b_{11} + \sum_{m=2}^{5}\sum_{n=2}^{5} b_{mn} + (\left|\sum_{n=2}^{5} b_{1n}\right| e^{j\mu} \cdot e^{j\phi_i})^* + \left|\sum_{n=2}^{5} b_{1n}\right| e^{j\mu} \cdot e^{j\phi_i} \\
&= b_{11} + \sum_{m=2}^{5}\sum_{n=2}^{5} b_{mn} + 2\left|\sum_{n=2}^{5} b_{1n}\right| \cos(\mu + \phi_i)
\end{aligned}
\tag{B.7}
$$

Obviously, the minimization of J_3 can be obtained for $\mu + \phi_i = -\pi + 2k\pi$ (k is an integer). Since $-\pi \le \mu < \pi$ and $-\pi < \phi_i < \pi$, thus

$$
\phi_i = \begin{cases} -\pi - \mu & (\mu \le 0) \\ \pi - \mu & (\mu > 0) \end{cases} \tag{B.8}
$$

So the closed-form solution to the estimate of ϕ_i is directly obtained by using the fast algorithm, which can significantly reduce the computational burden.

10. References

[1] P. A. Rosen, S. Hensley, I. R. Joughin, F. K. Li, S. N. Madsen, E. Rodrìguez and R. M. Goldstein. Synthetic Aperture Radar Interferometry. In: Proceeding of the IEEE, 2000,88(3):333-382.

[2] R. Bamler and P. Hartl. Synthetic aperture radar interferometry. Inverse Problem, 1998,14: R1-R54.

[3] Wei Xu and Ian Cumming. A Region-Growing Algorithm for InSAR Phase Unwrapping. IEEE Trans. On GRS, 1999, 37(1): 124-134.

[4] M. Costantini. A Novel Phase Unwrapping Method based on Network Programming. IEEE Trans. On GRS, 1998, 36(3): 813-831.

[5] R. M. Goldstein, H. A. Zebker, C. L. Werner.Satellite radar interferometry : two-dimensional phase unwrapping. Radio Sci ., 1988,23(4):713 -720.

[6] M. D. Pritt and J. S. Shipman.Least-Squares Two-Dimensional Phase Unwrapping Using FFT's. IEEE Trans. On GRS ,1994, 32(3):706-708.
[7] B. Zitova and J. Flusser. Image registration methods: A survey. Image Vis. Comput., 2003,21(11): 977-1000.
[8] A. Cole-Rhodes, K. L. Johnson, J. Le Moigne, and I. Zavorin. Multiresolution registration of remote sensing imagery by optimization of mutual information using a stochastic gradient. IEEE Trans. Image Process., 2003,12:1495-1511.
[9] X. Dai and S. Khorram. A feature-based image registration algorithm using improved chain-code representation combined with invariant moments. IEEE Trans. On GRS, 1999, 37(5):2351-2362.
[10] Harold S. Stone, Michael T. Orchard, Ee-Chien Chang, and Stephen A. Martucci. A Fast Direct Fourier-Based Algorithm for Subpixel Registration of Images. IEEE Trans. On GRS, 2001, 39(10):2235-2243.
[11] Eugenio Sansosti, Paolo Berardino, Michele Manunta, Francesco Serafino, and Gianfranco Fornaro. Geometrical SAR Image Registration. IEEE Trans. On GRS, 2006, 44(10):2861-2870.
[12] Francesco Serafino. SAR Image Coregistration Based on Isolated Point Scatterers. IEEE GRS Letters, 2006, 3(3):354-358.
[13] P. H. Eichel, D. C. Ghiglia, et al. Spotlight SAR Interferometry for Terrain Elevation Mapping and Interferometric Change Detection. Sandia National Labs Tech. Report, SAND93, December 1993, pp. 2539-2546.
[14] R. Lanari, G. Fornaro, et al. Generation of Digital Elevation Models by Using SIR-C/X-SAR Multifrequency Two-Pass Interferometry: The Etna Case Study. IEEE Trans. On GRS, 1996, 34(5):1097-1114.
[15] Jong-Sen Lee, Konstantinos P. Papathanassiou, et al. A New Technique for Noise Filtering of SAR Interferometric Phase Images. IEEE Trans. On GRS, 1998, 36(5):1456-1465.
[16] Ireneusz Baran, Mike P. Stewart, Bert M. Kampes, Zbigniew Perski, and Peter Lilly. A Modification to the Goldstein Radar Interferogram Filter. IEEE Trans. On GRS, 2003, 41(9):2114-2118.
[17] Nan Wu, Da-Zheng Feng, and Junxia Li. A locally adaptive filter of interferometric phase images. IEEE GRS Letters, 2006, 3(1):73-77.
[18] Qifeng Yu, Xia Yang, Sihua Fu, Xiaolin Liu, and Xiangyi Sun. An Adaptive Contoured Window Filter for Interferometric Synthetic Aperture Radar . IEEE GRS Letters, 2007, 4(1):23-26.
[19] Hai Li, Guisheng Liao. An estimation method for InSAR interferometric phase based on MMSE criterion. IEEE Trans On GRS , 2010,48(3):1457-1469.
[20] Hai Li, Zhenfang Li, Guisheng Liao, and Zheng Bao. An estimation method for InSAR interferometric phase combined with image auto-coregistration. Science in China, Series F, 2006, 49(3): 386-396.
[21] Zhenfang Li, Zheng Bao, Hai Li, and Guisheng Liao. Image Auto-Coregistration and InSAR Interferogram Estimation Using Joint Subspace Projection. IEEE Trans. On GRS ,2006,44(2):288-297.
[22] F. Lombardini, M. Montanari, and F. Gini. Reflectivity Estimation for Multibaseline Interferometric Radar Imaging of Layover Extended Sources. IEEE Trans On SP, 2003, 51(6): 1508-1519.
[23] F. Lombardini. Absolute Phase Retrieval in a Three-element Synthetic Aperture Radar Interferometer. In: Proceeding of the 1996 CIE Int. Conf. of Radar, Beijing, China, 1996, 309-312.

7

Airborne Passive Localization Method Based on Doppler-Phase Interference Measuring

Tao Yu
Shanghai Research Institute of Microwave Equipment, Shanghai
P. R. China

1. Introduction

Airborne passive localization with single station is a key technique which is quickly developing lately. If the exact localization to target can be realized in single mobile platform, its significance is self-evident. Because the information content obtained by single station is lesser, the difficulty would be obviously bigger relative to multi-station observation.

Phase interferometry is mainly in use for direction finding (DF). Moreover, Doppler shift may be used to determine the coordinate setting of target. As far as actual applications are concerned, these two methods are independently applied. Or, they firstly are respectively detected. And then, the obtaining data are syncretized.

Author researches the functional relation between Doppler shift and phase interferometry and presents some method which can syncretize these two methods directly from the equation of mathematical. Based on this fusion in physical relationship, we can obtain the Doppler shift by measuring phase. Or, we can determine the phase by measuring Doppler shift. This book describes these new methods as well as some applications in airborne passive localization system.

Section two describes the relationship between phase and Doppler shift. Firstly, the phase difference localization equation which is analogous to time difference localization equation is introduced by analyzing the phase in radial distance of target. And then, making use of the complementary relationship between the angle of advance at airborne platform and the angle of arrival for target, the approximate functional relation between Phase-difference and Doppler shift can be derived by combining DF formula based on phase interference and expression of Doppler shift. Where after, the calculating method of airborne Doppler rate is researched based on phase difference or Doppler frequency difference

Section three researches the problem of airborne Doppler direct range finding. Three solution methods are presented. First method is direct approximation for mathematical expression of Doppler rate. Second method is introducing differential transformation for Doppler shift equation based on angle rate. Third method is based on the specific value of Doppler rate between two adjacent detection nodes. On this basis, the airborne direct range finding formula based on phase measuring can be obtained by using the relationship between phase difference and Doppler shift.

In allusion to the problem to need solving integer ambiguity in available DF system based on phase interferometer, section four puts forward a DF method combining Doppler shift and phase interferometer. According to this method, the integer of wavelength in radial distant can be directly solved synthetically by making use of speed vector and Doppler shift as well as rate. Hence, the integer of wavelength in path length difference of radial distant between two adjacent antenna elements can be obtained. At the same time, the value less than one integer in path difference can be determined by using phase interferometer. As compared with available phase interferometer which firstly solves phase difference, this new method which firstly solves path difference has not phase ambiguity, nor it needs to limit baseline length.

Section fifthly presents a new airborne passive DF method using two orthogonal baselines based on the rate of direction cosine. This is a passive DF method which is only associated with Doppler frequency difference and is not associated with wavelength.

2. Relationship between phase and Doppler shift

2.1 Phase-detecting for distance

For single baseline interferometer as shown in fig.2-1, if the phase shift measured by descriminator, corresponding to the radial distance r_i , is ϕ_i , the expression of radial distance based on phase measuring is

$$r_i = \lambda(N_i + \frac{\phi_i}{2\pi}) \tag{2-1}$$

where: λ is wavelength; N_i the whole number of wavelength.

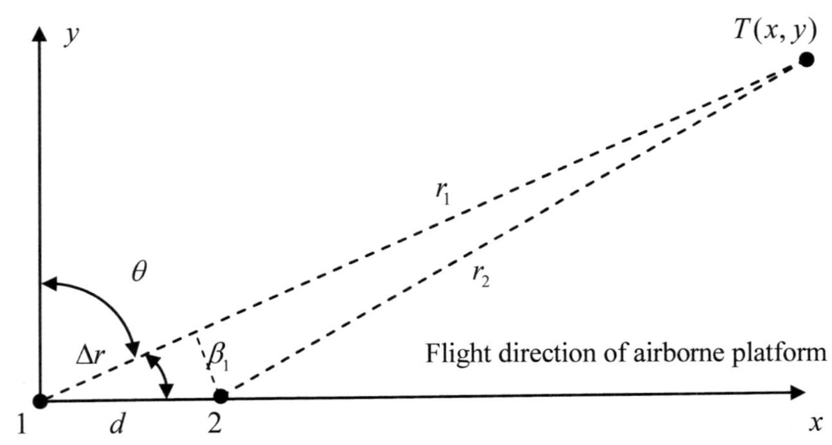

Fig. 2-1. Single baseline array on airborne platform.

Thus, the path length difference Δr of radial distance between two antenna elements can be determined by measuring for phase difference. The localization equation based on phase difference can be obtained whose formed expression is completely analogous to the localization equation based on time difference

$$\begin{aligned}\Delta r &= r_1 - r_2 \\ &= \lambda(N_1 - N_2 + \frac{\phi_1 - \phi_2}{2\pi}) \\ &= \lambda(\Delta N + \frac{\Delta\phi}{2\pi})\end{aligned} \quad (2\text{-}2)$$

where: $\Delta N = N_1 - N_2$ is the whole number of wavelength in path length difference; $\Delta\phi = \phi_1 - \phi_2$ phase difference.

Provided that incident wave from the some radiant is approximatively regarded as plane wave, the existing DF expression based on the principle of phase interferometry can be approximatively obtained due to sine theorem

$$\begin{aligned}\sin\theta &= \frac{\Delta r}{d} \\ &= \frac{\lambda}{d}(\Delta N + \frac{\Delta\phi}{2\pi})\end{aligned} \quad (2\text{-}3)$$

Or, by simply rearranging, the phase difference between two adjacent antennas can take the form

$$2\pi\Delta N + \Delta\phi = \frac{2\pi d}{\lambda}\sin\theta \quad (2\text{-}4)$$

where: θ is arrival angle of target signal; d baseline length between two antennas.

2.2 Phase difference detecting of Doppler shift

2.2.1 Recapitulation

Making use of the complementary relationship between the angle of advance and the angle of arrival, the approximate functional relation between phase-difference and Doppler shift can be obtained by combining the DF formula based on phase interference and the expression of Doppler shift. The analog calculation certificates the correctness of derived formula. The error analysis presents the measurement accuracy of Doppler shift obtained by phase-difference measuring. And calculation shows that the measuring error of Doppler shift obtained based on phase-difference measuring is lower if the detection error for integer of wavelength in path difference is not considered.

2.2.2 Primitive formula

As shown in fig.2-1, a single baseline array with two antenna elements is installed on airborne platform and the spacing of array is d . The direction of axis of baseline is parallel the axis of airborne platform. For the target T at stationary or low speed, the Doppler shift detected by airborne double channel measuring receiver in every element of single baseline interferometry is

$$\lambda f_{d1} = v\cos\beta_1 \tag{2-5}$$

where: f_{d1} is Doppler shift; v the flight speed of detection platform; β the angle of advance between radial distance and travelling direction of flight device.

According to the geometric relationship as shown fig.2-1 and the analysis results in a previous section, results in

$$\begin{aligned} \cos\beta_1 &= \sin(90^0 - \beta_1) \\ &= \sin\theta \approx \frac{\Delta r}{d} \\ &= \frac{\lambda}{d}(\Delta N + \frac{\Delta\phi}{2\pi}) \end{aligned} \tag{2-6}$$

Substituting (2-6) into the Doppler shift expression, we can obtain the calculating formula of Doppler shift based on phase difference measuring

$$\begin{aligned} f_{d1} &= \frac{v}{d}(\Delta N + \frac{\Delta\phi}{2\pi}) \\ &= \frac{v\Delta r}{d\lambda} \end{aligned} \tag{2-7}$$

According to above-mentioned derivation, we can directly written out the formula of Doppler frequency difference based on phase difference measuring by use of the array with double baseline in one-dimensional as shown in fig.2-2

$$\begin{aligned} \Delta f_d &= \frac{v}{d}\left[(\Delta N_1 + \frac{\Delta\phi_1}{2\pi}) - (\Delta N_2 + \frac{\Delta\phi_2}{2\pi})\right] \\ &= \frac{v}{d}\left[(\Delta N_1 - \Delta N_2) + \frac{1}{2\pi}(\Delta\phi_1 - \Delta\phi_2)\right] \end{aligned} \tag{2-8}$$

where: $\Delta N_i = N_i - N_{i+1}$; $\Delta\phi_i = \phi_i - \phi_{i+1}$.

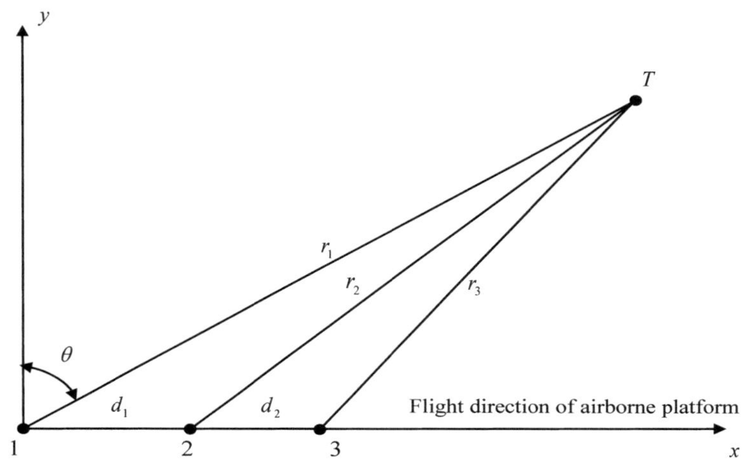

Fig. 2-2. One-dimensional array with double baseline.

2.2.3 Computational error

In order to verify the calculating formula of Doppler shift based on phase difference measuring, we make the analog calculation by replacing measured value with theoretical value. Firstly, we preset following parameter: radial distance r_1, wavelength λ, baseline length d, the flight speed of flight device v. And then, we make the arrival angle θ change in prescribed domain. Hence, the rest radial distance as well as path difference can be obtained according to circular function relationship. Simultaneity, the theoretical value of Doppler shift corresponding to every radial distance can be calculated by Doppler shift formula.

On this basis, the Doppler shift based on phase measuring is computed by Eq.(2-7) and then the relative calculation error is obtained by compared with the theoretical value

$$\varepsilon_f = \left| \frac{f_{di} - f_{dai}}{f_{di}} \right| \times 100\% \tag{2-9}$$

where: subscript a expresses calculating value.

Without the notice, the adopted parameter is as follows: $r_1 = 100\ km$, $v = 100\ m/s$, $d = 5\lambda$, $\lambda = 0.0375\ m$.

Fig.2-3 depicts the calculating error curve of Doppler shift f_{d1} when the baseline length is different. The mathematical simulation shows that calculating error is independent of flight speed and directly related to wavelength and inversely related to radial distance.

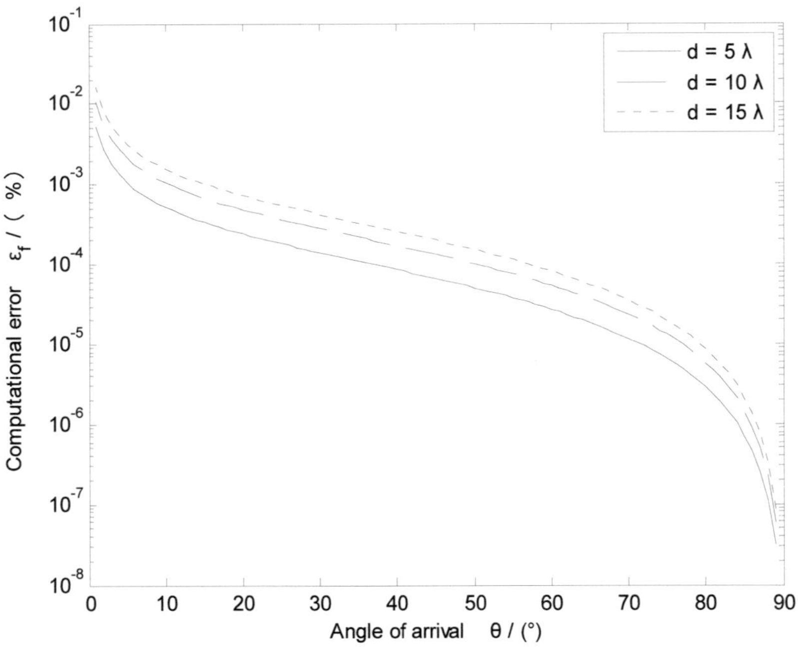

Fig. 2-3. Computational error of Doppler shift with different baseline lengths.

2.2.4 Precision analysis

According to error estimation theory, the measuring error of Doppler shift produced by the measuring error of phase difference, flight speed and integer of wavelength in path difference is

$$\begin{aligned} \sigma &= \sqrt{\left(\frac{\partial f_{d1}}{\partial \Delta\phi}\sigma_{\phi}\right)^2+\left(\frac{\partial f_{d1}}{\partial v}\sigma_{v}\right)^2+\left(\frac{\partial f_{d1}}{\partial \Delta N}\sigma_{n}\right)^2} \\ &= \sqrt{\left(\frac{v}{2\pi d}\sigma_{\phi}\right)^2+\left(\frac{\Delta r}{\lambda d}\sigma_{v}\right)^2+\left(\frac{v}{d}\sigma_{n}\right)^2} \end{aligned} \qquad (2\text{-}10)$$

where: σ_{ϕ}, σ_{v} and σ_{n} is respectively the mean-root-square error measuring phase difference, flight speed and integer of wavelength in path difference. In where, unit for σ_{ϕ} is radian. Without the notice, the value of mean-root-square error is respectively as follows: $\sigma_{\phi}=\pi/180^0(=1^0)$, $\sigma_{v}=1\ m/s$, $\sigma_{n}=1$.

It can be seen from (2-10) that the Doppler measuring error produced only by measuring error of phase difference is just a constant term. If $\sigma_{\phi}=\pi/180^0$ (radian), σ value will be less than $2Hz$ making use of the parameter presented in previous section. The Doppler measuring error produced only by integer of wavelength in path difference is also a constant term. Because the speed is usually larger than baseline length for Airborne applications, the Doppler measuring error will be prodigious if the measuring for integer of wavelength in path difference has error. According to the same reason, the larger measuring error of Doppler shift is also produced by the error of speed measuring in axle direction of baseline.

Fig.2-4 depicts the Doppler measuring error with different baseline length. The analysis shows that the measuring error in integer of wavelength has prodigious influence for measuring error of Doppler shift. Also, the change for baseline length is also considerable sensitive.

Fig.2-5 illustrates the change curve versus wavelength for Doppler shift error when the integer of wavelength in path difference can be accurately detected. And the analyzer shows that the measuring error is not associated baseline and the influence produced by change of flight speed is also lesser.

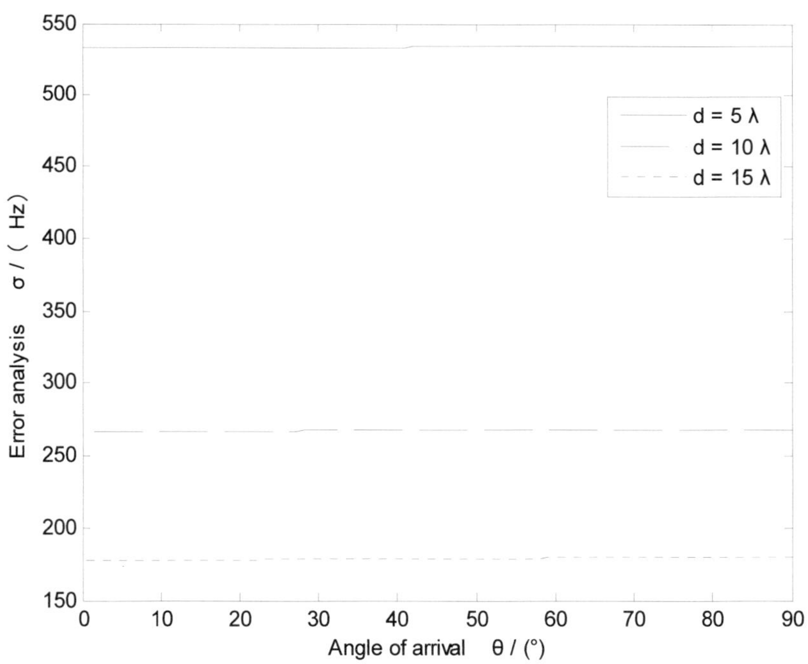

Fig. 2-4. Measurement error of Doppler shift.

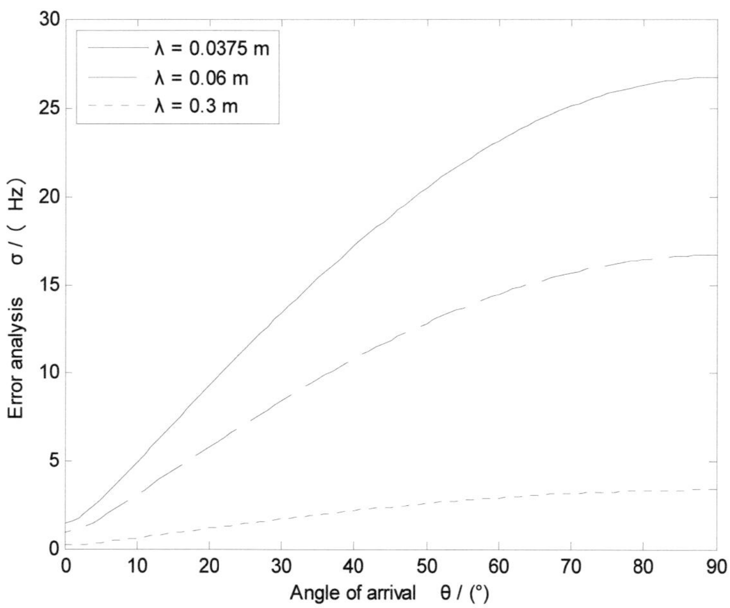

Fig. 2-5. Measurement error of Doppler shift without integer error.

More, relative calculation error of Doppler shift is analyzed by perfect differential, that is

$$df_{d1} = \frac{\partial f_{d1}}{\partial \Delta\phi} d\Delta\phi_i + \frac{\partial f_{d1}}{\partial v} dv + \frac{\partial f_{d1}}{\partial \Delta N} d\Delta N \tag{2-11}$$

When the error of observing variable is zero-mean and unaided reciprocally, the relative calculation error of Doppler shift is

$$\begin{aligned} \sigma_r &= \left|\frac{df_{di}}{f_{di}}\right| \\ &= \frac{1}{f_{di}}\left(\left|\frac{\partial f_{di}}{\partial \Delta\phi_i}\sigma_\phi\right| + \left|\frac{\partial f_{d1}}{\partial v}\sigma_v\right| + \left|\frac{\partial f_{d1}}{\partial \Delta N}\sigma_n\right|\right) \\ &= \frac{\lambda d}{v\Delta r}\left(\left|\frac{v}{2\pi d}\sigma_\phi\right| + \left|\frac{\Delta r}{\lambda d}\sigma_v\right| + \left|\frac{v}{d}\sigma_n\right|\right) \end{aligned} \tag{2-12}$$

The advantage making the expression which contains phase difference transform the expression which includes path difference Δr is that the analysis and calculation for integer and phase difference can be avoided. Fig.2-6 illustrates the change curve versus baseline length for relative calculation error of Doppler shift when the measuring for integer is without error. The calculation shows that the curvilinear change is basically not associated with flight speed and wavelength. The relative calculation error is inversely relative to baseline length. It can be seen that the relative calculation error of Doppler shift is less than 1.5% after the angle of arrival is bigger than 10^0

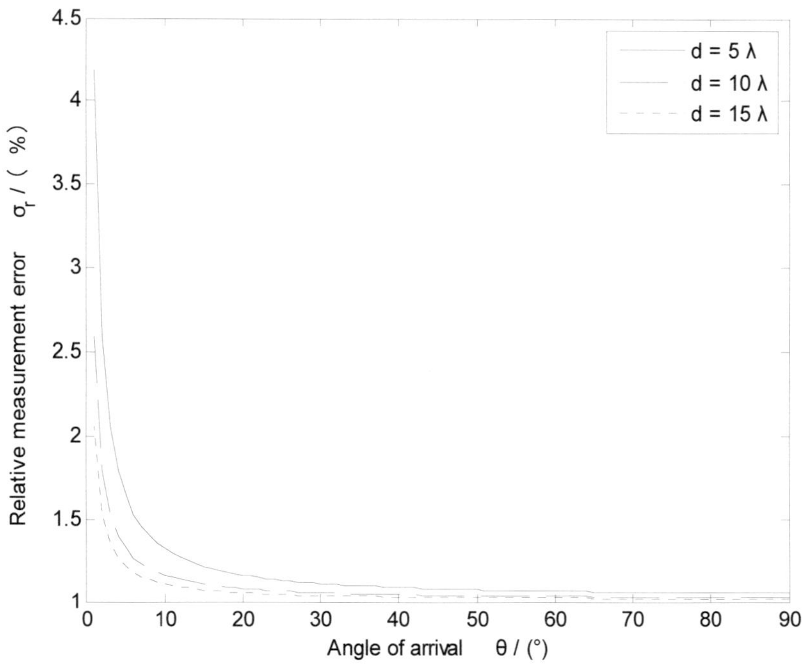

Fig. 2-6. Relative measurement error of Doppler shift.

2.3 Airborne Doppler changing rate obtaining by frequency difference or phase difference

2.3.1 Recapitulation

According to localization theory, the radial distance between measuring platform and measured target can be directly obtained based on Doppler changing rate equation. But, in fact, the localization method based on Doppler rate of change is not the classical method in current target localization for electronic warfare. A main reason is that measuring for change rate of Doppler shift is also relative difficult at present. At the same time, because Doppler rate of change is directly concerned with the tangential velocity, the problem related to direct range finding using Doppler rate equation is that the detecting system has to measure the angle of advance between the traveling direction of detection platform and the radial distance to target, along with measuring Doppler rate. Thereout, the localization equation can be solved. Hence, the localization method based on Doppler rate of change can be completed with other localization method at present.

In addition, for localization and tracker system, it is extraordinary valuable to obtain the Doppler changing rate of received signal in order to estimate the state and position of target. At present, Doppler rate of change is mainly obtained by estimating the frequency variation of received signal based on the principle that Doppler rate of change is the same in mathematical analysis as the rate of carrier frequency of received signal. These estimative algorithms are not only associated with modulation mode, but they are also complicated.

On the basis of researching Doppler direct ranging, this section presents a method determining Doppler changing rate only by detecting Doppler shift value of received signal at some discrete node. Here, an analytic method solving Doppler changing rate only by detecting the frequency difference of Doppler shift or received signal is presented based on azimuth rate. This new method is not only straightforward in detection mode; it is also succinct on expression. More important characteristic is lying in:

1. The Doppler rate of change can be directly obtained when the wavelength of measured signal is yet unknown.
2. It is wholly not associated with light speed. And this is completely helpful for realizing high-accuracy measuring. The existing analyzing indicates that the result of error analysis will become very bad if the resolving result is relative to light speed.

2.3.2 Derivation

According to the geometric relationship as shown in fig.2-1, we have as following the approximate representation of circular function

$$\sin\beta_1 \approx \frac{\sqrt{d^2 - \Delta r^2}}{d} \tag{2-13}$$

$$\cos\beta_1 \approx \frac{\Delta r}{d} \tag{2-14}$$

By applying differential transformation, we can make a deformation to the Doppler shift equation

$$\begin{aligned} \lambda f_d = v\cos\beta_1 &= \frac{v}{\omega}\frac{\partial \sin\beta_1}{\partial t} \\ &= -\frac{v\Delta r}{\omega d\sqrt{d^2 - \Delta r^2}}\frac{\partial \Delta r}{\partial t} \\ &= \frac{v\Delta r \lambda \Delta f_d}{\omega d\sqrt{d^2 - \Delta r^2}} \end{aligned} \tag{2-15}$$

where: $\Delta f_d = f_{d2} - f_{d1}$ is the frequency difference of Doppler shift; $\omega = \partial\beta / \partial t$ angular velocity.

Making use of circular function and by rearranging, we get

$$\omega f_d = \frac{v}{d}\Delta f_d ctg\beta_1 \tag{2-16}$$

Further, the Eq.(2-16) can be deformed by use of angular velocity $\omega = \frac{v_t}{r}$ and expression $\lambda f_d = v\cos\beta$

$$\frac{v^2 \sin^2 \beta_1}{\lambda r_1} = \frac{v}{d} \Delta f_d \tag{2-17}$$

In where: $v_t = v \sin \beta_1$ is tangential velocity.

In fact, when the flight device is uniform motion, the left-hand component of equation is namely basic expression of Doppler rate of change

$$\dot{f}_d = \frac{v^2 \sin^2 \beta}{\lambda r_1}$$

So we obtain the calculation formula of Doppler rate only based on measuring Doppler frequency difference

$$\dot{f}_d = \frac{v}{d} \Delta f_d \tag{2-18}$$

According to the mathematical definition of Doppler rate

$$\dot{f}_d = \Delta f_d / \Delta t$$

There results:

$$\frac{v}{d} = \frac{1}{\Delta t}$$

It can be seen that the time variation of detecting Doppler shift can be equivalently expressed by specific value between flight distance and flight speed of detection platform. Hence, we also prove that Doppler changing rate is only associated with flight speed and it is wholly not associated with light speed.

Again, the frequency difference of Doppler shift can is also obtained from one of radiation frequency

$$\begin{aligned} \Delta f_d &= f_{d1} - f_{d2} \\ &= f_{t1} - f_{t2} = \Delta f_t \end{aligned}$$

where: f_{ti} is measured value of signal.

So Doppler changing rate can be determined by the actual measurement to frequency difference of signal.

As soon as we substitute the expression (2-8) of Doppler frequency difference based on phase difference measuring into (2-18), we can obtain the formula of Doppler changing rate based on phase difference

$$\dot{f}_d = \left(\frac{v}{d}\right)^2 \left[(\Delta N_2 - \Delta N_1) + \frac{1}{2\pi} (\Delta\phi_2 - \Delta\phi_1) \right] \tag{2-19}$$

2.3.3 Analog calculation

In order to verify the accuracy of analysis formula of Doppler rate based on measuring frequency difference, the mathematical simulation is applied with measured value replaced by theoretical value. The wavelength λ, radial distance r_1, the flight speed of flight device v and time interval Δt are set beforehand and making the angle of advance β_1 change in prescribed domain. Hence, the rest radial distance and angle of advance can be obtained according to geometric relationship as shown in fig.2-1. Thus, the theoretical value of Doppler shift and Doppler rate of change can be calculated corresponding to every radial distance.

On this basis, Doppler changing rate is computed by Eq.(2-18) and then the relative calculation error is obtained by compared with the theoretical value.

Without the notice, the adopted parameter is as follows: $\lambda = 0.25\ m$, $r = 100\ km$, $v = 100\ m/s$, $\Delta t = 5\ s$.

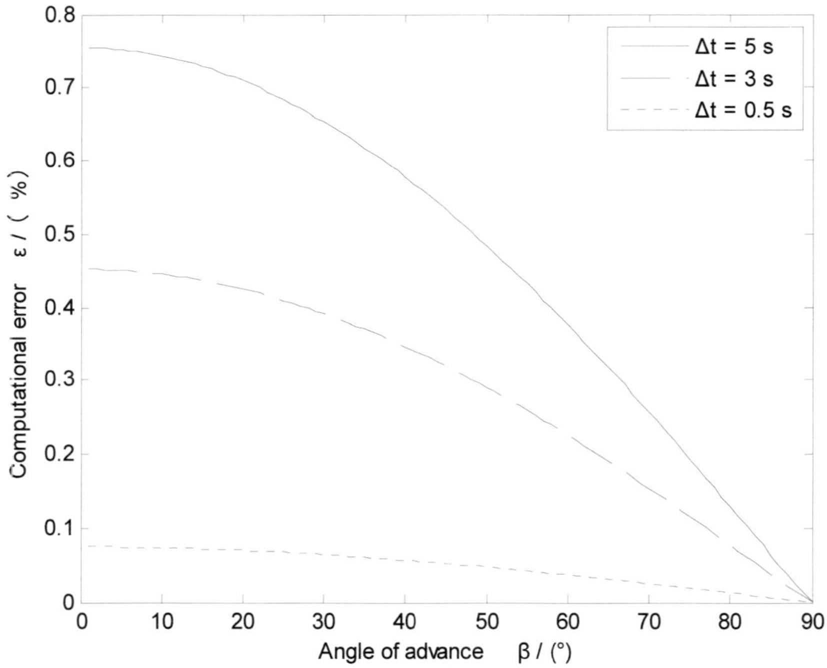

Fig. 2-7. Error curve of Doppler rate against advancing angle in different detecting times.

Fig. 2-7 depicts the relative calculation error curve against the angle of advance for Doppler changing rate when the detecting time interval is different. It can be seen that the error will be conspicuously augmentation if the time interval is too long. Fig.2-8 depicts the relative calculation error is inversely relative to radial distance.

The mathematical simulation shown yet that relative calculation error are independent of flight speed and wavelength. The relative calculation error curve is smoother as compared to the curve given by existing literature. And the behavior is basically identical.

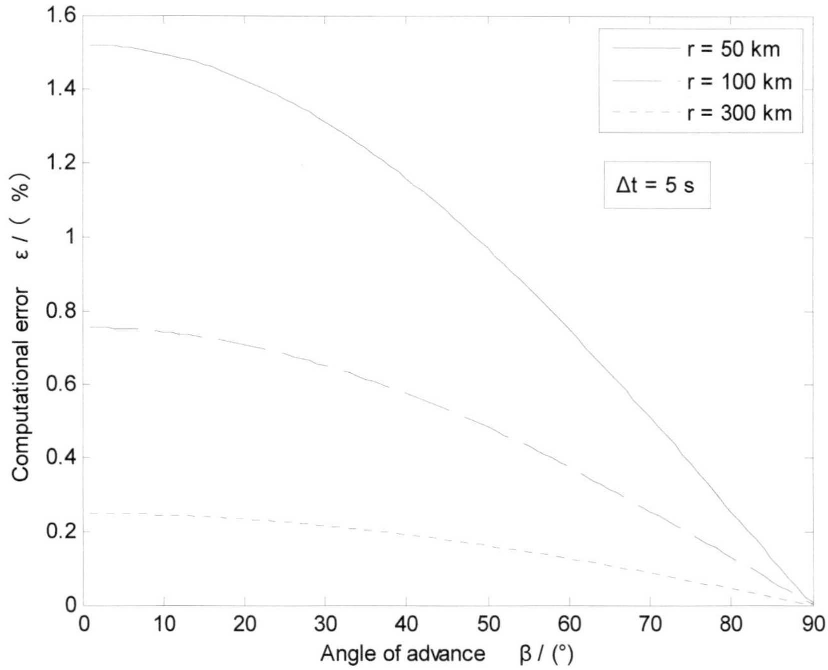

Fig. 2-8. Relative calculation error curve of Doppler changing rate against advancing angle in different radial distance

3. Airborne range finding method

3.1 Principle of Doppler direct range finding

3.1.1 Recapitulation

According to mathematical definition, we can realize airborne single-station ranging only based on Doppler shift measurement by approximatively dealing with Doppler changing rate equation. On this basis, the expressing problem of average for Doppler changing rate in a time interval is researched. On condition that airborne measuring station is uniform flight along linear motion, the analysis shows that the average value of Doppler changing rate is directly related to the product of tangential velocity at two terminals and is inversely relative to radial distance at end position in a time interval. And the analog calculation verifies that the better ranging result can be obtained only from this average expression. As contrasted to existing method, the ranging method derived in this text requires neither to detect Doppler changing rate directly nor to use other localization methods.

3.1.2 Basic range finding formula

Provided that airborne platform is uniform motion along straight line as shown in fig.3-1, the Doppler changing rate equation between detection platform and measured target is

$$\dot{f}_d = \frac{v_t^2}{\lambda \cdot r} \tag{3-1}$$

From the viewpoint of basic mathematical definition, during a time interval Δt, the Doppler changing rate can be approximatively expressed by measured value of Doppler frequency difference between two detection nodes

$$\begin{aligned} \dot{f}_d &= \frac{\Delta f_d}{\Delta t} \\ &= \frac{f_{d2} - f_{d1}}{\Delta t} \end{aligned} \tag{3-2}$$

where: Δt is a time interval.

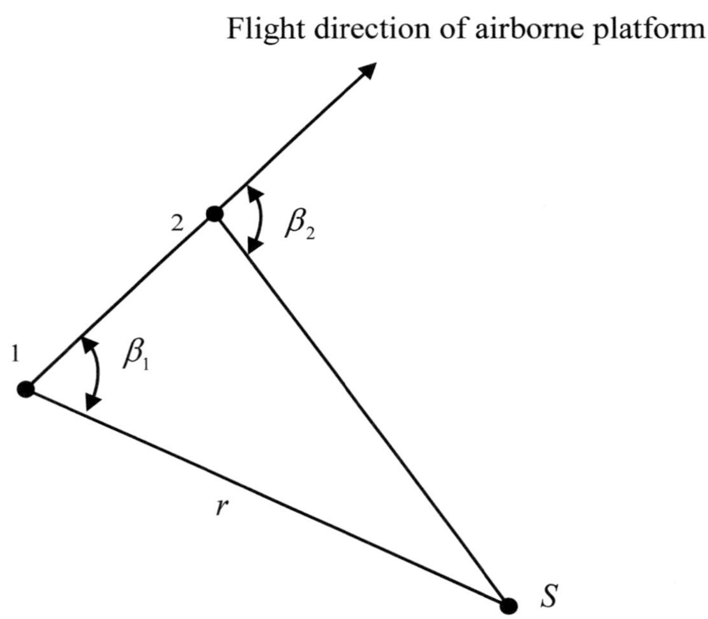

Fig. 3-1. Geometric relationship used for mobile ranging system.

Synthetically applying above-named two equations and speed vector equation $v^2 = v_r^2 + v_t^2$ as well as the relationship between radial speed and Doppler shift $v_r = \lambda f_d$, the basic range finding formula can be obtained

$$r = \frac{v_t^2 \Delta t}{\lambda \Delta f_d} = \frac{(v^2 - \lambda^2 f_d^2)\Delta t}{\lambda |\Delta f_d|} \tag{3-3}$$

3.1.3 Improvement of measuring accuracy

In fact, Doppler changing rate equation only describes the change of Doppler shift in a certain moment or at a certain point. It can not shows the average change of Doppler shift within a time interval. Later-day research has proved that the speed value derived according to basic physical definition is adverse to one from plane geometry. For speed change within a time interval, the speed value obtaining by plane geometry relationship is more exact.

The Doppler frequency difference can be formulated as

$$\lambda \Delta f_d = v(\cos \beta_2 - \cos \beta_1) \tag{3-4}$$

According to the relationship between interior angle and exterior angle $\beta_2 = \beta_1 + \Delta\beta$ as shown in fig.3-1, using approximate expression $\cos \Delta\beta \approx 1$ and $\sin \Delta\beta \approx \Delta\beta$, we have after rearranging for (3-4)

$$\begin{aligned} \lambda \Delta f_d &= v[\cos(\beta_1 + \Delta\beta) - \cos \beta_1] \\ &= v(\cos \beta_1 \cos \Delta\beta - \sin \beta_1 \sin \Delta\beta - \cos \beta_1) \\ &\approx -v\Delta\beta \sin \beta_1 \end{aligned} \tag{3-5}$$

Substituting (3-5) into (3-2) it follows that

$$\begin{aligned} \dot{f}_d &= \frac{\Delta f_d}{\Delta t} \\ &\approx -\frac{v\Delta\beta \sin \beta_1}{\lambda \Delta t} \\ &\approx -\frac{v_{t1}\omega}{\lambda} \end{aligned} \tag{3-6}$$

The angular velocity ω is associated with the rotary radius vector r, that is

$$r\omega = v_t \tag{3-7}$$

As a result

$$\frac{\Delta f_d}{\Delta t} \approx -\frac{v_{t1} v_{t2}}{\lambda r} \tag{3-8}$$

The result shows that the average of Doppler changing rate within a certain time is directly related to the product of tangential velocity at two-terminal of flight distance, that is directly related to geometric mean value at two-terminal of flight distance. And it is inversely related to radial distance at terminal position. Right now, the ranging finding equation whose accuracy can be improved can be obtained after expressing tangential velocity in term of speed vector equation and Doppler speed equation

$$\begin{aligned} r &= \frac{v_{t1} \cdot v_{t2} \cdot \Delta t}{\lambda(f_{d1} - f_{d2})} \\ &= \frac{\sqrt{v^2 - \lambda^2 f_{d1}^2}\sqrt{v^2 - \lambda^2 f_{d2}^2}\Delta t}{\lambda(f_{d1} - f_{d2})} \end{aligned} \tag{3-9}$$

3.1.4 Analog verification

In order to validate the accuracy of range finding, the analog calculation is done by replacing measured value with theoretical value. Preassigning following parameter: wavelength λ, radial distance r, flight speed v, flight time Δt or spacing d, and making β_1 continuously change within preassigned interval, the radial distance and angle of advance in other nodes can be calculated in turn according to the geometric relationship as shown in fig. 3-1. Hence, the theoretical value of Doppler shift f_{di} corresponding to every radial distance can be computed according to Doppler shift equation.

On this basis, the range can be calculated according to (3-9). The relative calculation error can be obtained by compared with the theoretical value.

Without the notice, the adopted parameter is as follows: $r = 100\ km$, $v = 100\ m/s$, $\Delta t = 5\ s$, $\Delta t = d/v$ (in where: $d = 1000m$).

Fig.3-2 is the relative calculation error curve again the angle of advance for ranging calculation with different flight time, time interval in graphs has been transformed into mobile distance d. It can be seen that the calculation value obtained from (3-9) has more accuracy if the time interval is less.

Analog calculation proves that the relative calculation error is not associated with wavelength and movement speed of airborne platform. Fig.3-3 presents the comparison of error result between (3-3) and (3-9). Apparently, the calculation accuracy after mean processing has biggish improvement.

In processing analog calculation, we must advert that Doppler frequency difference in denominator of ranging formula must take absolute value. The correct result can be also obtained without absolute value if the operation order of Doppler shift in two detection nodes can be determined according to derivation process presented in 3.1.2 section.

3.2 Direct ranging method based on angle rate

3.2.1 Recapitulation

An airborne Doppler direct ranging method with single baseline is presented. Based on the principle that the angle rate can be determined by Doppler frequency difference between

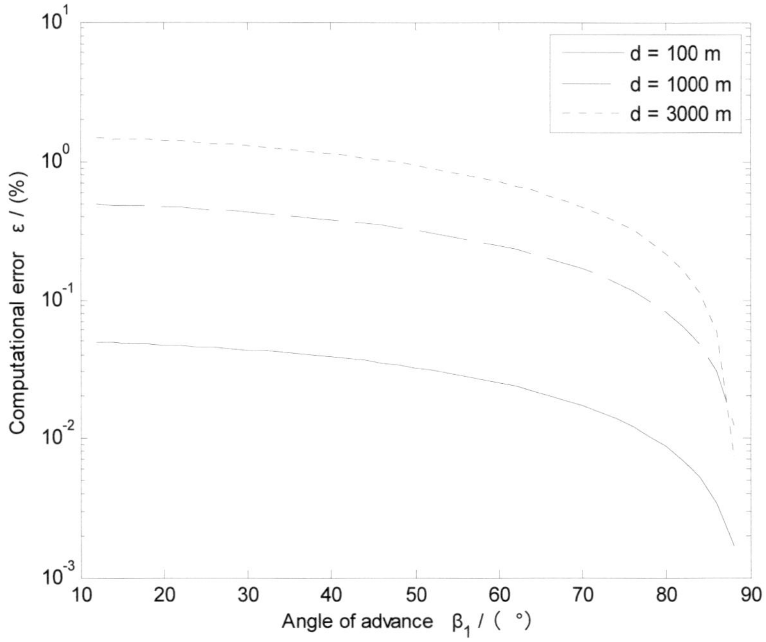

Fig. 3-2 Relative calculation error curve against advancing angle for radial distance with different displacement.

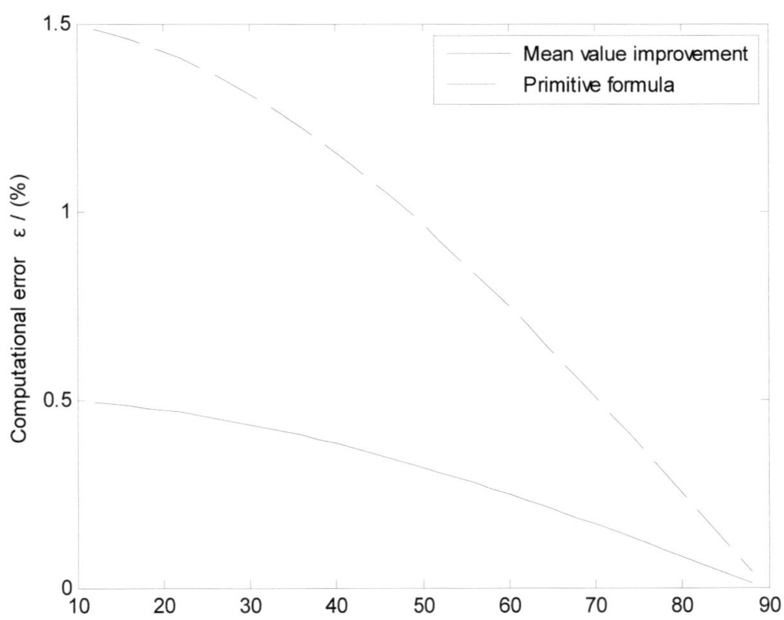

Fig. 3-3. Improvement of calculation accuracy after mean processing.

two antennas, the radial distance from measured target to detection platform can be derived making use of angular velocity that is obtained by differential deformation after introducing sine angle rate of change into Doppler shift equation. The analog calculation verifies that derived formula is correct. Moreover, it is proofed that the direct ranging method based on angle rate of change is equivalent to the one based on Doppler rate of change. As a comparison, the direct ranging method based on angle rate of change has following characteristics:

1. Only requiring a measurement
2. Independent of time measurement
3. It needs not assume that the detecting platform must be uniform motion.

3.2.2 Primitive formula

As shown in fig.3-4, a single baseline array with two antenna elements is installed on airborne platform and the spacing of array is d . The direction of axis of baseline is parallel the axis of airborne platform. For fixed or low speed target T , the Doppler shift detected by airborne double channel measuring receiver in no.1 antenna element is

$$\lambda f_{d1} = v\cos\beta_1 \tag{3-10}$$

By differential transmutation, the Doppler shift is expressed in terms of the sine rate of angle of advance

$$\lambda f_{d1} = \frac{v}{\omega}\frac{\mathrm{d}\sin\beta_1}{\mathrm{d}t} \tag{3-11}$$

where: $\omega = v_t/r_1$.

Regarding the incident signal from target as parallel wave, according to the geometric relationship in diagram, we approximatively have

$$\sin\beta_1 \approx \sqrt{d^2 - \Delta r^2}\Big/d \tag{3-12}$$

Again

$$\dot{r}_i = v_{ri} = \lambda f_{di} \tag{3-13}$$

Resulting in

$$\begin{aligned}\lambda f_{d1} &= \frac{v}{\omega}\frac{\mathrm{d}\sin\beta_1}{\mathrm{d}t} \\ &= \frac{v}{\omega}\frac{\Delta r}{d}\frac{\lambda\Delta f_d}{\sqrt{d^2 - \Delta r^2}}\end{aligned} \tag{3-14}$$

where: $\Delta f_d = f_{d1} - f_{d2}$.

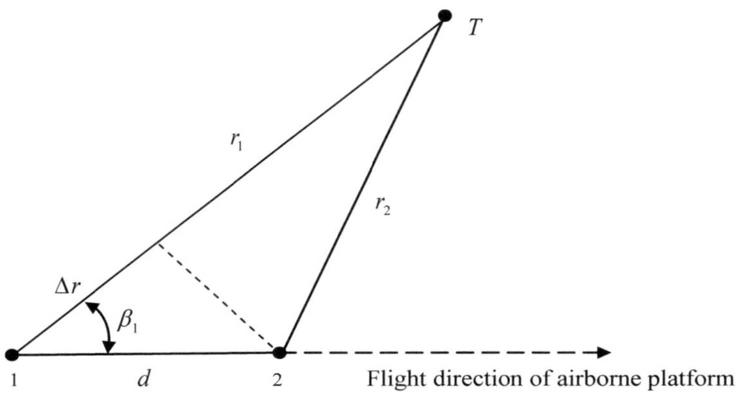

Fig. 3-4. Geometric relationship used for airborne Doppler direct ranging.

Because of

$$ctg\beta = \Delta r \big/ \sqrt{d^2 - \Delta r^2} \tag{3-15}$$

By rearrangement, the equation can be expressed

$$\omega d \sin \beta_1 = \lambda \Delta f_d \tag{3-16}$$

By substituting $\omega = v_t / r_1$ into (3-16), we obtain

$$r_1 = \frac{d v_t \sin \beta_1}{\lambda \Delta f_d} = \frac{d\left[v^2 - (\lambda f_{d1})^2 \right]}{\lambda v \left| \Delta f_d \right|} \tag{3-17}$$

The simulation analysis shows that the frequency difference in denominator must introduce absolute value sign.

As soon as the Doppler shift expression based on phase difference measurement is substituted into the Doppler ranging equation derived based on angle change rate, according to the geometric relationship as shown fig.2-2, we can obtain the airborne ranging formula only using phase shift measuring

$$r_1 = \frac{d_1 \left[1 - \frac{\lambda^2}{d_1^2} (\Delta N_1 + \frac{\Delta \phi_1}{2\pi})^2 \right]}{\lambda \left| \frac{1}{d_1} (\Delta N_1 + \frac{\Delta \phi_1}{2\pi}) - \frac{1}{d_2} (\Delta N_2 + \frac{\Delta \phi_2}{2\pi}) \right|} \tag{3-18}$$

3.2.3 Error between computational and theoretical value

Based on the equivalence in computation, the ranging formula (3-18) can be transformed into as follows form

$$r = \frac{d_1\left[1 - \frac{\Delta r_1^2}{d_1^2}\right]}{\left|\frac{\Delta r_1}{d_1} - \frac{\Delta r_2}{d_2}\right|} \tag{3-19}$$

Preassigning following parameter: radial distance r and baseline length d, and making the angle of arrival $\theta = 90^0 - \beta_1$ continuously change within preassigned interval. The other radial distance can be in turn solved making use of cosine law. Then, the path difference is obtained. With that, the radial distance of target can be calculated by (3-19) and the relative calculation error ε between computational and preassigning value can be obtained by comparison

$$\varepsilon = \frac{|X - X_a|}{X} \times 100\% \tag{3-20}$$

where: X and X_a respectively express computational and preassigning value.

The benefit using (3-19) in analog calculation is that the analysis for integer of wavelength and phase difference value can be avoided.

Fig.3-5 shows the error curve of radial distance between computational and preassigning value in which the total length of baseline is different when the angle of arrival is linear variation from 0^0 to 90^0. Its basic feature is that the longer is the baseline length, the greater is the computational error, namely, the computational error is inversely proportional to radial distance.

The basic parameters adopting in simulation analysis is: $r_1 = 50\ \ km$ and $d = d_1 = d_2$.

3.3 Direct ranging method based on Doppler rate of change

3.3.1 Recapitulation

Under the condition that detection platform is uniform motion along straight line and continuing multipoint detection, on the one hand, the ratio of Doppler changing rate between two adjacent detecting nodes can be expressed as the cubic of specific value of tangential velocities making use of circular function relationship. On the other hand, making use of the formula of Doppler rate of change based on measuring Doppler frequency difference, the ratio of Doppler changing rate between two adjacent detecting nodes can be also expressed as the specific value of Doppler frequency difference. On this basis, a direct ranging formula based on Doppler shift as well as frequency difference can be obtained from the identical relation of speed vector.

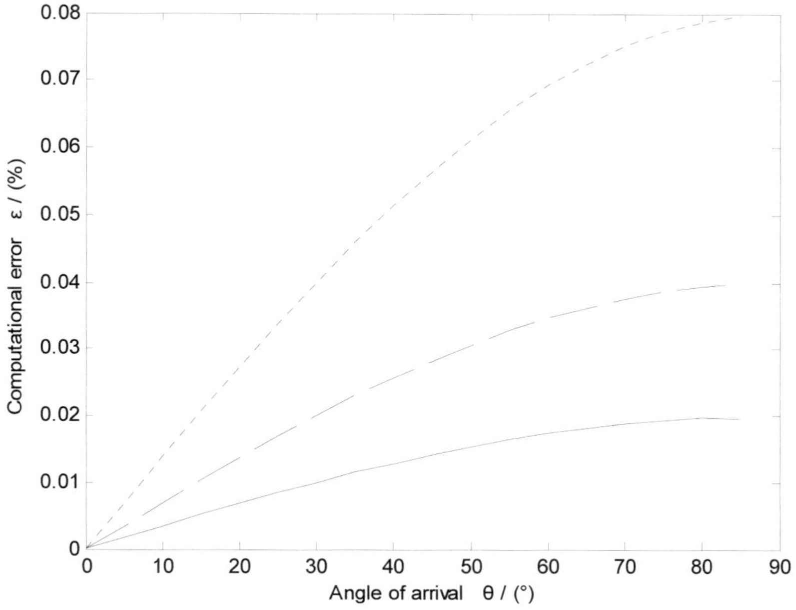

Fig. 3-5. Computational error of ranging formula with different baseline length.

3.3.2 Ratio of Doppler changing rate between two adjacent detecting nodes

As shown in fig. 3-6, provided that the detection platform which is uniform motion along straight line detects the signal of target by fixed cycle and carries out at least three continuing measurement respectively to earth-fixed target. The expression of Doppler changing rate at per node is

$$\dot{f}_{di} = \frac{v_{ti}^2}{\lambda r_i} \quad (i = 1,2,3) \tag{3-21}$$

The ratio of Doppler changing rate between two adjacent detecting nodes is

$$q = \frac{\dot{f}_{d2}}{\dot{f}_{d1}} = \frac{r_1}{r_2}\frac{v_{t2}^2}{v_{t1}^2} \tag{3-22}$$

The ratio of radial distance between two adjacent detecting nodes according to sine theorem is

$$\frac{r_{i+1}}{r_i} = \frac{\sin\beta_i}{\sin\beta_{i+1}} = \frac{v\sin\beta_i}{v\sin\beta_{i+1}} = \frac{v_{ti}}{v_{t(i+1)}} \tag{3-23}$$

Namely, in the case of uniform motion, the ratio of radial distance between two adjacent detecting nodes equals the ratio of tangential velocity. Substituting Eq.(3-23) into (3-22) gives

$$q = \frac{\dot{f}_{d2}}{\dot{f}_{d1}} = \frac{v_{t2}^3}{v_{t1}^3} \tag{3-24}$$

Hence, when detection platform is uniform motion, the ratio of Doppler change rate between two adjacent detecting nodes will equal the cubic of specific value of tangential velocities on two adjacent nodes. Introducing the Doppler change rate expression (2-18) based on Doppler frequency difference, the specific value of Doppler change rate between two adjacent detecting nodes can be also written as

$$q = \frac{d_1}{d_2}\frac{\Delta f_{d2}}{\Delta f_{d1}} \tag{3-25}$$

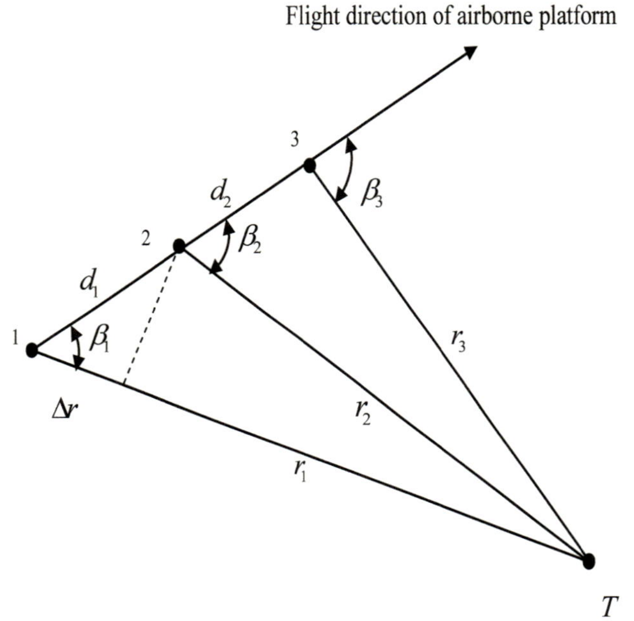

Fig. 3-6. Geometric model used for analyzing Doppler passive localization of moving single station.

3.3.3 Solution of radial distance

According to the velocity component at every node in mobile path of the platform, we have identical relation about speed

$$v^2 = v_{r2}^2 + v_{t2}^2 = v_{r1}^2 + v_{t1}^2 \tag{3-26}$$

Rearranging yield

$$v_{r2}^2 - v_{r1}^2 = v_{t1}^2 - v_{t2}^2 \tag{3-27}$$

Then, separately substituting the Doppler shift which is associated with radial velocity and the Doppler changing rate which is associated with tangential velocity as well as their ratio into (3-27) gives

$$\lambda(f_{d2}^2 - f_{d1}^2) = r_2 \dot{f}_{d2}(u^{-1} - 1) \tag{3-28}$$

where: $u = \sqrt[3]{q^2} = \sqrt[3]{\left(\frac{d_1}{d_2}\frac{\Delta f_{d2}}{\Delta f_{d1}}\right)^2}$.

Hence, we can obtain the calculating formula of radial distance

$$\begin{aligned} r_2 &= \frac{\lambda(f_{d2}^2 - f_{d1}^2)}{(u^{-1} - 1)\dot{f}_{d2}} \\ &= \frac{\lambda d_2 \Delta f_{d1}(f_{d1} + f_{d2})}{(u^{-1} - 1)v\Delta f_{d2}} \end{aligned} \tag{3-29}$$

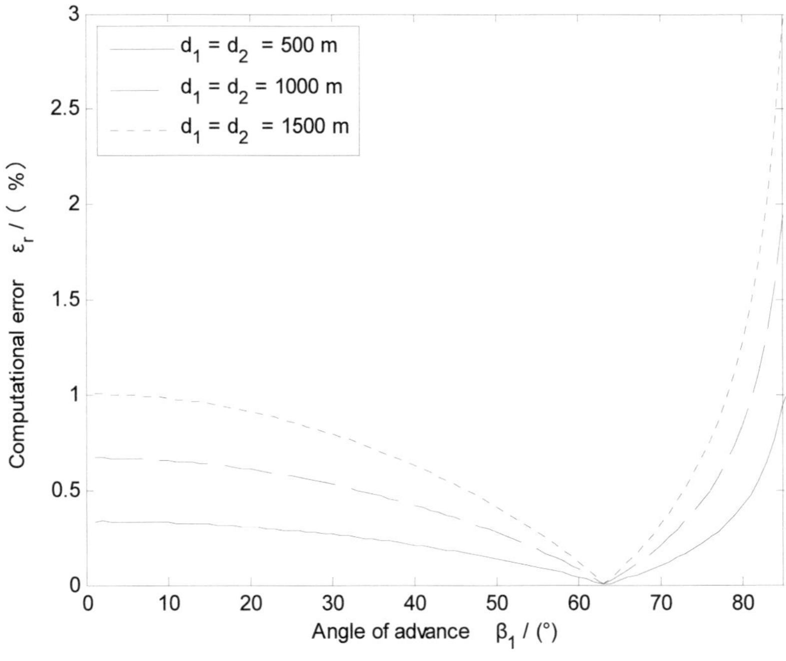

Fig. 3-7. Computational error of ranging formula in equally spaced detection.

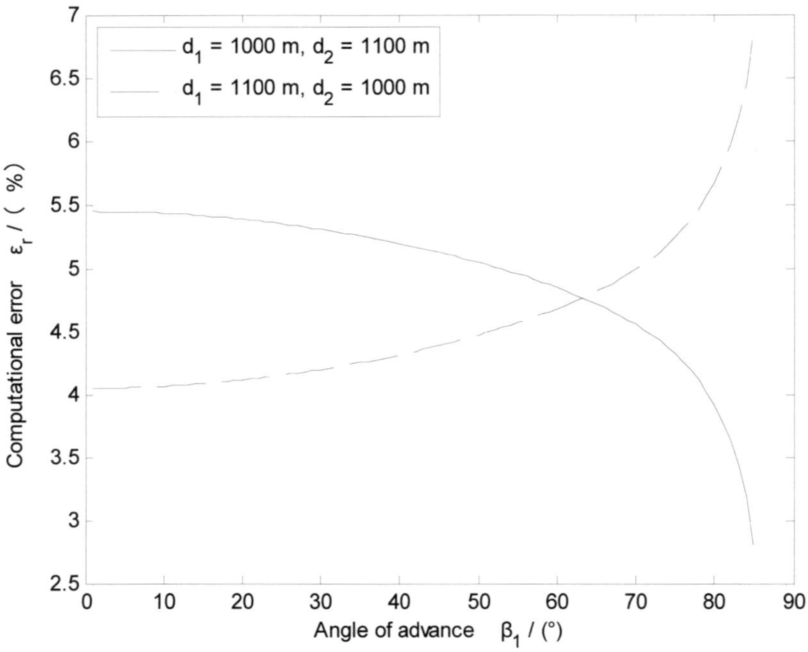

Fig. 3-8. Computational error of ranging formula in unequal spaced detection.

According to calculation error formula between calculated value and theoretical value, fig.3-7 shows the computational error curve between computational and theoretical value when the platform is movement in equal interval. Fig.3-8 is not equal interval. It can be seen that the formula has best accuracy when two adjacent flight distances is equal. If the distance is not equal, the error is larger. There is divergence phenomenon when the angle of advance goes to 90^0. The simulation calculation shows that the error is not connected with the flight speed of mobile platform and wavelength of measured signal.

The basic parameters adopting in simulation analysis is: $v = 100\ m/s$, $r_1 = 100\ km$ and $\lambda = 0.25\ m$.

4. Airborne DF method without ambiguity based on Doppler as well as rate

4.1 Recapitulation

The phase interferometry is a direction finding method with better measurement accuracy. It is widely used for active and passive detection system. But for single baseline phase interferometry, there is the contradiction between accuracy of direction finding and maximum angle without ambiguity. To solve this problem, existing method is to utilize multi-baseline system including the method combining long baselines with short ones and algorithm resolving phase ambiguity with multi-baseline.

In actual application, the method combining long baselines with short ones have two limitations. In fact, corresponding baselines will also become extremely small since wavelength is very short for high-frequency signal. This moment, not only the antenna element must be made to do very small, but also very high demand is put forward for antenna arrangement. It will bring about coupled between antennas and bring down antenna gain. At the same time, higher demand will be required for measurement accuracy of interferometer. For algorithm resolving phase ambiguity with multi-baseline, the computing amount is heavy due to demanding multidimensional integer search.

The study presented in this section shows that applying a single baseline array in airborne platform can realize the high-accuracy DF without the ambiguity of viewing angle after combining with Doppler information. Only from the viewpoint of principle of measurement, selection for baseline length is arbitrary. So it is more suitable for carrying out detection operation in a broadband.

The analysis discovers that the integer of wavelength in radial distance can be directly obtained compositely making use of the velocity vector equation and Doppler shift as well as Doppler changing rate equation. From this, the integer of wavelength in path length difference of radial distance between two adjacent antenna elements can be determined. Further, the value less than a wavelength in path length difference can be obtained by phase difference measurement. As compared with now existing interferometry firstly determining phase difference, this sort of direction finding method associating with Doppler and phase difference and firstly determining path length difference does not exist phase ambiguity nor require restricting baseline length.

4.2 Analytical derivation

According to derivation in section three, in fact, we can obtain two relations from the identical relation about speed

$$\lambda(f_{d1}^2 - f_{d2}^2) = r_1 \dot{f}_{d1}(u-1) \tag{4-1}$$

$$\lambda(f_{d1}^2 - f_{d2}^2) = r_2 \dot{f}_{d2}(1-\frac{1}{u}) \tag{4-2}$$

Thus, we can obtain integer values of wavelength about two radial distances

$$N_1 = \text{int}\left[\frac{r_1}{\lambda}\right] = \text{int}\left[\frac{f_{d1}^2 - f_{d2}^2}{\dot{f}_{d1}(u-1)}\right] \tag{4-3}$$

$$N_2 = \text{int}\left[\frac{r_2}{\lambda}\right] = \text{int}\left[\frac{(f_{d1}^2 - f_{d2}^2)u}{\dot{f}_{d2}(u-1)}\right] \tag{4-4}$$

Substituting these expressions into the existing DF expression based on the principle of phase interferometry, we can obtain the airborne DF formula based on phase shift and Doppler shift as well as its rate

$$\sin\theta = \frac{\Delta r}{L} = \frac{\lambda}{L}\left(\mathrm{int}\left[\frac{f_{d1}^2 - f_{d2}^2}{\dot{f}_{d1}(u-1)} \right] - \mathrm{int}\left[\frac{(f_{d1}^2 - f_{d2}^2)u}{\dot{f}_{d2}(u-1)} \right] + \frac{\Delta\phi}{2\pi} \right) \tag{4-5}$$

5. Airborne passive DF with orthogonal baseline

5.1 Recapitulation

At present, main methods applicable to airborne DF have the amplitude comparison and phase interferometry, etc. The measuring precision of direction-finding system based on amplitude comparison have always had to suffer the biggish influence conduced by incompatible from antenna and reception channel of measuring receiver. Moreover, phase interferometry needs to solve phase ambiguity.

This section presents a Doppler DF method applicable to airborne based on the direction cosine change rate.

If three antenna units are divided into two set and two baselines are placed at right angles to each other, in which the direction of one baseline is parallel to the actual flight direction of air vehicle, the sine and cosine function of target bearing respectively in two baseline directions can be simultaneously obtained according to the analysis principle of the direction cosine change rate. The angle measurement formula only based on Doppler frequency difference can be derived after eliminating the unknown parameters including angular velocity and wavelength by the specific value of two circular functions. The analog calculation shows that the relative calculation error is in direct proportion to the baseline length provided that the incident wave is parallel in derivation. Furthermore, the derived formula has irregularity in airborne axis direction. But the error analysis depicts that the measurement accuracy is in direct proportion to the baseline length. Moreover, the measurement accuracy when the azimuth angle is minor can be usefully enhanced by changing the specific value between two baselines. Since the new method is not associated with wavelength, the direction finding only based on Doppler frequency difference will be more adapted to passive sounding as compared with phase interference method.

5.2 Derivation

If we can use the direction cosine change rate for single baseline array, the incidence angle of measured signal can be expressed as the function depending on the Doppler frequency difference and on the angular velocity and on the baseline length and on the wavelength. According to this result, the study discovery that the airborne direction finding only based on Doppler frequency difference can be realized by making use of the direction cosine change rate for orthogonal double-baseline. Firstly, a planar array with L-shape is structured by use of three antennas. Then, the sine and cosine circular function with regard to the incidence angle of measured signal can be simultaneously obtained due to the direction cosine change rate. Moreover, the unknown angular velocity and the wavelength can be eliminated by the specific value of the sine and cosine circular function. Thus, the derived tangent angle is only associated with known Doppler frequency difference and the baseline length.

As shown in fig.5-1, three antenna units arrange in L form in horizontal plane. Two baselines are placed at right angles to each other. In which, the direction of one baseline is parallel to axes of aerocraft. On condition that measured target is motionless or low speed motion, the Doppler shift received by airborne receiver in every antenna unit is

$$\lambda f_{di} = v\cos\theta_i \tag{5-1}$$

Approximately upon condition that the incident wave is parallel, according to the direction cosine change rate, we have

$$\begin{aligned}\frac{\partial\cos\theta_1}{\partial t} &= \frac{\partial}{\partial t}\left(\frac{\Delta r_1}{d_1}\right)\\ &= \frac{\dot{r}_1-\dot{r}_2}{d_1}\\ &= \frac{\lambda}{d_1}(f_{d1}-f_{d2})\end{aligned} \tag{5-2}$$

Hence, making use of two antenna units whose baseline is parallel to the axis of aerial vehicle can obtain the sinusoidal triangle function with regard to relative azimuth depending on the Doppler frequency difference and on the angular velocity and on the base length

$$\begin{aligned}\sin\theta_1 &= -\frac{1}{\omega_\theta}\frac{\partial\cos\theta_1}{\partial t}\\ &= -\frac{1}{\omega_\theta}\frac{\partial}{\partial t}\left(\frac{\Delta r_1}{d_1}\right)\\ &= -\frac{\lambda\Delta f_{d1}}{\omega_\theta d_1}\end{aligned} \tag{5-3}$$

where: $\Delta f_{d1} = f_{d1} - f_{d2}$ is Doppler frequency difference; $\omega_\theta = \dfrac{v\sin\theta_1}{r_1}$ angular velocity ; d_1 baseline length parallel to the axis of aerial vehicle..

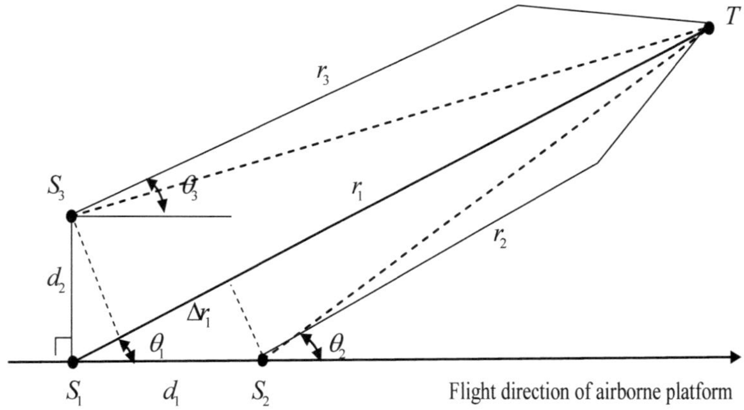

Fig. 5-1. Airborne passive direction finding array with L form.

Further, making use of two antenna units whose baseline is at right angles to the axis of aerial vehicle can obtain the cosine triangle function with regard to relative azimuth depending on the Doppler frequency difference and on the angular velocity and on the base length

$$\cos\theta_1 = \sin(90-\theta_1)$$

$$= -\frac{1}{\omega_\theta}\frac{\partial\cos(90-\theta_1)}{\partial t} \tag{5-4}$$

$$= -\frac{1}{\omega_\theta}\frac{\partial}{\partial t}\left(\frac{\Delta r_2}{d_2}\right) = -\frac{\lambda \Delta f_{d2}}{\omega_\theta d_2}$$

where: $\Delta f_{d2} = f_{d1} - f_{d3}$ is Doppler frequency difference; d_2 baseline length at right angles to the axis of aerial vehicle.

By way of the specific value, the tangential triangle function only depending on the Doppler frequency difference and on the baseline length can be obtained

$$tg\theta_1 = \frac{\sin\theta_1}{\cos\theta_1}$$

$$= \frac{d_2}{d_1}\frac{\Delta f_{d1}}{\Delta f_{d2}} \tag{5-5}$$

The relative azimuth between aerial vehicle and measured target is

$$\theta_1 = tg^{-1}\left[\frac{d_2}{d_1}\frac{\Delta f_{d1}}{\Delta f_{d2}}\right] \tag{5-6}$$

This analytic function in form is analogous to the formula of the amplitude comparison and phase interference method. Moreover, it is also not connected with wavelength.

5.3 Simulation analysis

We make the analog verification by replacing measured value with theoretical value. Firstly, presetting following parameter: radial distance r_1 and base length d_i as well as wavelength and speed, the theoretical value of the rest radial distance and azimuth angle can be computed by making azimuth angle continuous change in specified interval. Hence, we can obtain the theoretical value of the Doppler shift corresponding to each radial distance. After that, the value of azimuth angle can be calculated by Eq.(5-6) and the relative calculation error can be obtained by comparison with the theoretical value.

Because the simulation analysis of the relative calculation error is not associated with both wavelength and speed, there is not the specified value of wavelength and speed.

Fig.5-2 and fig.5-3 depict the relative calculation error curve with different baseline and radial distance. When azimuth angle tends to zero, the derived formula has irregularity. Obviously, the farther is the object distance, or the shorter is the baseline length, the smaller is the relative calculation error of formula. There occurs this occurrence provided that the incident wave is parallel in derivation.

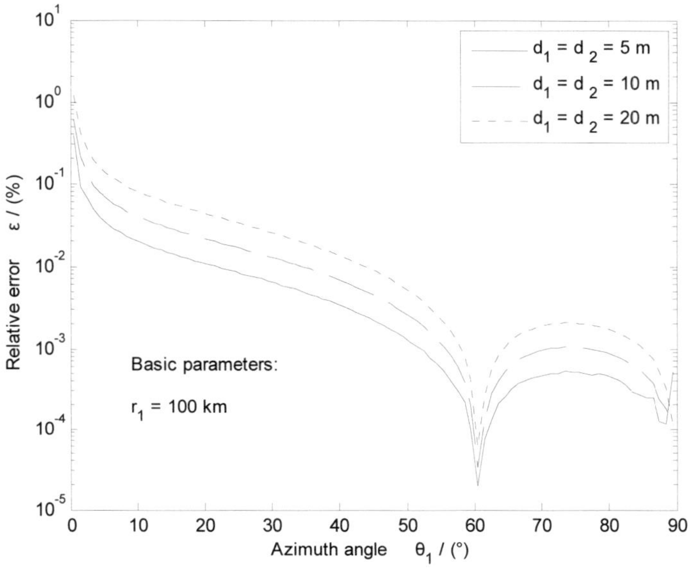

Fig. 5-2. Relative calculation error: different baseline lengths.

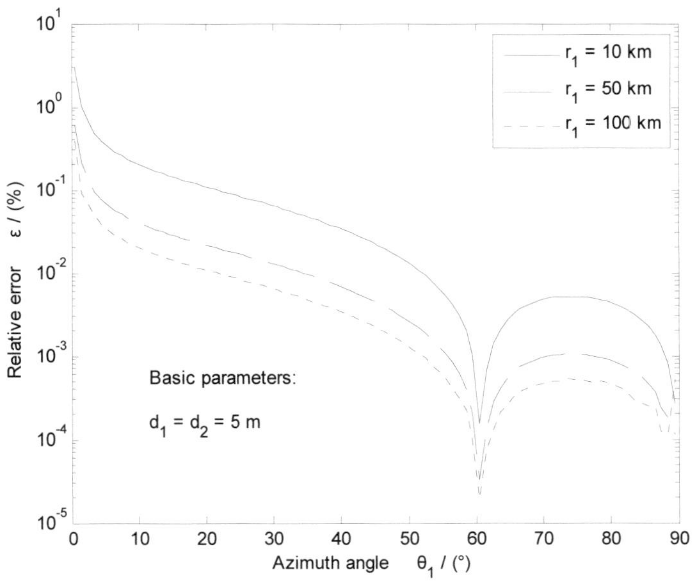

Fig. 5-3. Relative calculation error: different radial distances.

5.4 Error analysis

On condition that the locating error of baseline is neglected, making differential for Doppler frequency difference gives

$$\frac{\partial\theta_1}{\partial\Delta f_{d1}} = \frac{1}{1+A^2}\frac{d_2}{d_1\Delta f_{d2}} \tag{5-7}$$

$$\frac{\partial\theta_1}{\partial\Delta f_{d2}} = \left|\frac{1}{1+A^2}\frac{d_2\Delta f_{d1}}{d_1\Delta f_{d2}^2}\right| \tag{5-8}$$

where: $A = \frac{d_2}{d_1}\frac{\Delta f_{d1}}{\Delta f_{d2}}$

According to error estimation theory, the overall error produced by the measurement of Doppler frequency difference is

$$\sigma = \sigma_f \sqrt{\sum_{i=1}^{n=2} \frac{\partial\theta_i}{\partial\Delta f_{di}}} \tag{5-9}$$

where: σ_f is the mean-root-square error measuring Doppler frequency difference.

Fig.5-4 shows that the error curve falls exponentially. When azimuth angle tends to zero, there is obviously the blind zone for direction finding. And the accuracy is best when azimuth angle is close to right angle. The analysis shows that the measuring error can usefully decrease if the length of two baselines is simultaneously increased. For example, as soon as the length of two baselines can be increased to 15 meter, the maximum angle measuring error is less than 2^0 after azimuth angle is bigger than 10^0.

Fig.5-5 depicts that the measurement error when the azimuth angle is less than 30^0 can be quickly reduced by changing the specific value between two baselines.

Although the relative calculation error is not connected with wavelength, the calculation shows that the measurement accuracy is connected with wavelength. Fig.5-6 shows that the measurement accuracy is inversely proportional to the wavelength.

Moreover, though the relative calculation error is not connected with the flight speed, the calculation shows that the measurement accuracy is connected with the flight speed. Fig.5-7 shows that the measurement accuracy is in direct proportional to the flight speed.

5.5 Summarize briefly

Since the new method presented in this text is not associated with wavelength, the direction finding only based on Doppler frequency difference will be more adapted to passive sounding as compared with phase interference method.

The multi-objective passive localization and wideband operation from single airborne observer is a key and difficulty subject for modern electron reconnaissance. Because the

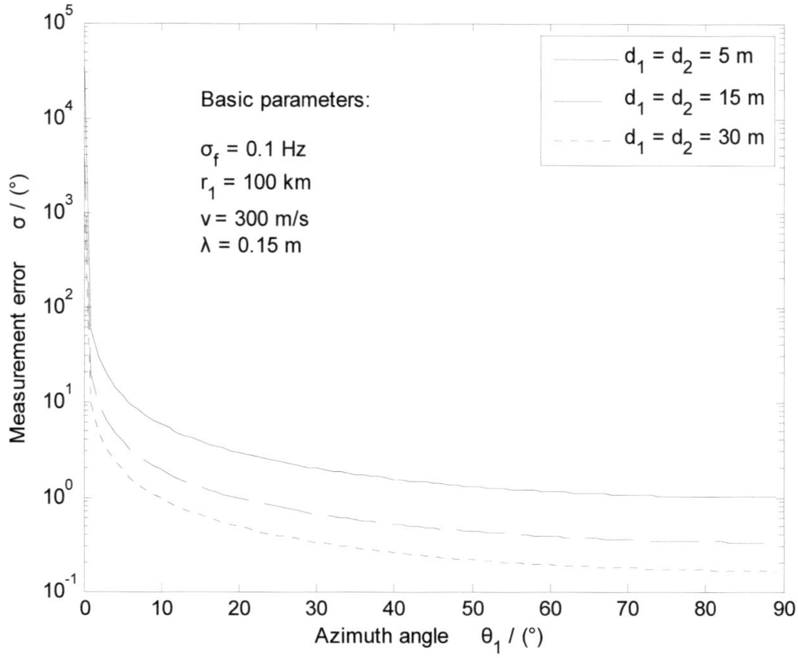

Fig. 5-4. Measurement errors: different baseline lengths.

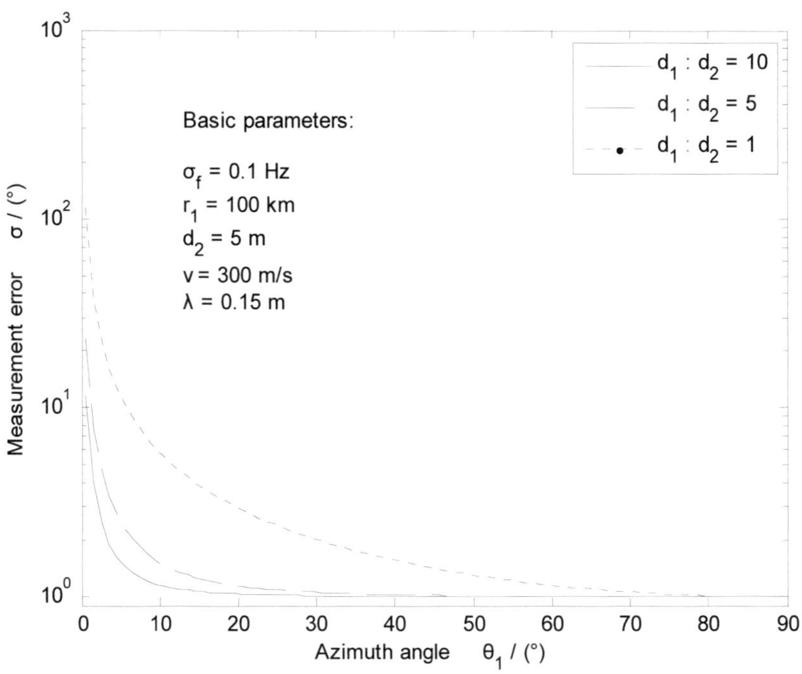

Fig. 5-5. Measurement errors: different ratio between two baseline lengths.

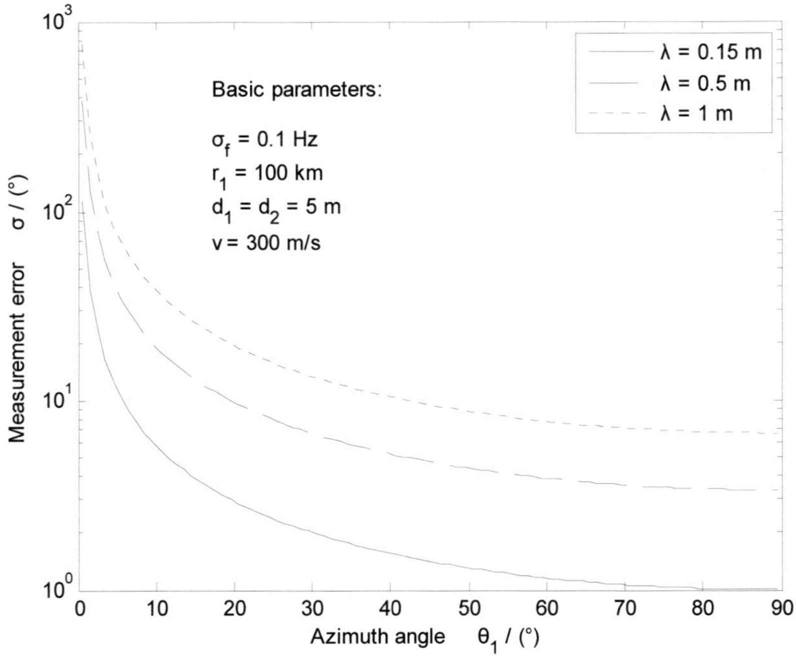

Fig. 5-6. Measurement errors: different wavelengths.

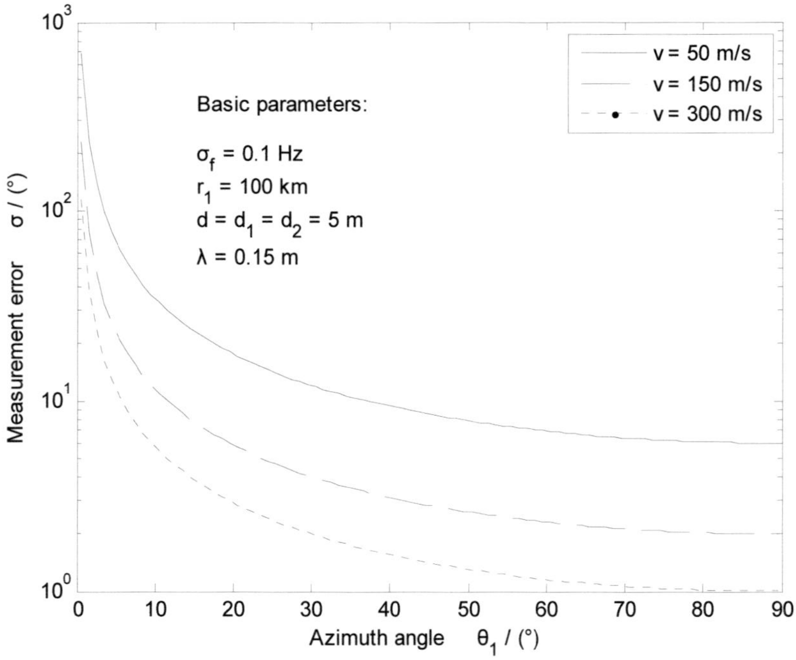

Fig. 5-7. Measurement errors: different flight speeds.

direction finding method presented in this text is only associated with Doppler frequency difference, that is, which is not associated with the setting up the wavelength and baseline length, this method is very applicable to wideband operation.

At the same time, this method is also very applicable to the electron reconnaissance for multi-objective from single airborne observer because the information of frequency measurement is advantageous for signal sorting and recognition.

6. Conclusions

Available localization method based on Doppler as well as its rate of change uses the coordinate variation in rectangular coordinate system for analyzing the angle of advance in polar coordinate system and uses geometric projection for analyzing the speed vector. Because the Doppler shift is the function of position and motion state of target, there are 2n unknown number in n-dimensional plane. For the localization of single station, in order to resolve the position and motion state of target, 2n nonlinear equations must be set up according to the measured value obtained during n measuring cycle. The solution process is comparative complicated and the analytic result can not be obtained.

The method of Doppler direct ranging changes such status. Recently, research done by author shows that the direct range finding for target with analytic form can be realized making use of Doppler frequency difference during a time interval from the viewpoint of the basic definition of Doppler rate. More, the direct ranging formula whose characteristic is better can be obtained by applying angle rate.

Though the auctorial research result has proves that the single station which is moving can realize fast ranging for fixed or slow speed target making use of Doppler shift as well as its rate of change. However, the measurement accuracy for carrier frequency in passive localization is just $\sigma_f = 10^{-4} f$ based on actual technique level. This measuring ability restricts the application of Doppler passive localization technique in engineering. As a rule, the value of Doppler frequency difference may be less than Hz since the baseline length of antenna array is shorter in single moving platform. This conduces that the demand for measurement accuracy of frequency difference is very rigor.

In allusion to such difficulty, author researches the phase detection method for Doppler shift and presents the airborne passive phase interference ranging method based on the principle of Doppler direct ranging. The other one which is being explored by author is the DF method combining Doppler with phase interference in which the path length difference is firstly determined by directly solving the integer of wavelength in path length difference based on measuring Doppler as well as its rate. Though these methods are also imperfect, elementary result has shown that the measuring method of Doppler-phase interference will help to development of airborne passive localization technique as well as correlation technique.

As soon as the phase shift can directly integrate with the Doppler frequency shift in physical relationship, more novel method of localization can be appeared. The mutual equivalent conversion between phase shift and Doppler shift means that the method obtained by a kind of measurement technology can be quickly expanded to another measurement technology.

Hence, the research for passive localization by applying the functional relation between phase shift and Doppler shift may be a more meaningful work.

7. References

Abatzoglou T. Fast maximum likelihood joint estimation of frequency and frequency rate. *IEEE Trans. On Aerospace and Electronic Systems*, AES-Vol.22, No.6, (1986), pp.708-715, ISSN 0018-9251

An Xiao-jun. Research on DF Based on Improved Phase Interferometer. *Radio Engineering of China*, Vol.39, No.03, (2009), pp.59-61, ISSN 1003-3106

Diao ming, Wang Yue. Research of passive location based on the Doppler changing rate. *Systems Engineering and Electronics*, Vol28, No.5, (2006), pp. 696-698, ISSN 1001-506X

Djuric Petar M and Kay Steven M. Parameter estimation of chirp signals. *IEEE Trans. on Acoustics, Speech and Signal Processing*, Vol.38, No.12, (1990), pp. 2118-2126, ISSN 1053-587X

Ernest J. Ambiguity resolution in interferometer. *IEEE Trans. Aerospace and Electronic Systems*, Vol.117, No.6, (1981), pp.766-780, ISSN 0018-9251

Feng Daowang & Zhou Yiyu & Li Zonghua. A Fast and Accurate Estimation for Doppler Rate-of-Change with the Coherent Pulse Train. *Signal Processing*, Vol.20, No.1, (2004), pp.40-43, ISSN 1003-0530

Gong Xiang-yi & Yuan Jun-quan & Su Ling-hua. A Multi-pare Unwrap Ambiguity of Interferometer Array for Estimation of Direction of Arrival. *Journal of Electronics & Information Technology*, Vol.28, No.1, (2006), pp.55-59, ISSN 1009-5896

Gong Xiang-yi & Huang Fu-kan & Yuan Jun-quan. A New Algorithm for Estimation of Direction of Arrival Based on the Second-Order Difference of Phase of Interferometer Array. *Acta Electronica Sinica*,.Vol.33, No.3, (2005), pp.444-446, ISSN 0372-2112

Hua Yang. Research into the Single-station Passive Location for Electronic Reconnaissance UAV. *Shipboard Electronic Countermeasure*, Vol.32, No.2, (2008), pp.14-17, ISSN 1673-9167

Li Yong & Zhao Guo-wei & Li Tao. Algorithm of Solving Interferometer Phase Difference Ambiguity by Airborne Single Position. *Chinese Journal of Sensors and Actuators*, Nol.19, No.6, (2006), pp.2600-2602, ISSN 1004-1699

Li Zonghua, Xiao Yuqin, Zhou Yiyu, Sun Zhongkang. Single-observer passive location and tracking algorithms using frequency and spatial measurements. *Systems Engineering and Electronics*, Vol.26, No.5, (2004), pp. 613-616, ISSN 1001-506X

Li Zonghua,Xiao Yuqin,Zhou Yiyu,et al. Single-observer passive location and tracking algorithms using frequency and spatial measurements. *Systems Engineering and Electronics*, Vol.26, No.5, (2004), pp. 613-616, ISSN 1001-506X

Liao Ping & Yang Zhong-hai & Jiang Dao-an. A Single Observer Passive Location Algorithm Based on Probability in Multi-target Environment, *Telecommunication Engineering*. Vol.46, No.1, (2006), pp.45-49, ISSN 1001-896X

Lin Yi-meng & Liu Yu & Zhang Ying-nan. Algorithm of Direction Finding for Broadband Digital Signal. *Journal of Nanjing University of Aeronautics & Astronautics*, Vol.37, No.3, (2005), pp.335-340, ISSN 1005-2615

Lu Xiaomei. Overview on the Technology of the Single Airborne Station Passive Location, *Shipboard Electronic Countermeasure*, Vol.26, No.3, (2003), pp:20-23, ISSN 1673-9167

McCormick W S, Tsui J B Y, BakkieV L. A Noise Insensitive Solution to an Ambiguity Problem in Spectral Estimation. *IEEE Trans on Aerospace and Electronic Systems*, Vol.25, No.5, (1989), pp.729-732, ISSN 0018-9251

Messer H,Singal G. On the achievable DF accuracy of two kinds of active interferometers. *IEEE Trans. on Aerospace and Electronic Systems*, Vol.32, No.3, (1996), pp.1158-1164, ISSN 0018-9251

Poisel,Richard A. (2008). Electronic warfare target location methods. Translated by Qu Xiao-xu. *Electronic Industry Press,* ISBN 978-7-121-06326-8, Beijing, China

Sun Zhongkang, Guo Fucheng, Feng Daowang. (2008). Passive location and tracking technology by single observer. *National Defense Industry Press*, ISBN 978-7-118-05817-8, Beijing, China.

Sundaram K R, Ranjan K M. Modulo conversion method for estimation the direction of arrival. *IEEE Trans. on Aerospace and Electronic Systems*, Vol.36, No.4, (2000), pp.1391-1396, ISSN 0018-9251

Wang Qiang. A Research on Passive Location and Tracking Technology of a Single Airborne Observer. National University of Deefnse Technology, Changsha, Hunan. PR.China. (Nov. 2004).

Wang Zhi-rong. On High Precision DOA Estimation. *Radio Engineering of China*,.Vol.37, No.11, (2007), pp.24-25, ISSN 1003-3106

Wei Xing & Wan Jian-wei & Huang Fu-kan. Study of Passive Location System Based on Multi Base-line Interferometers. *Modern Radar*,.Vol.29, No.5, (2007), pp.22-25,35, ISSN 1004-7859

Xu Yao-wei & Sun Zhong-kang. Passive Location of Fixed Emitter Using Phase Rate of Change. *Systems Engineering and Electronics*, Vol.21, No.3, (1999), pp. 34-37, ISSN 1001-506X

Yang Yue-lun. Summary of New Airborne Passive Detection and Location Technologies, *Shipboard Electronic Countermeasure*, Vol.34, No.3, (2011), pp.14-17+56, ISSN 1673-9167

Yu Chun1ai, Wan Jianwei, Zhan Ronghui. An Estimation Algorithm for Doppler Frequency Rate-of-Change with PCM Coherent Pulse Train. *Journal of Electronics & Information Technology*, Vol.30, No.10, (2008), pp. 2303-2306, ISSN 1009-5896

Yu Tao, An Analytic Method for Doppler Changing Rate Based on Frequency Measurement[C]. 2010 International Conference on Communications and Intelligence Information Security. Beijing, China, December 17-19, 2010

Yu Tao. A new airborne passive DF method only based on frequency difference. *Advanced materials research,* Vols.219-220, (2011), pp.846-850. ISSN 1022-6680

Yu Tao. Airborne Direction Finding Method Based on Doppler-Phase Measuremen. *Frontiers of Electrical and Electronic Engineering in China*, Vol.5, No.4, pp.493-495, ISSN 1673-3584

Yu Yang. The Improvement of Airborne DF Adjust Method Based on Four-channel Amplitude Comparison. *Electronic Information Warfare Technology*, Vol.22, No.3, (2007),pp.24-26,39, ISSN 1674-2230

Zhang Gangbing & Liu Yu & Liu Zongmin. Unwrapping Phase Ambiguity Algorithm Based on Baseline Ratio. *Journal of Nanjing University of Aeronautics & Astronautics*, Vol.40, No.5, (2008), pp.665-669, ISSN 1005-2615

Zhao Yefu & Li Jinhua. (2001).The Radio Tracking and Instrumentation System. *National Defense Industry Press*, ISBN 7-118-02418-X, Beijing, China

Zhou Ya-qiang & Huang Fu-kan. Solving ambiguity problem of digitized multi-baseline interferometer under noisy circumstance. *Journal of China Institute of Communications*,.Vol.26, No.8, (2005), pp.16-21, ISSN 1000-436X

Zhou Zhen,Wang Geng-chen. Passive locating and tracking of maneuvering targets from single airborne observer. *Electronics Optics & Control*, Vol.15, No.3, (2008), pp.60-63, ISSN 1671-637X

8

Experiences in Boreal Forest Stem Volume Estimation from Multitemporal C-Band InSAR

Jan Askne[1] and Maurizio Santoro[2]
[1]*Chalmers University of Technology,*
[2]*Gamma Remote Sensing,*
[1]*Sweden*
[2]*Switzerland*

1. Introduction

During the last two decades synthetic aperture radar (SAR) and interferometric synthetic aperture radar (InSAR) have become important tools for airborne and satellite remote sensing. SAR delivers a high resolution image of the radar backscatter and by means of InSAR a second image from almost the same orbit is combined with the first one to provide the relative phase difference and stability/coherence of the backscattered signal. Principles of InSAR can be found in various review articles (Bamler and Hartl, 1998; Massonnet and Feigl, 1998; Rott, 2009). This presentation will focus on the use of InSAR observations for investigation of forest stem volume or above ground biomass.

Satellite methods are important to determine forest above-ground biomass on a global scale in order to follow changes in a consistent manner over long periods of time as part of climate modeling. Methods to determine forest stem volume are also important for the economical management of forest areas. For these applications many remote sensing techniques have been investigated. Our interest will focus on a radar method. In contrast to methods based on optical images, a semi-empirical, physically-based relation between measurements and stem volume/biomass can be established because microwaves penetrate the forest canopy to a certain extent and the backscattered signal is therefore modulated by the forest structural properties.

With the early C-band SAR systems, the European Remote Sensing Satellites ERS-1 and ERS-2, and in particular the observations during the tandem period with one day interval between the observations, large amounts of InSAR data became available, which allowed the possibility to retrieve forest parameters. One argument against C-band for forest applications has been that the backscatter is originating from the top layer of the forest with little relation to the major components of the biomass, such as the stem and then saturate for low biomass values (Imhoff, 1995). However, this changed with repeat pass InSAR data, and coherence can under certain conditions provide high accuracy stem volume estimates up to the stem volumes available in the test areas studied below, i.e. up to 539 m^3/ha or approximately 265 tons/ha. Microwave penetration into forests is related to the vegetation properties including the gaps in the vegetation. Since the density of boreal forests is often low this is a fundamental property for the application of C-band and higher frequencies, where gaps as small as some wavelengths will allow penetration and result in a contrast between the relatively high ground coherence, and the relatively low canopy coherence induced by wind effects on the canopy. In addition, part of the coherence, the volume decorrelation, which increases with baseline length, is dependent on forest height. This means that C-band tandem coherence measurements include information about forest density as well as height, thus being related to the above ground stem volume/biomass.

This paper will describe methods and results giving insight into the applicability of C-band InSAR coherence data for retrieval of stem volume in boreal forests. The semi-empirical model, the Interferometric Water Cloud Model, IWCM, relating the InSAR coherence to forest parameters will be presented. The possibility to determine or reduce the number of parameters in the IWCM will be discussed, e.g. the possibility to express the mean tree height as a function of stem volume in form of an allometric relation. For this, a large number of measurements from ground surveys in Sweden will be used. The model parameters which vary with the meteorological conditions will in the first step be determined by training the model by means of known forest sites. In a second step, a method for the estimation of the model parameters without such training sites will be presented.

The goal is to investigate the possible accuracy of C-band InSAR for retrieval of forest stem volume and thus biomass by means of forest sites with accurate *in situ* data, Brattåker, Remningstorp, Tuusula, and Kättböle. Since the usefulness of a semi-empirical model is related to the possibility to validate the model and demonstrate its accuracy, four different test sites with different properties have been studied, and results are presented. As part of the goal, IWCM is developed in various ways, and new results derived.

2. Model for InSAR data from a forest, the Interferometric Water Cloud Model

A model describing the relation between the remote sensing observables and the stem volume should present a fairly simple formulation if the aim is to invert it to retrieve stem volume from InSAR coherence values. The Interferometric Water Cloud Model, IWCM, which is extensively presented elsewhere (Askne et al., 1997; Santoro et al., 2002; Askne et al., 2003), assumes a volume above ground of randomly distributed scatterers with gaps (within canopy as well as between trees), and takes into account the backscatter and coherence variations, decorrelation of the ground and the vegetation layer including the volume decorrelation. The canopy cover or the areafill η represents the relative percentage of the area covered by vegetation. Leaf area index and gap fraction are associated aspects (Nilson, 1999). The backscatter and coherence will be described in terms of the attenuation back and forth through the partially covering vegetation layer, $exp(-\alpha h)$, but also in terms of an empirically related variation with stem volume (Pulliainen et al., 1994), $exp(-\beta V)$. The forest backscatter components also determine the weight of the statistically independent coherence components.

$$\sigma^0_{for} = \eta\left[\sigma^0_{gr}e^{-\alpha h} + \sigma^0_{veg}\left(1 - e^{-\alpha h}\right)\right] + (1 - \eta)\sigma^0_{gr} = \sigma^0_{gr}e^{-\beta V} + \sigma^0_{veg}\left(1 - e^{-\beta V}\right) \quad (1)$$

$$\gamma_{for} = \eta\left[\gamma_{gr}\sigma^0_{gr}e^{-\alpha h} + \gamma_{veg}\gamma_{vol}\sigma^0_{veg}\left(1 - e^{-\alpha h}\right)\right] + (1 - \eta)\gamma_{gr}\sigma^0_{gr} + \varepsilon_{noise} = \gamma_{gr}\sigma^0_{gr}e^{-\beta V} + \gamma_{veg}\gamma_{vol}\sigma^0_{veg}\left(1 - e^{-\beta V}\right) + \varepsilon_{noise} \quad (2)$$

σ_{gr}^{0} is the backscatter from ground, σ_{veg}^{0} the backscatter from the vegetation layer, γ_{gr} and γ_{veg} represent the temporal decorrelation of the ground scatterers and the vegetation scatterers. If the variation of scattering and stability with height is only determined by the attenuation, the volume decorrelation is determined by

$$\gamma_{vol} = \frac{\int_0^{h(V)} e^{-\alpha z'} * e^{-jK(B_n)z'} dz'}{\int_0^{h(V)} e^{-\alpha z'} dz'} = \frac{\alpha}{\alpha - jK(B_n)} \frac{e^{-jK(B_n)} - e^{-\alpha h(V)}}{1 - e^{-\alpha h(V)}} \tag{3}$$

where $K=4\pi B_n/\lambda R sin\theta$. α and $h(V)$ are assumed to be known. α was shown to be ≈ 1 dB/m during winter conditions and 2 dB/m during summer conditions (Santoro et al., 2007b), however, the possibility to determine α from the measurements will be studied here. To reduce the number of forest parameters, the height is parameterized as a function of stem volume, $h(V)$. This relationship is assumed to be given by an allometric relation between height and stem volume, see below.

For the two expressions in (1) and (2) to agree, a demand is obtained that the areafill factor $\eta(V)$ is given by a relation between α and β

$$\eta(V) = \frac{1 - e^{-\beta V}}{1 - e^{-\alpha h(V)}} \tag{4}$$

Traditionally an extinction coefficient, κ, is used to define the attenuation through a homogeneous vegetation layer. From above $\frac{2\kappa h(V)}{cos\theta_i} = \beta V$ or with (4)

$$\kappa(V) = -\frac{cos\theta_i}{2h(V)} \ln[1 - \eta(V)(1 - e^{-\alpha h(V)})] \tag{5}$$

illustrating how the extinction coefficient is dependent on stem volume, height, areafill, as well as temperature, humidity etc. through α.

3. Forest properties influencing the model

The formulation of the IWCM in Eq. (1) and (2) highlights that the modeled coherence depends on a number of parameters related to the scattering properties of the forest and the forest structure. Scope of this Section is to investigate the relationship among some of these parameters. α and $h(V)$ will be discussed, and also the relation between stem volume and biomass, since above ground biomass is an important parameter for climate modeling.

3.1 Areafill and attenuation

The areafill is the fraction of ground covered by the vegetation, looking in the direction of the satellite. The areafill can be observed by different methods, and some results are given in Fig. 1a. The three observed areas are showing the same trend.

α is the attenuation factor per unit height of the radar wave going back and forth the idealized vegetation layer. If α is made dependent of β in the way described by Eq. (6), an areafill variation is obtained in line with the observations in Kättböle. When β is varying from 0.002 to 0.007 in steps of 0.001 as shown in Fig 1b, the range of values of the associated areafill is marked by the vertical lines in Fig. 1a. α has normally a small effect on the final retrieval result and a fixed value of 0.23 can be used for winter conditions and 0.46 for summer conditions. Such a fixed value is needed if not the areafill information is available.

$$\alpha(\beta) = -\frac{1}{7.4}\ln\left(1 - \frac{\beta}{0.008}\right) \quad (6)$$

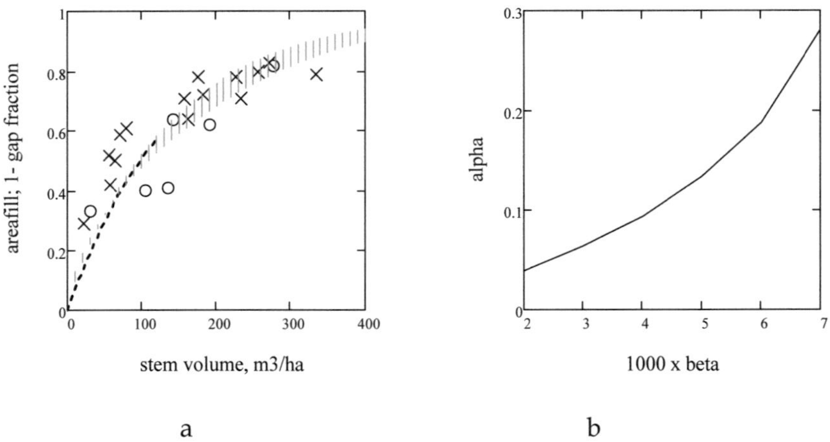

Fig. 1. Observations of areafill/gap fraction, x based on hemispherical photography of Kättböle stands (Santoro et al., 2002), o based on results in (Nilson, 1999) from a neighboring area; dashed line: regression line from MODIS observations in Northern Sweden (Västerbotten) (Cartus et al., 2010). Note that the uncertainties in the observations are large. Vertical lines are illustrating range of values of areafill variations for $0.002 < \beta < 0.007$ in Eq. (6).

3.2 Allometric relation for h(V)

The IWCM describes the forest coherence as a function of two forest parameters: stem volume and height. Since we are primarily after stem volume, we investigated the possibility to replace height with a function related to stem volume. The volume decorrelation is dependent on the forest height and increasing with baseline although typically of lower importance than temporal decorrelation. For this reason, a simple allometric relation for *h(V)* has been considered. Replacement of the actual height with an allometric equation assumes that this does not introduce significant errors in the modeled coherence. In (Askne et al., 1997) an expression based on a very limited dataset was given, but this expression has also been found to be applicable for other areas investigated below using InSAR data, although an allometric relation can be expected to be dependent on forest type, growing conditions etc.

$$h(V)=(2.44\ V)^{0.46} \quad (7)$$

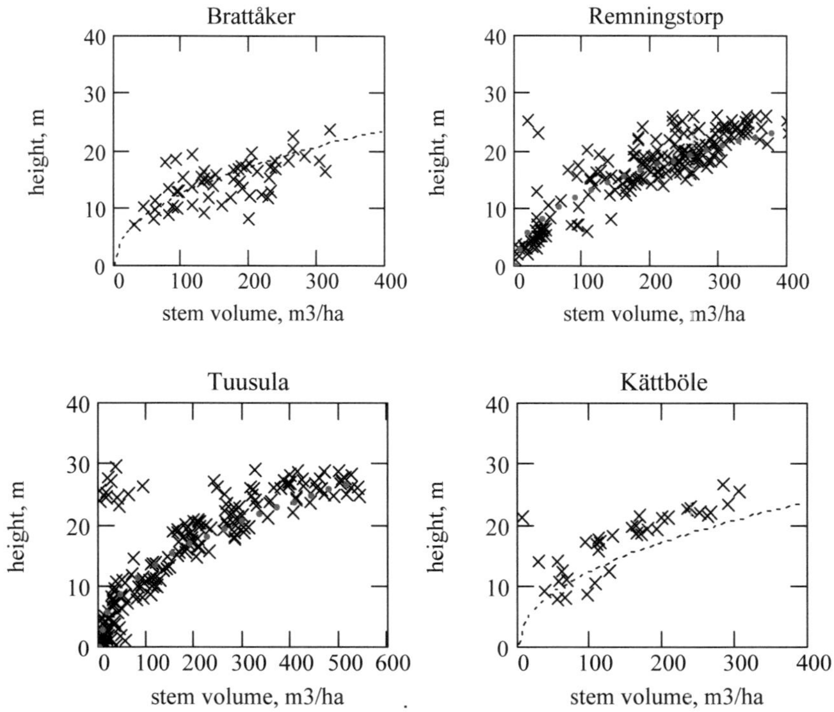

Fig. 2. Illustrating forest properties for investigated test sites in the form of the relation between height and stem volume (stand averages) together with the allometric relation in Eq. (7), dotted curve.

Height versus stem volume is illustrated in Fig. 2 for the test sites studied below. From the figures some anomalies associated with e.g. seed trees and thinning practice can be noticed.

In order to investigate the allometric relation for more general conditions, a large number of statistically sampled measurements from the Swedish National Forest Inventory (courtesy Anders Lundström and Göran Kempe) have been used. In total, forest parameters from 3046 plots (in productive forest) with 10 m diameter (permanent plots) or 7 m diameter (temporary plots) have been used. The measurements took place between 2002 and 2006. The measurements are presented for four areas of Sweden, Skaraborg County (N 57° 38′ - 59° 02′, O 12° 35′ - 14° 35′), Uppsala County (N 59° 28′ - 60° 38′, O 16° 47′ - 18° 39′) and Västerbotten County (N 63° 30′ - 65° 49′, O 14° 54′ - 21° 45′). Västerbotten County is divided in two parts, the coast land and the inland.

The measured heights and stem volumes for the plots are illustrated in Fig. 3 by the dots. The scatter plots have been limited to heights up to 30 m and stem volumes up to 600 m^3/ha (only eight stands fall outside these plots). Since coniferous are dominating at the test sites, plots with more than 70% coniferous have been marked with circles. For each region, Eq. (5) has also been included as reference. In Fig. 4 the regression function for each region has been included and only a slight difference between the curves relative the spread of values around the regression curves is seen, as shown in Fig. 2 and 3.

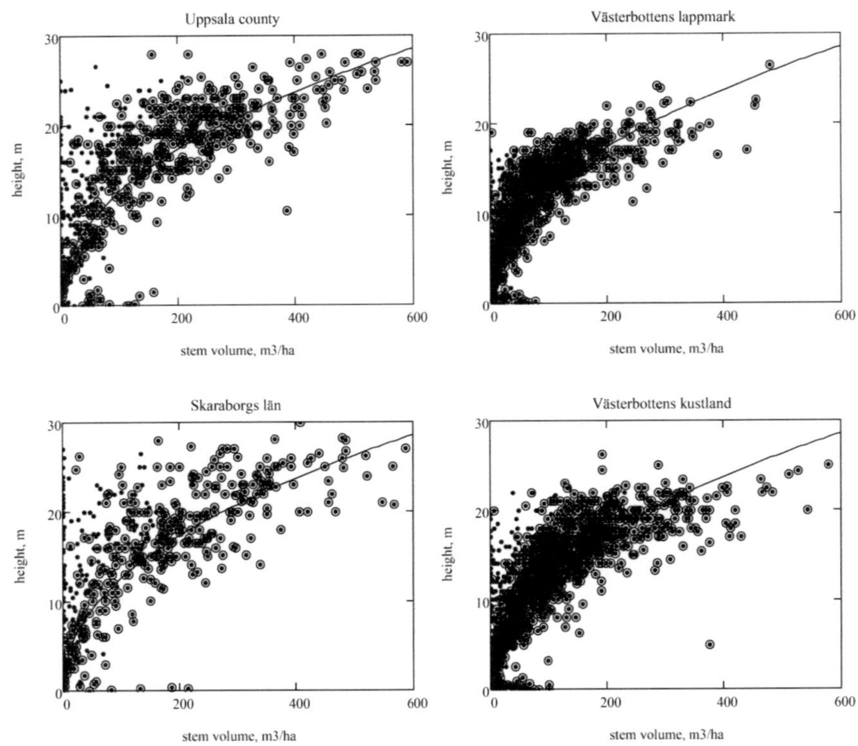

Fig. 3. Plot wise values for height and stem volume. All plots included in the dataset are marked by black dots, and those with > 70% coniferous are marked with an added circle. The black curve represents Eq. (7), while the regression curves to the observations are included in Fig. 4.

The rather constant allometric relation between height and stem volume for the investigated areas in Sweden confirmed the plausibility of the assumption on replacing the true height with Eq. (7) in the IWCM. The deviation between the allometric $h(V)$ relation from the *in situ* observations means an uncertainty in the volume decorrelation term of IWCM. For Uppsala county, illustrated in Fig. 3, and for a perpendicular baseline of 200 m, this would imply a coherence shift of up to 10% (or 0.04) with a maximum for stem volumes around 200 m^3/ha, adding to the model errors from other sources.

Since the analysis is based on National Forest Inventory plots, and then typical for the Swedish landscape, the heterogeneity of the forest explains the large spread around the allometric expression. In turn, this indicates that retrieval of stem volume based on an allometric equation and height estimated for example with a remote sensing technique might be affected by significant uncertainty. For example, if only the height had been measured, e.g. by PolInSAR, and the allometric relation for the county of Uppsala, had been used to determine the stem volume in this county, this would result in a relative Root Mean Square Error (RMSE) of 51% and a mean value of 219 m^3/ha instead of 192 m^3/ha obtained from *in situ* measurements of stem volume. Therefore, if the goal is to determine stem volume or biomass, it is not sufficient to determine the height by remote sensing, but also the forest density has to be estimated.

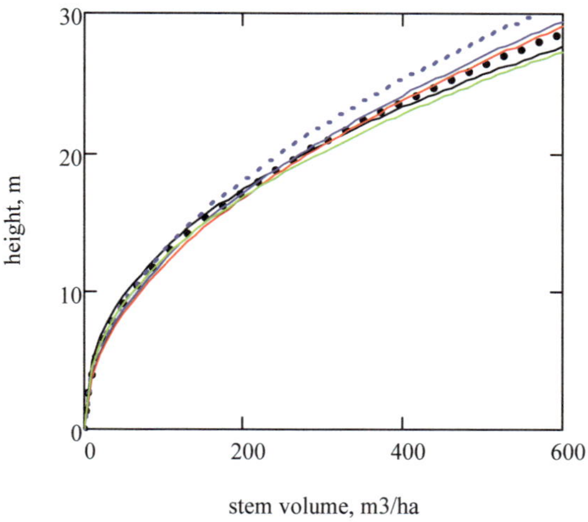

Fig. 4. Regression curves for the different data sets: black solid line: all investigated areas (3046 plots); black dotted line: allometric expression $h(V)=(2.44\ V)^{0.46}$. With only those with >70% coniferous are included: green solid line: Skaraborgs County (343 plots); red solid line: Uppsala County (455 plots); blue dotted line: Vasterbotten County inland (855 plots); blue solid line Vasterbotten County coast land (1070 plots)

It is important to remark that the allometric relation *h(V)* is area-dependent and may be valid for the managed forests in Scandinavia. For the area of the former USSR, yield tables derived from inventory data can be used to relate height and stem volume. The relationship depends on the productivity, see (Santoro et al., 2007b) Eq. (7) can be applied for forests with high relative stocking representing a more managed type of forest, while stands with low relative stocking are characterized by larger tree heights. For German forests presented in (Schober, 1995; Mette et al., 2004) the *h(V)* relation is quite different from Swedish forests.

3.3 Relation between stem volume and biomass

Stem volume has been the property of interest for forest administration purposes and for the Swedish NFI. However for climate modeling applications the above ground biomass is of interest. The International Panel on Climate Change (IPCC) guidelines for national greenhouse gas inventories (IPCC, 2006) give biomass expansion factors for boreal forest depending on growing stock level. For pine, it is 0.45-0.58 and for spruce 0.45-0.605 in both cases for a growing stock level > 100 m^3/ha. The factors are multiplied by the growing stock in m^3/ha to obtain the above ground biomass in tons/ha. (Formally the boreal climatic zone is defined by $\leq$ 3 months at a temperature > 10°C but still seems most relevant for the test sites.) There is more specific information on the relation between stem volume and biomass for two sites, Remningstorp and Krycklan (35 km north Brattåker). For Remningstorp, stem volume and above ground biomass have been calculated from tree level by means of allometric equations according to (Näslund, 1947) for stem volume and (Marklund, 1988) for above ground biomass while for Krycklan the updated HEUREKA expressions have been used (ESA, 2010). It is concluded that the conversion factor between stem volume and above ground biomass is 0.47, (regression formula 0.47 x stem volume +12.7) and that stem volume and above ground biomass are highly correlated (0.98).

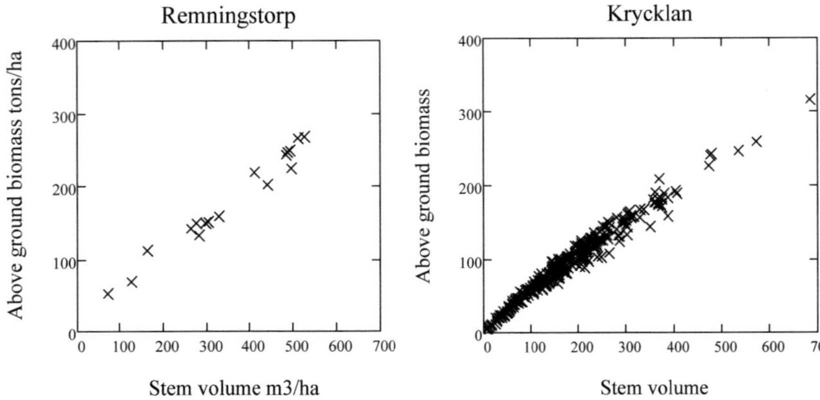

Fig. 5. Relation between above ground biomass, tons/ha, and stem volume, m3/ha (correlation 0.99, slope 0.471 for Remningstorp, correlation 0.98, slope 0.465 for Krycklan) Data courtesy: Gustaf Sandberg for Remningstorp, BioSAR 2008 campaign for Krycklan.

4. Test sites for stem volume retrieval

For testing the retrieval accuracy of C-band InSAR data from ERS-1/ERS-2 tandem period four forest areas have been investigated. The areas, Brattåker, Remningstorp, Tuusula, and Kättböle, see Fig. 6, were selected based on the availability of accurate *in situ* data, and represent examples of boreal forest in northern Europe - two of them at the southern edge, and Remningstorp just south of the edge, see Fig. 6. They are managed by different organizations/companies. All are covered mainly with typical boreal coniferous species, i.e. Scots pine (*Pinus sylvestris*) and Norway spruce (*Picea abies*); but some deciduous species are also present, the commonest being birch (*Betula pendula*). The areas are divided in stands consisting of relatively homogeneous forest properties and site conditions.

Brattåker (lat. 64°16′ N; long. 19°33′ E) is a forest research park located in northern Sweden and managed by the Swedish forest company Holmen Skog AB. It covers approximately 6000 ha. Elevations in the test site range from 160 to 400 m above sea level and slopes up to 12° are common. For one stand the mean slope angle is 21°. The *in situ* data included 54 stands carefully inventoried (at the 10% level) in 2000. Based on growth factors the status for the year of acquisition of the ERS data, i.e., 1996, was estimated. However possible thinning etc could have decreased the accuracy of the *in situ* data.

Tuusula (lat. 60° 25′N long. 25° 1′E) consists of 210 stands of which eight stands were smaller than the SAR image pixel size and thus too small for estimates of backscatter and coherence. Elevations in the test site ranged from 35 to 90 m above se level. *In situ* data were collected in summer 1997 as part of the EUFORA project (Hallikainen et al., 1997). The *in situ* data together with InSAR data been analyzed in several papers (Santoro et al., 1999; Pulliainen et al., 2003; Engdahl et al., 2004; Askne and Santoro, 2005). It resulted that 37 of the 202 measured stands could be labeled as being suitable for retrieval of stem volume because of being large and homogeneous (Santoro et al., 1999).

Kättböle (lat. 59° 59′ N long. 17°7′ E) is a forest estate covering 550 ha, with relatively flat topography where elevation ranges between 75 and 110 m above sea level. Accurate *in situ*

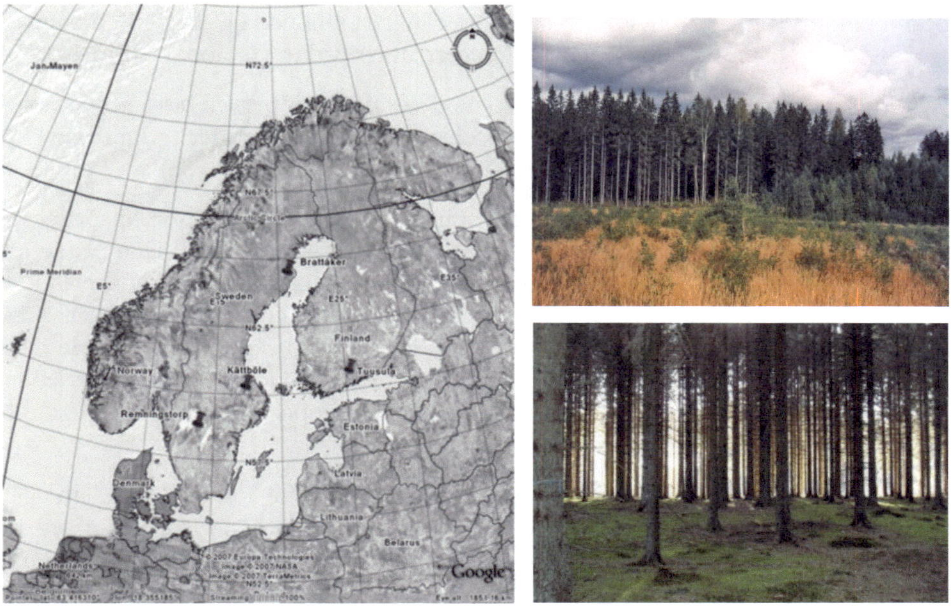

Fig. 6. Map with test sites marked (courtesy Google) and forest photos from Tuusula and from Remningstorp (courtesy Leif Eriksson) to the right.

data were collected 1995 and 1996 and those from 1995 corrected by growth factors. 42 stands were studied. This test site was analyzed in (Fransson et al., 2001; Santoro et al., 2002)

Remningstorp (lat. 58°30′ N; long. 13°40′ E) is a 1200 ha forest estate run by Skogssällskapet AB. This is a site used in many remote sensing verification experiments (Holmgren, 2003; Magnusson and Fransson, 2004a; ESA, 2008; Sandberg et al., 2011). Elevations in the test site range from 120 to 145 m above sea level. The *in situ* data included 105 stands inventoried at the 10% level. The inventory took place during 1997-2002. To decrease the uncertainties in the *in situ* values associated with unknown thinnings etc, stands that were inventoried not later than 2000 were considered and growth factors were applied to estimate the status in 1996.

5. Retrieval of stem volume from the traditional model training

The retrieval of stem volume requires that the model parameter, unknown *a priori*, are estimated first. The traditional estimation approach relies on measurements of stem volume, coherence and backscatter forming a model training dataset. The parameters are then estimated with non-linear least squares regression techniques (Santoro et al., 2002). In Eq. (1) and (2), σ_{gr}, σ_{veg}, γ_{gr}, γ_{veg}, and β are estimated. Although α can also change depending on the seasonal condition, in the analysis it was instead fixed to 0.23 (i.e., 1 dB/m).

While the inversion of Eq. (1) to retrieve stem volume from a backscatter measurement is straightforward, the inversion of Eq. (2) cannot be expressed in a closed form and a numerical approach is required (Santoro et al., 2002). Finally, if multiple measurements of coherence and backscatter are available, a multi-temporal estimate of stem volume based on a linear combination of individual estimates can be obtained. Details of the inversion process including the multitemporal averaging are described in (Santoro et al., 2002).

Modelling and retrieval using the traditional model training approach have been carried out using stand-wise averages of stem volume, coherence and backscatter to limit the effect of pixel-wise noise.

For assessing the retrieval accuracy of stem volume, the stands were sorted for increasing stem volume and divided in two groups by taking each second stand. One of the groups was used for training and one for testing. The groups were interchanged and mean results determined. For Remningstorp, however, all stands were used for training as well as testing. For improved statistical significance, cross-validation can be used, but the simple methods have been chosen due to a relatively large number of stands and also in the conviction that the errors are mainly due to the variability of forest properties and the environmental conditions which will vary from site to site.

For each site, the coherence distribution versus stem volume for the "best" image (i.e., characterized by the highest retrieval accuracy) is illustrated in Fig. 7. Results of the InSAR observations are summarized in Table 1 together with baselines and environmental conditions for the three best image pairs. First will some comments related to each of the test sites be given and then general experience from all the test sites. Which observations can be assumed to be most suitable for retrieval of stem volume from repeat-pass coherence will be discussed below.

The retrieval accuracy is here reported in form of the relative Root Mean Square Error (RMSEr), i.e., the RMSE relative to the true mean stem volume. Typically a value between 10% and 20% is considered acceptable by forest inventories. Larger uncertainties can be accepted by the carbon and climate communities.

Brattåker: Five pairs from the winter season were selected but two were only partly overlapping with the forest site and only three image pairs were used. Of these, for the two image pairs acquired in 1996 the temperature varied close to zero degrees during the day between acquisitions causing decorrelation due to changes of the dielectric and structural properties of the snow layer. The topography varies over the area more than the other investigated areas for which the topography was mainly negligible; however, no effect due to the topography was detected on the retrieval accuracy. For a multitemporal combination RMSEr = 33% was obtained.

Remningstorp: Five image pairs, acquired when the temperature is close to the freezing point, were available. The maximum temperature during the day in between acquisitions was in all cases above 0° C, causing decorrelation due to changes in the snow layer. The retrieval accuracy was related to stand size and quality of the *in situ* data. For 34 stands including at least 24 pixels or at least 1.5 ha large, a multitemporal RMSEr of 32% was obtained. For 30 stands including at least 18 pixels and with inventory data collected not later than 1997, a multitemporal RMSEr of 27% was obtained.

Tuusula: For this site 15 image pairs were available of which 4 with temperatures below zero. By applying a multitemporal combination RMSEr = 20% is obtained for 37 selected stands. In (Askne and Santoro, 2005) all 202 stands were considered and RMSEr = 58%. However, when stands less than 4 pixels were excluded and those with *in situ* stem volume less than 100 m^3/ha RMSEr = 26% (85 stands) is obtained. This explains that the high RMSEr for all 202 stands originates from small stands and stands with small stem volume. For larger stem volumes one might have expected larger errors associated with saturation, but from the analysis of the stands in Tuusula including stands up to 539 m^3/ha (≈ 265 tons/ha) this was not seen, cf. Fig. 6. Similar investigations of other test sites gave the same tendency.

Kättböle: For this site nine image pairs were available of which two with temperatures below zero. The multitemporal combination of all nine pairs resulted in RMSEr = 19% for 42 stands.

In total 34 ERS-1/2 tandem pairs covering the four test sites have been investigated. Forest properties of 338 forest stands have been determined by foresters, for two sites (Kättböle and Tuusula) within one year and for the other two within two to four years time difference relative the ERS tandem observations. The analysis illustrates the importance of having *in situ* data close in time with the acquisitions. The topography, mostly gentle, did not seem to have much effect on the retrieval accuracy.

The Interferometric Water Cloud Model was found to well describe the observations, under all environmental conditions and interferometric geometries. The retrieval results have illustrated that the accuracy is related to:

- **Weather conditions**: Best with temperatures consistently a few degrees below the freezing point, a dry snow layer on the ground (characterizing high ground coherence) and a wind corresponding to an at least moderate breeze (characterizing low vegetation coherence)
- **Stem volume**: small stem volumes present larger relative retrieval errors than large stem volumes, no saturation limit identified - RMSEr decreases with increasing stem volume
- **Stand size**: at least 1.5 ha (large and homogeneous stands give best agreement with *in situ* data)
- **Baseline of image pair**: Of the 34 image pairs studied seven had baselines in the range 200 - 250 m. Five of these belong to the best twelve image pairs in Table 1.

With the northern location of Brattåker one would have expected suitable environmental conditions to occur relatively often, still no acquisitions were obtained under optimal conditions. This is probably related to the limited number of acquisitions in this area., The winter season 1995/96 included 120 days with temperature < 0° C of which 70 days with wind speed at least 3 m/s and a snow layer or 35 days with wind speed at least 4 m/s and a snow layer. For the southernmost, hemi-boreal site, Remningstorp, as a mean per winter season based on observations from 1993 - 1998 78 days had temperature < 0° C of which 52 days with wind speed at least 3 m/s (of which 33 with a snow layer) or 11 days with wind speed at least 4 m/s (6 also with a snow layer). If the acquisition plan for ERS-1 and ERS-2 had taken into account the conditions for optimal meteorological conditions for forest stem volume retrieval, it is expected that it would have been possible to obtain a reasonable number of images with strong sensitivity of the coherence to stem volume each winter season in the boreal zone.

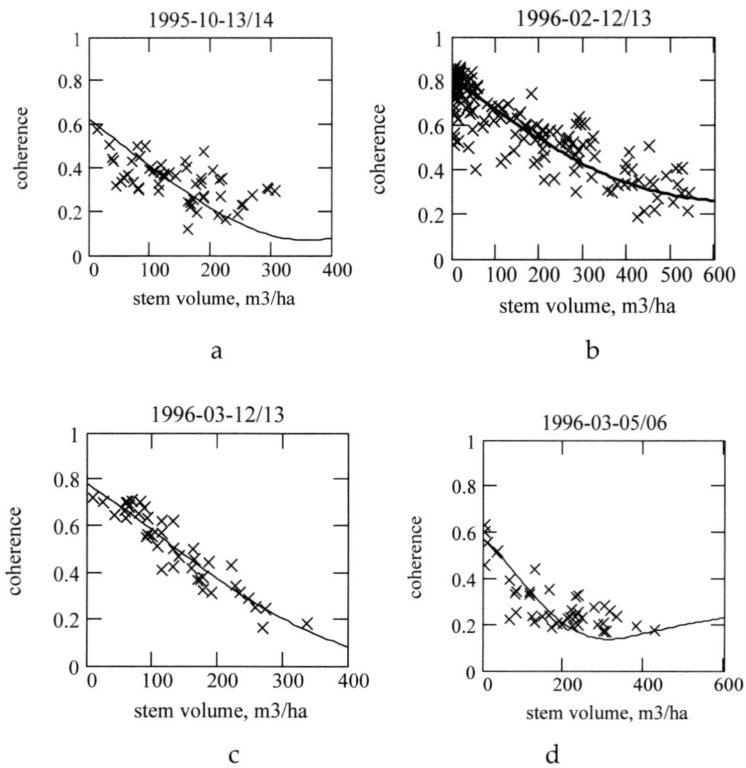

Fig. 7. Measured and modeled coherence as a function of stem volume for each of the best image pairs from a) Brattåker; b) Tuusula; c) Kättböle and d) Remningstorp.

6. Consistency plots

To analyze the properties of a multitemporal set of coherence and backscatter values, the so called "consistency plots" can be investigated, the plots of two observations of the same area from different dates versus each other. The plots show to what extent the observed quantity for the same pixels change consistently or randomly and then if the variations can be described by a model. With n observations there are $\frac{1}{2} n(n-1)$ different combinations. As an example Fig. 8 illustrates the observations from two coherence images of forest pixels in the Kättböle area and also the corresponding ERS-1 backscatter observations. If the temporal coherence had been exactly the same at the two occasions the points had fallen more or less along a straight line (depending on baseline). Lowest coherence would correspond to the forest with highest stem volume and highest coherence to bare surface. From Fig. 8 it can be concluded that there has been a certain consistent change in the coherence observations and in addition a spread due to noise and temporal variation of forest properties. For the backscatter there is no or very little trend due to the mostly weak sensitivity of the backscatter upon stem volume. From the consistency plot the density of observation values can be determined and the corresponding matrix values presented as a contour plot or "density plot" (2D-histogram).

Date	Baseline m	Temp. °C	Wind speed m/s	Snow depth cm	Prec mm	RMSEr %
Brattåker, stem volume:< 305 m³/ha, multitemporal RMSEr = 33%						
1995-10-13/14	219.7	+5.4/+8.4	5/10	0/0	0/0	42
1996-03-03/04	228.5	-2.3/-2.5	3/3	36/40	0/0	38
1996-03-17/18	84.3	-1.7/-2.6	3/3	33/33	0.1/0	44
Tuusula, stem volume: < 539 m³/ha, multitemporal RMSEr = 58% For 37 large, homogeneous stands: 20%, for 85 stands > 4 pixels and V > 100 m3/ha: 26%						
1996-02-12/13	85	-14.9/-10.8	5/1	16/19	1.4/0	60
1996-03-02/03	76	-4.2/-5.5	4/3	32/32	0.1/0	74
1995-10-14/15	220	+8.6/+6.2	6/1	0/0	0/0	82
Kättböle, stem volume: < 335 m³/ha, multitemporal RMSEr = 18%						
1996-03-12/13	219.2	-4.9/-4.4	2.5/2.5	10/10	0/0	19.8
1996-03-17/18	65.6	-1.7/-2.8	1.7/0.9	10/10	0/0	38.0
1996-04-21/22	-54.0	+12.8/+12.5	2.3/1.3	0/0	0/0	31.6
Remningstorp, stem volume: < 494 m³/ha, multitemporal RMSEr = (32 or) 27%						
1996-03-05/06	250.2	-0.6/-1.2	1/1	8/7	0/2.3	46
1996-04-11/12	-89.1	-3.1/-3.1	5/5	0/0	0/0	45
1996-04-27/28	-84.5	+2.4/+6.6	3/5	0/0	0.8/0	44

Table 1. List of the three best image pairs in terms of relative RMSEr, weather conditions and retrieval statistics. Weather conditions refer to the time of acquisition of the SAR data.

For backscatter, in contrast to coherence, the density plot is roughly circular and there is no evident sensitivity to different forest properties.

For the Tuusula area it was noted (Askne and Santoro, 2007) that the large and homogeneous stands resulted in the high retrieval accuracy while small stands and those with small stem volume resulted in low retrieval accuracy. Part of the small retrieval accuracy could be related to errors in the *in situ* observations, but the question is if the coherence observations could provide some indication about the properties of forest stands and the possible accuracy that can be obtained by studies of the consistency plots. For this purpose the "outlier concept" was introduced in (Askne and Santoro, 2007), as based on a measure of the spread of coherence values between different image pairs. High outlier values are related to pixels located in the outer contours of the consistency plots related to the coherence values for the same pixel from two coherence images. It was shown in (Askne and Santoro, 2007) that the highest stem volume retrieval accuracy is obtained for large stem volumes, large stand areas and for small outlier values. All of these properties can then be estimated from remote sensing, and pixels for which the stem volume accuracy can be suspected to be low can be identified, cf Fig. 9.

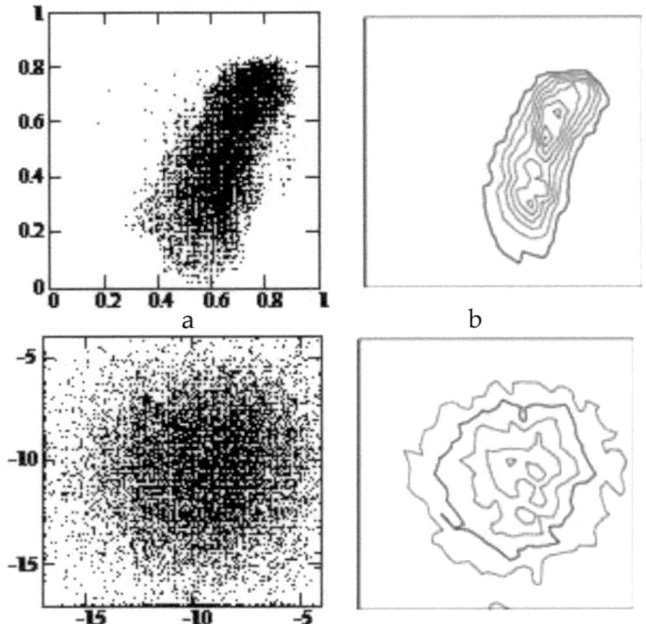

Fig. 8. Illustrating coherence observations from Kattböle from 12/13 March 1996 versus 16/17 April, a and b, and the corresponding ERS-1 backscatter in dB, c and d. Consistency plots showing pair of pixel values, a and c, and the corresponding density plots illustrating the density of points, b and d (with the same scale as in a and c).

The reason for the low accuracy could be aspects related to the forest properties, properties not related to stem volume having a relatively high and varying influence. The major conclusion is that high accuracy of stem volume for a forest stand can be retrieved only when the pixels show up in the consistency plots with little spread, i.e. low outlier value.

7. Retrieval of stem volume with modeling without training stands

There is always a need for reference sites for verification of the accuracy of satellite data. However the need for training data for the interpretation is an issue since such data have to be available on distances related to the spatial scale of weather changes. This is a problem for all remote sensing methods sensitive to environmental properties.

A proposal for an automatic training method of ERS tandem coherence and JERS backscatter data was presented in (Wagner et al., 2003). A matching technique between different frames was used such that the 10th and 90th percentiles of the histograms within the overlap areas of two frames were matched to those of the image with the greater frame coverage. The technique was combined with an exponential coherence model for interpretation of the observations of stem volumes in three classes up to 80 m^3/ha and a fourth class above.

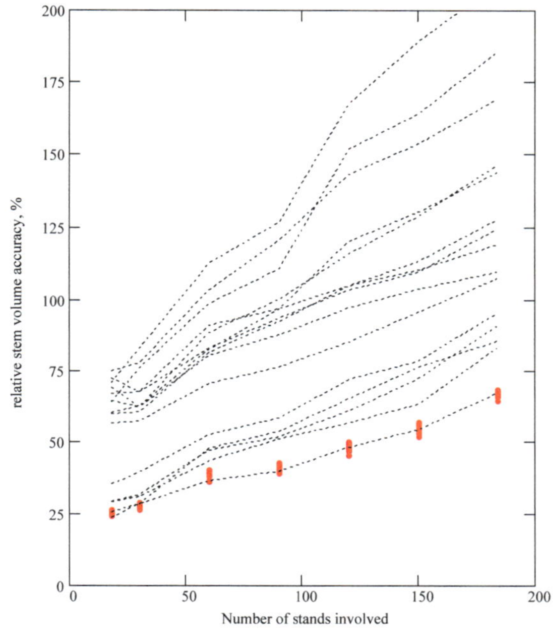

Fig. 9. Results based on 18 stands used for training and 10 up to 180 for testing with increasing outlier value. The dotted lines illustrate the RMSEr of the individual pairs. The solid red intervals illustrate the RMSE when multitemporal analysis is used combining up to 15 image pairs. From (Askne and Santoro, 2007).

A method using the IWCM for interpretation of ERS tandem coherence data without ground reference data was proposed in (Cartus et al., 2007; Cartus et al., 2011). The IWCM parameters were determined by taking statistics of coherence and backscatter of open areas and dense vegetated areas, identified by means of the MODIS Vegetation Continuous Fields, VCF, tree canopy cover product. A fixed value of 0.006 is used for β.

Another method based on multitemporal and multibaseline ERS tandem data was presented in (Askne and Santoro, 2009). The technique is based on consistency plots and density plots introduced in § 6. The majority of forest pixels cluster along a central "ridge" while those more affected by noise and temporal decorrelation will be located in the external parts. It was also shown in § 6 that in the case of the Tuusula data set those forest plots with data along the ridge are those most accurately described by the model, while those consistently further away from the ridge result in lower accuracy. In other words, it is demanded that the model parameters should be such that modeled coherence will follow the ridge, when the stem volume varies.

7.1 Kättböle

The method is simplest to demonstrate for a dataset with relatively limited noise and temporal variability, and the Kättböle test case will be used. It is assumed that *a priori* information is available of the highest stem volumes in the region of interest, from e.g. NFI information. If there is no information on typical areafill variations a value for α corresponding to 1 dB/m can also be assumed. For simplicity it will also be assumed in this analysis that the pixels studied already have been classified as forest pixels. However it has been shown, see e.g. (Wegmüller and Werner, 1995; Dammert et al., 1999; Engdahl and Hyyppä, 2003) that InSAR can be used to differentiate between forest and non-forest pixels, and that the entire classification process then can be based on ERS tandem data.

In Fig. 10 the coherence values for the forest pixels associated with the four coherence image pairs for the Kättböle area (12.5 m pixels) are shown in consistency plots. The IWCM parameters now have to be determined such that, when co-plotted, the IWCM curves for the four cases go through the ridges of the six consistency plots. For Kättböle the areafill variation with stem volume is known and a β- dependent attenuation according to Eq. (6) is

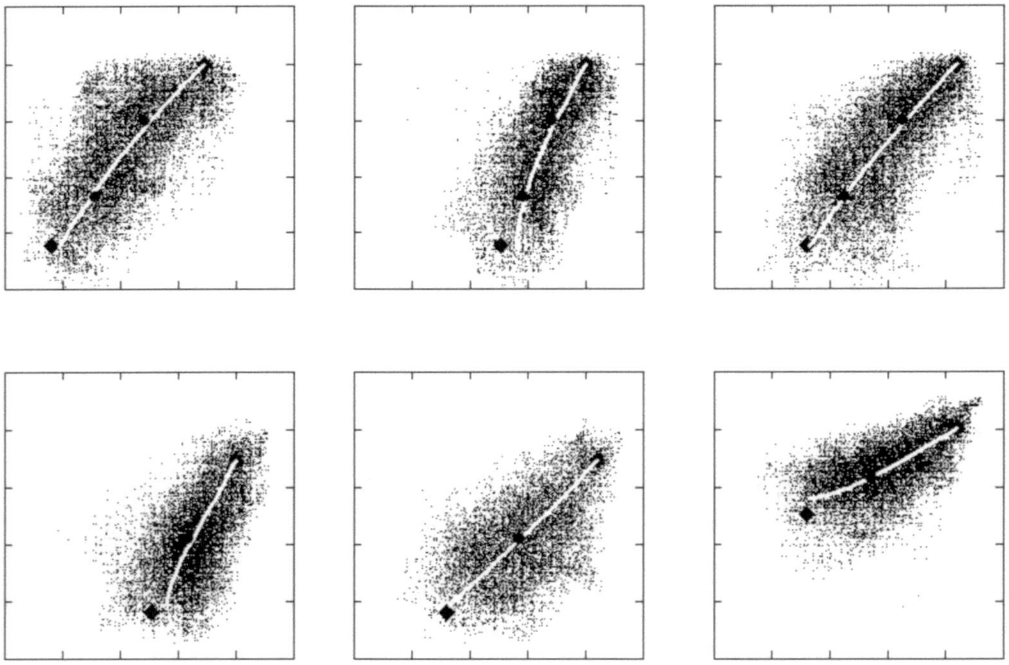

Fig. 10. Illustrating coherence observations and model variations for the four coherence image pairs for the Kättböle area, denoted $\gamma 1$, $\gamma 2$, $\gamma 3$, and $\gamma 4$, of observations for the Kättböle area. In upper row $\gamma 1$ versus $\gamma 2$, $\gamma 1$ versus $\gamma 3$, $\gamma 1$ versus $\gamma 4$, and lower row: $\gamma 2$ versus $\gamma 3$, $\gamma 2$ versus $\gamma 4$, and $\gamma 3$ versus $\gamma 4$. The model curves spans the interval 0 - 378 m^3/ha. Reference points are marked along the ridge, γ_{gr} , γm, and γ_{dv}, see text.

used. From the coherence distributions of each image it is first determined what pixel fall in the upper 10% and the lower 15% (more noise influences the lower coherence values) and these are defined as potential pixels representing ground and dense vegetation respectively. Those pixels in common for all image pairs are then defined as ground and dense vegetation pixels. From these pixels the average coherence and backscatter are determined, σ_{gr} and γ_{gr}, and values for "dense vegetation", σ_{dv} and γ_{dv}.

The model parameters σ_{veg} and γ_{veg} represent the backscatter and temporal coherence for an infinitely opaque forest and to derive these from σ_{dv} and γ_{dv} model corrections have to be made by means of an assumption on the stem volume corresponding to dense vegetation, V_{dv}. The stem volume distribution has reached 80% of the cumulative stem volume histogram typical for the region (e.g. determined from NFI-plots) at 315 m^3/ha, cf. Fig. 3, and a 20% higher value or 378 m^3/ha is here considered as typical for V_{dv}. Since the backscatter is normally saturated we may assume that $\sigma_{veg} \approx \sigma_{dv}$. To obtain γ_{veg} Eq. (5) has to be solved for each coherence image

$$|\gamma(V_{dv}, \gamma_{veg}, \beta)| - \gamma_{dv} \leq \varepsilon \tag{8}$$

starting from guess-values, to be discussed below. ε is the uncertainty in the coherence value of the manually determined point, say ε = 0.02. When the baseline is such that IWCM has a minimum in the coherence versus stem volume plane, then the meaning of γ_{dv} changes, see § 7.2.

The demand now is that the model curve should follow the "ridge" identified in the density or consistency plots. The ridge is identified by manual inspection and one or two points along the ridge may be identified, γm_{pair}, for each one of the coherence pairs, see Fig. 10. These points are associated with unknown values for the stem volume, say V1 and V2. Together with the four unknown β-values there are six unknowns, beside Eq. (8) regarding γ_{veg}, with demands like $\gamma 1(V1,\beta 1) - \gamma m1_1 < \varepsilon$, $\gamma 2(V1,\beta 2) - \gamma m1_2 < \varepsilon$, $\gamma 3(V1,\beta 3) - \gamma m1_3 < \varepsilon$, and $\gamma 4(V1,\beta 4) - \gamma m1_4 < \varepsilon$. Four other conditions are related to V2. Minimizing these expressions with a Levenberg-Marquardt algorithm starting from some initial guess values, expressions for the β:s are obtained. Since the algorithm is sensitive to the initial guess the process is initiated by assuming $\gamma_{veg} = \gamma_{dv}$ and a manual variation of the β-values over a range 0.003 – 0.007 in order to make the model curves follow as closely as possible to the "ridges". Also γ_{veg} may be adjusted.

Once stem volume has been retrieved for each individual coherence observation, the individual estimates are combined to form the multi-temporal stem volume value. The multitemporal stem volume value is determined from the individual image results using relative weights determined by the coherence contrast in each image between the values for ground and for dense vegetation, and by the percentage of observations that fall within this range.

The results from the traditional training method were reported in Table 1 and the results without training are presented in Table 2 for the case illustrated in Fig. 10. The derived β-values are 0.0034, 0.0043, 0.0060, and 0.0045. R^2 is the square of the correlation between estimated stem volumes and *in situ* values, RMSEr is the relative root mean square error and the Bias stands for the difference between the estimated mean volume for all stands and the corresponding *in situ* value. All steps in the process are based on pixel values, but the

	IWCM	IWCM	Exp
Vegetation attenuation, α	1 dB/m	variable	
R^2	0.91	0.91	0.91
RMESr, multitemporal	17%	17%	18%
Bias	0.3 m^3/ha	-3.5 m^3/ha	-1.6 m^3/ha
RMSEr 1996-03-12/13	22.4	22.9	27.0
RMSEr 1996-03-17/18	33.4	33.2	34.6
RMSEr 1996-04-16/17	34.8	34.4	36.5
RMSEr 1996-04-21/22	31.5	31.5	31.7

Table 2. Statistical measures on the accuracy of stem volume estimates using no training data based on the Kättböle area illustrated in Fig. 8b with 42 stands. The first column refers to results obtained with the IWCM and constant α value. The second column refers to results obtained with the IWCM and α based on area-fill and β estimates. The third column refers to results obtained with a simplified IWCM model that does not take into account volume decorrelation.

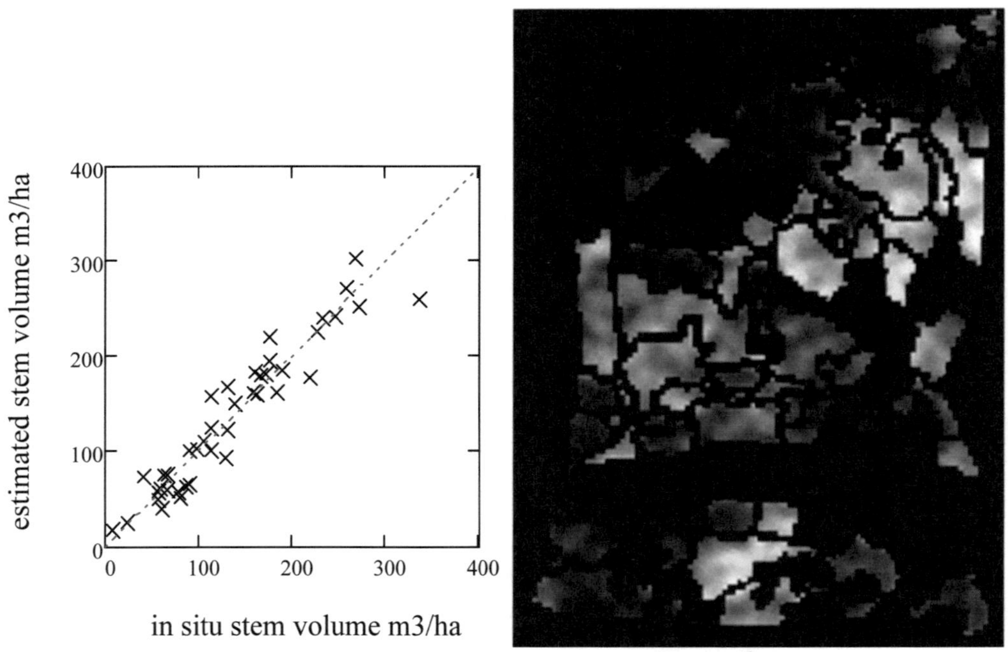

a b

Fig. 11. a: Stand averages of stem volume compared with *in situ* values; b: Gray scale image of stem volume variation in Kättböle site (stand boundaries marked). The image is normalized such that white corresponds to the highest estimated stem volume of 378 m3/ha in a single pixel.

comparison with *in situ* data is done for stand averages. Results for α fixed to 0.23 corresponding to 1 dB/m are given as well as β-dependent α-values. The values of α in the last case are 0.31, 0.45, 0.80, and 0.48 dB/m. This is corresponding to an extinction coefficient between 0.1 and 0.2 dB/m for a 22 m vegetation. The temperature varied between - 5° and

+13° C with the lowest extinction at -5° C. The results are in line with observations reported in (Treuhaft et al., 1996). We conclude that the results without training stands are in line with those derived by means of training stands.

The stand averaged multitemporal estimate is compared with the *in situ* values of stem volume in Fig. 11a and an image of the estimated stem volume variation in the test area is illustrated in 11b for possible comparison with laser scanning data in the future.

In (Cartus et al., 2011) it was argued that the performance of the IWCM suffers from errors under certain environmental conditions that reduce the sensitivity of the coherence to stem volume. An exponential model, corresponding to $\gamma_{vol} \equiv 1$, appeared to be more suitable to describe the coherence as a function of stem volume in such cases. For completeness, we compared the retrieval performance of the IWCM and the simple exponential model, cf column 3 in Table 2. The results were similar except for pair 1 with a baseline of 229 m. The reason is that IWCM tends to an exponential form when the baseline is zero, whereas there is a difference when volume decorrelation is introduced by the long baseline.

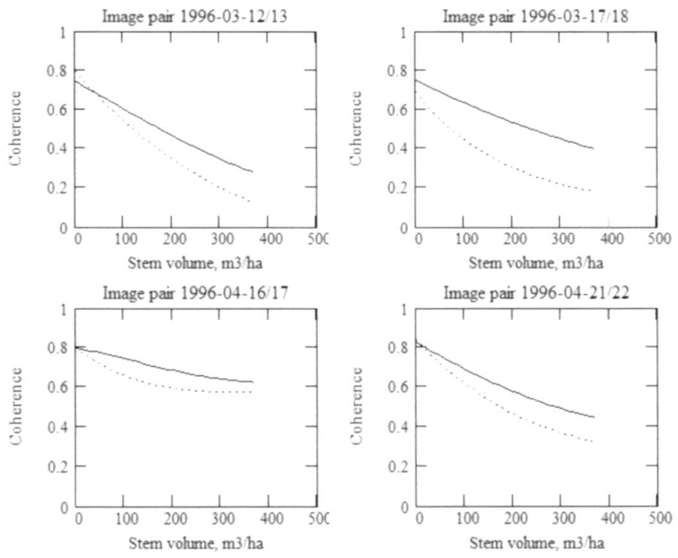

Fig. 12. Model curves for the four coherence images centered on the test site of Kättböle. Solid line for model realizations determined for an area 25 km away from Kättböle, and dotted line model realizations for the Kättböle area. The model spans the interval 0 - 378 m^3/ha.

To test whether the retrieval approach based on consistency plots is adaptive to variations of the coherence due to different environmental conditions across the coherence image, we looked at an area covered by the same ERS scenes, 25 km SE of the Kättböle site. In an earlier analysis (Santoro et al., 2002), it was noted that there were spatial variations of the coherence as a consequence of possible differences of the environmental conditions with respect to the Kättböle area. From the consistency plots for the new area it is realized that the model curve, see Fig. 12, (solid line) is shifted from the model curve derived for the Kättböle area (dotted line). These changes could be interpreted as caused by a lower wind speed in the new area.

The results indicate the need of a spatially adaptive method to retrieve stem volume from coherence data to avoid significant retrieval errors by using a single realization of the coherence model.

7.2 Remningstorp

For Remningstorp the environmental conditions were not as optimal as for Kättböle (see Table 1). Fig. 13 shows the consistency plots in the case of four images and 100 stands, 39 being smaller than 10 pixels. The curves are concentrated around low coherence values, and the structure is less clear compared to the consistency plots shown in Fig. 10 for Kättböle.

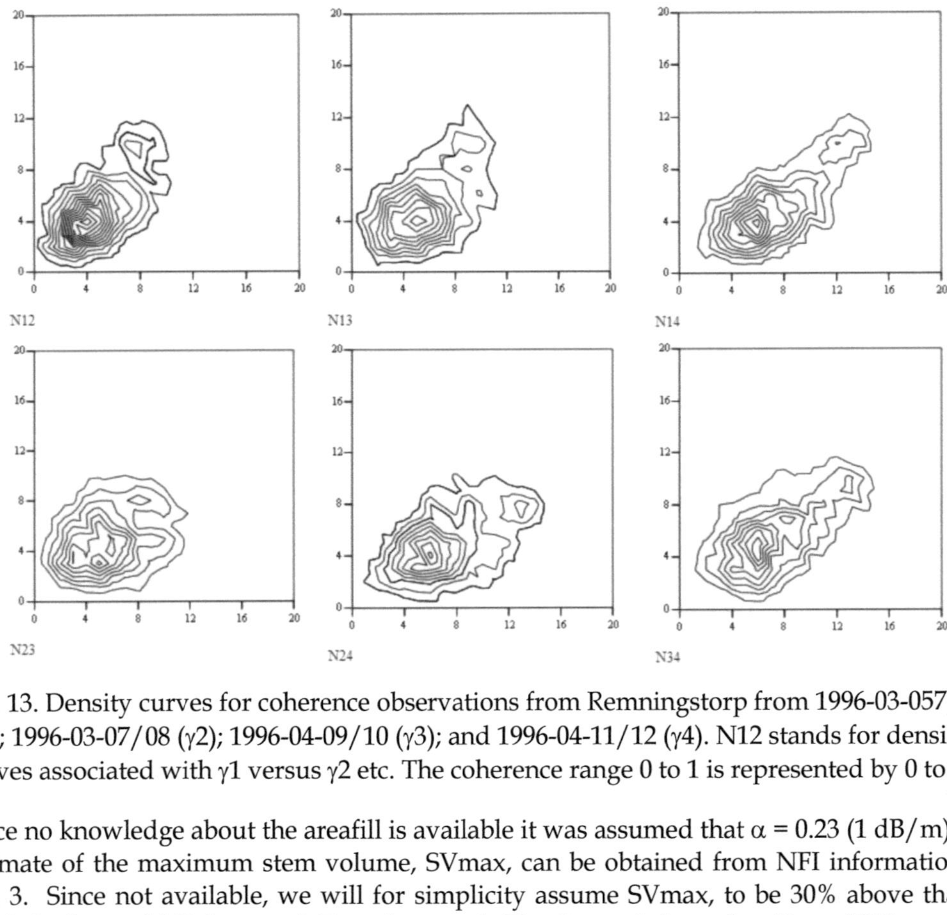

Fig. 13. Density curves for coherence observations from Remningstorp from 1996-03-057/06 ($\gamma1$); 1996-03-07/08 ($\gamma2$); 1996-04-09/10 ($\gamma3$); and 1996-04-11/12 ($\gamma4$). N12 stands for density curves associated with $\gamma1$ versus $\gamma2$ etc. The coherence range 0 to 1 is represented by 0 to 20.

Since no knowledge about the areafill is available it was assumed that $\alpha = 0.23$ (1 dB/m). An estimate of the maximum stem volume, SVmax, can be obtained from NFI information, cf Fig. 3. Since not available, we will for simplicity assume SVmax, to be 30% above that in Kättböle (located 200 km north Remningstorp). The first pair has a baseline of 250 m and a coherence minimum, which means that the low coherence value we have identified as γ_{dv} represents this minimum, and γ_{veg} is larger than otherwise expected. This is realized when fitting the model curves to the consistency plots, see Fig. 14. For this reason the exponential model may seem easier to fit and was also used. For the 61 stands > 10 pixels and for four images an RMSEr of 27.3% is obtained using IWCM and 27.4% using the exponential model. The result is not sensitive to the assumed SVmax; RMSEr ≤ 30% for 350 ≤ SVmax ≤ 650 m^3/ha. The insensitivity to SVmax is in line with results in (Cartus et al., 2011).

The stand averaged multitemporal estimate is compared with the *in situ* values of stem volume in Fig. 15a and an image of the estimated stem volume variation in the test area is illustrated in 15b.

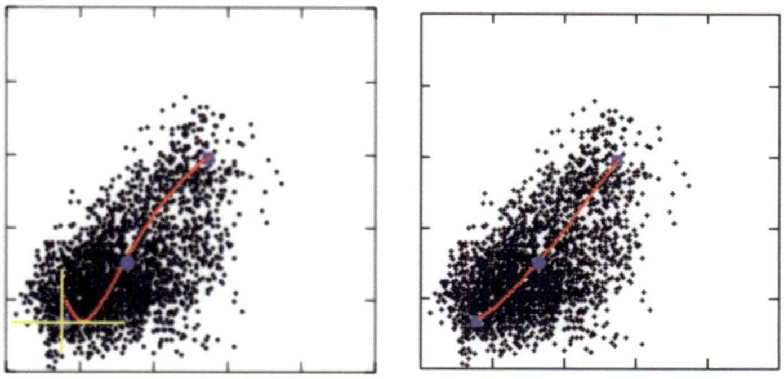

a b

Fig. 14. Illustrating consistency plots for coherence pair 1 and 3, cf Fig. 13, and co-plotted model curve (red lines) for stem volumes from 0 to SVmax. a. IWCM, b. exponential model. The yellow lines, crossing in the point γ_{dv}, indicate the levels for the lowest coherence levels for these coherence pairs when the stem volume is ≤ SVmax. The blue points represent γ_{gr}, γm, and γ_{dv}.

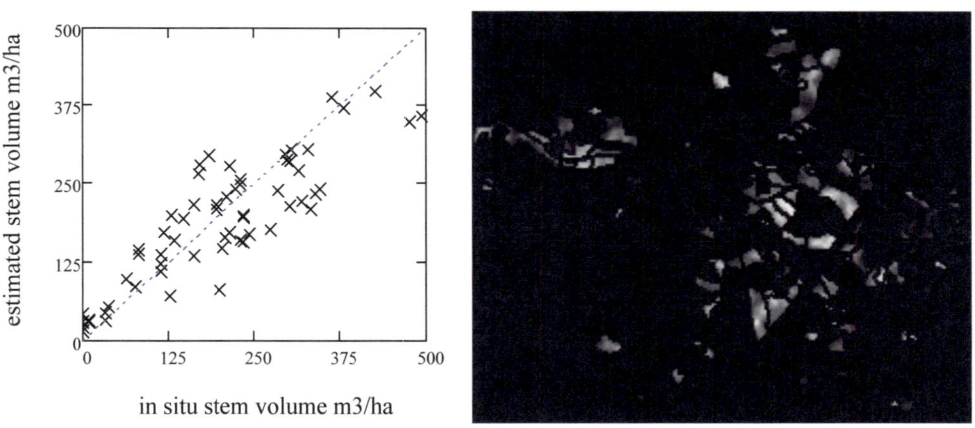

a b

Fig. 15. a: Stand averages of stem volume for 61 stands in Remningstorp compared with *in situ* values. b: a pixel based map over the Remningstorp stands (white corresponds to maximum estimated stem volume of 491 m^3/ha).

8. Conclusions and future outlook

A large dataset of ERS tandem InSAR data has been investigated with focus on the accuracy by which stem volume can be estimated in boreal forest. For this purpose data from four test sites with *in situ* data of high quality have been used. Since the accuracy is related to the environmental conditions the analysis has been concentrated on winter-time images, which are considered to give the best conditions for high retrieval accuracy (sub zero temperature, dry snow layer, moderate breeze). Although individual images may have relatively high RMSEr, a multitemporal RMSEr of 20% may be obtained for such conditions. In case of less favorable environmental conditions with temperature conditions close to zero degrees, but also the *in situ* data less accurate, the RMSEr is in the 30% range. Stands with small stem volumes (< 100 m^3/ha) showed the largest relative retrieval errors, whereas mature forests showed the highest accuracy when investigated for Tuusula, while this was not as clear for Kättböle, where large area was most important for high accuracy. Since the ultimate goal is to have a method that allows retrieving stem volume for large areas it has to be taken into account that the environmental conditions, e.g. wind speed, can vary on a relatively small scale. For this reason a method without the need for *in situ* data for estimating model parameters has been introduced. In the investigated cases, it has been demonstrated that the estimation of stem volume is characterized by the same accuracy as when the model is trained by means of *in situ* measurements. The technique is based on a method to analyze multitemporal data by means of consistency plots. Such plots could also illustrate for what forest stands the coherence observations are less stable and the expected accuracy worse than for those closer to the ridge in the plots.

Observations using other sensors have been performed for the same test sites as studied here, cf. (Smith and Ulander, 2000; Fransson et al., 2001; Fransson et al., 2004; Magnusson and Fransson, 2004b; Magnusson and Fransson, 2004a). Looking at individual acquisitions, optical images (SPOT and Landsat) are reported to have RMSEr 24 - 38%. In the other end of the frequency range there is CARABAS, an airborne VHF-SAR, with much higher spatial resolution and with RMSEr 19 - 30% excluding stands with small stem volume. Other sensors are based on accurate height observations to determine stem volume and for laser scanning from helicopter the RMSEr for Remningstorp is reported to 19 - 26% on plot level and 11% on stand level (Holmgren, 2003). For standard methods like photo-interpretation, one can find figures in the literature indicating accuracies such as 26 - 39%. The accuracy of *in situ* observations may vary 10 - 34% with a mean of 21.4% cf. (Haara and Korhonen, 2004). Polarimetric SAR interferometry is a technique estimating forest height from the volume decorrelation at different polarizations assuming the temporal decorrelation to be negligible or small (Cloude and Papathanassiou, 1998; Papathanassiou and Cloude, 2001). Flights were made over Remningstorp in 2007 and an L- and P-band as part of the ESA project BIOSAR 2007 (ESA, 2008). For L-band data, the best results reported to our knowledge were obtained using HV-polarized backscatter, giving an RMSEr between 31% and 46%. Using HV- or HH-polarized P-band backscatter or both HV and HH, the RMSEr:s were between 18 and 27% (Sandberg et al, 2011).

Values on RMSEr for other sensors given above apply for individual images and can be expected to improve by multitemporal acquisitions when such are suitable. However, thanks to the multitemporal character of satellite radar acquisitions, C-band repeat pass interferometry becomes an interesting alternative for estimation of stem volume and biomass above ground in boreal forests over large areas with potential of accuracy comparable to standard *in situ* methods, without the need for training data and without showing saturation in the relative accuracy for the investigated areas covering up to 539 m^3/ha or ≈ 265 tons/ha.

For large area studies using ERS tandem coherence the best reported result is given in (Santoro et al., 2007b) obtaining an RMSEr of 30% for two of the ten acquired coherence pairs. (Wagner et al., 2003) used tandem coherence and an exponential model in combination with JERS-1 backscatter to determine stem volume in Siberia. In (Drezet et al., 2006; Drezet, 2007) a set of coherence images over UK with a good coherence range was used. Instead of biomass or stem volume, age was used as variable, and results cannot be compared. Finally, (Cartus et al., 2011) used a large number of images in order to cover large areas in China and Siberia. Environmental unstable conditions caused problems and IWCM, used without training data, but with MODIS VCF data as complement, was in many cases not successful and an exponential model was often preferred. It is concluded that, although a high accuracy in stem volume or biomass estimation can be obtained by short-repeat C-band InSAR data, large scale investigations have had complications due to limitations in the number of data acquisitions for optimal environmental conditions and due to limited access to accurate *in situ* data for verification of the results.

An important aspect is that the result is strongly dependent on the short-repeat pass and on timeliness of acquisitions, which puts demands on the acquisition program and also on a short repeat cycle of such a spaceborne interferometric configuration. Currently there are no plans for configurations similar to the ERS-1/2 tandem period. The ERS-ENVISAT configuration offered some possibility for short repeat cycle (Santoro et al., 2007a) but the very long baseline requirement was a major complication because of the strong volume decorrelation. The 12 day repeat cycle of Sentinel-1A is decreased to 6 days with Sentinel-1B and the temporal decorrelation can be acceptable, since the ERS-1 3-day period was showing good coherence up to at least a 6-day interval for winter conditions (Smith et al., 1996). An interesting alternative may be found using X-band missions with more than one satellite such as Tandem-X or COSMO-SkyMed in pursuit mode. Typically the coherence is decreasing with higher frequencies, but, with the possibility to have two acquisitions within seconds, the ground coherence (contributing through gaps in the canopy) is expected to be very close to one while the vegetation layer should still decorrelate because of foliage movement induced by wind even in a few seconds. The extinction at X-band is higher than at C-band and the effective vegetation layer causing volume decorrelation is thinner, but still the volume decorrelation will add to the temporal decorrelation. This should mean a possibility for high contrast between ground and dense vegetation and therefore a potential for high accuracy of stem volume estimates.

9. Acknowledgment

The ERS tandem pairs obtained as an ESA Cat-1 project, ID 3121 and *in situ* data for the Swedish test sites obtained by the help of Dr Johan Fransson, Swedish Univ. of Agricultural Sciences are gratefully acknowledged.

10. References

Askne, J., P. Dammert, L. M. H. Ulander & G. Smith (1997). C-band repeat-pass interferometric SAR observations of forest. *IEEE Trans. Geosci. Remote Sensing,* Vol. 35, No. 1, pp. 25-35.

Askne, J. & M. Santoro (2005). Multitemporal repeat pass SAR interferometry of boreal forest. *IEEE Trans. Geosci. Remote Sensing,* Vol. 43, No. 6, pp. 1219-1228.

Askne, J. & M. Santoro (2007). Selection of forest stands for stem volume retrieval from stable ERS tandem InSAR observations. *IEEE Geosc. Remote Sensing Letters,* Vol. 4, No. 1, pp. 46-50.

Askne, J., M. Santoro, G. Smith & J. Fransson (2003). Multi-temporal repeat pass SAR interferometry of boreal forests. *IEEE Trans. Geosci. Remote Sensing,* Vol. 41, No. 7, pp. 1540-1550.

Bamler, R. & P. Hartl (1998). Synthetic aperture radar interferometry. *Inverse problems,* Vol. 14, No. pp. R1-R54.

Cartus, O., M. Santoro & C. Schmullius (2010). C-band intensity-based growing stock volume estimates versus the MODIS vegetation continuous fields tree canopy cover: Does C-band see more than canopy cover?, Proceedings of *ESA Living Planet Symposium, Bergen, Norway, 2010*

Cartus, O., M. Santoro, C. Schmullius & Z. Li (2011). Large area forest stem volume mapping in the boreal zone using synergy of ERS-1/2 tandem coherence and MODIS vegetation continuous fields. *Remote Sensing of Environment,* Vol. 115, No. pp. 931–943.

Cartus, O., M. Santoro, C. Schmullius, Y. Pang & Z. Li (2007). Creation of large area forest biomass maps for NE China using ERS-1/2 tandem coherence, Proceedings of *Envisat Symposium, Montreux, CH, 2007*

Cloude, S. R. & K. P. Papathanassiou (1998). Polarimetric SAR interferometry. *IEEE Trans. Geosci. Remote Sensing,* Vol. 36, No. 5, pp. 1551-1565.

Dammert, P. B. G., J. Askne & S. Kühlmann (1999). Unsupervised segmentation of multi-temporal interferometric SAR images. *IEEE Trans. on Geosci. Remote Sensing,* Vol. 37, No. 5, pp. 2259-2271.

Drezet, P. M. L., S. Quegan & (2006). Environmental effects on the interferometric repeat-pass coherence of forests *IEEE Transactions on Geoscience and Remote Sensing,* Vol. 44, No. 4, pp. 825-837.

Drezet, P. M. L., Quegan, S. (2007). Satellite-based radar mapping of British forest age and Net Ecosystem Exchange using ERS tandem coherence. *Forest Ecology and Management,* Vol. 238, No. pp. 65-80.

Engdahl, M., J. Pulliainen & M. Hallikainen (2004). Boreal forest coherence-based measures of interferometric pair suitability for operational stem volume retrieval. *IEEE Geosc. Remote Sensing Letters,* Vol. 1, No. 3, pp. 228-231.

Engdahl, M. E. & J. M. Hyyppä (2003). Land-cover classification using multitemporal ERS-1/2 InSAR data. *IEEE Trans. Geosci. Remote Sensing,* Vol. 41, No. 7, pp. 1620-1628.

ESA (2008). BIOSAR 2007.
http://earth.esa.int/campaigns/DOC/biosar_finalreports_nosynthesis.pdf,

ESA (2010). BIOSAR 2008.
http://earth.esa.int/campaigns/DOC/BIOSAR2_final_report.pdf,

Fransson, J. E. S., G. Smith, J. Askne & H. Olsson (2001). Stem volume estimation in boreal forests using ERS-1/2 coherence and SPOT XS optical data. *Int. J. Remote Sensing,* Vol. 22, No. 14, pp. 2777-2791.

Fransson , J. E. S., G. Smith, F. Walter, A. Gustavsson & L. M. H. Ulander (2004). Estimation of forest stem volume in sloping terrain using CARABAS-II VHF SAR data. *Can. J. Remote Sensing,* Vol. 39, No. 4, pp. 651-660.

Haara, A. & K. T. Korhonen (2004). Kuvioittaisen arvioinnin luotettavuus (Reliability of stand-wise forest inventory data, in Finnish). *Metsätieteen aikakauskirja, publ. Metsäntutkimuslaitos and Suomen Metsätieteellinen Seura,* Vol. 4/2004, No. pp. 489–508.

Hallikainen, M., H. Hyyppä, J. T. Koskinen, M. Roschier & P. Ahola (1997). EUFORA Campaign Plan, Version 2. Helsinki Univ. Technol., Lab. Space Technol., Helsinki.

Holmgren, J. (2003). Estimation of forest variables using airborne laser scanning. Forest Resource Management and Geomatics. Swedish University of Agricultural Sciences, Umeå, Sweden. PhD thesis, 43 pages plus 5 articles.

Imhoff, M. (1995). Radar backscatter and biomass saturation: ramifications for global biomass inventory. *IEEE Transactions on Geoscience and Remote Sensing,* Vol. 33, No. 2, pp. 511-518.

IPCC (2006). 2006 IPCC Guidelines for National Greenhouse Gas Inventories. http://www.ipcc-nggip.iges.or.jp/public/2006gl/pdf/4_Volume4/V4_04_Ch4_Forest_Land.pdf, 4 Agriculture, Forestry and Other Land Use, 83 pages.

Magnusson, M. & E. S. Fransson (2004a). Combining airborne CARABAS-II VHF SAR data and optical SPOT-4 satellite data for estimation of forest stem volume. *Can. J. Remote Sensing,* Vol. 30, No. pp. 661-670.

Magnusson, M. & E. S. Fransson (2004b). Combining CARABAS-II VHF SAR and Landsat TM satellite data for estimation of forest stem volume, Proceedings of *IGARSS'04, 0-7803-8743-0, Anchorage, Alaska.*

Marklund, L. G. (1988). Biomassafunktioner för tall, gran och björk i Sverige. Institutionen för skogstaxering, Sveriges lantbruksuniversitet, Umeå, Sweden. Rapport 45, in Swedish:

Massonnet, D. & K. L. Feigl (1998). Radar interferometry and its application to changes in the earth's surface. *Reviews of Geophysics,* Vol. 36, No. 4, pp. 441-500.

Mette, T., K. Papathanassiou, I. Hajnsek, H. Pretzsch & P. Biber (2004). Applying a common allometric equation to convert forest height from Pol-InSAR data to forest biomass, Proceedings of *IGARSS'04, 0-7803-8743-0, Anchorage, Alaska, 2004*

Nilson, T. (1999). Inversion of gap frequency data in forest stands. *Agricultural and Forest Meteorology,* Vol. 98-99, No. pp. 437-448.

Näslund, M. (1947). Funktioner och tabeller för kubering av stående träd. Tall, gran och björk i södra Sverige samt hela landet. Meddelande från Statens skogsforskningsinstitut. 36 (3) in Swedish:

Papathanassiou, K. P. & S. R. Cloude (2001). Single-baseline polarimetric SAR interferometry. *IEEE Trans. Geosci. Remote Sensing,* Vol. 39, No. 11, pp. 2352-2363.

Pulliainen, J., M. Engdahl & M. Hallikainen (2003). Feasibility of multi-temporal interferometric SAR data for stand-level estimation of boreal forest stem volume. *Remote Sens. Environ.,* Vol. 85, No. pp. 397-409.

Pulliainen, J. T., K. Heiska, J. Hyyppä & M. T. Hallikainen (1994). Backscattering properties of boreal forests at the C-and X-band. *IEEE Trans. Geoscience and Remote Sensing,* Vol. 32, No. 5, pp. 1041-1050.

Rott, H. (2009). Advances in interferometric synthetic aperture radar (InSAR) in earth system science. *Progress in Physical Geography,* Vol. 33, No. 6, pp. 769-791.

Sandberg, G., L. M. H. Ulander, J. E. S. Fransson, J. Holmgren & T. L. Toan (2011). L- and P-band backscatter intensity for biomass retrieval in hemiboreal forest. *Remote Sensing of Environment,* Vol. 115, Issue 11, pp 2874–2886.

Santoro, M., J. Askne, P. B. G. Dammert, J. E. S. Fransson & G. Smith (1999). Retrieval of biomass in boreal forest from multi-temporal ERS-1/2 interferometry, Proceedings of *FRINGE'99, Liège, Belgium, 1999*

Santoro, M., J. Askne, G. Smith & J. E. S. Fransson (2002). Stem volume retrieval in boreal forests from ERS-1/2 interferometry. *Remote Sens. Environ.,* Vol. 81, No. 1, pp. 19-35.

Santoro, M., J. Askne, U. Wegmüller & C. Werner (2007a). Observations, modelling and applications of ERS-ENVISAT coherence over land surfaces. *IEEE Trans. Geosci. Remote Sensing,* Vol. 45, No. 8, pp. 2600-2611.

Santoro, M., A. Shvidenko, I. McCallum, J. Askne & C. Schmullius (2007b). Properties of ERS-1/2 coherence in the Siberian boreal forest and implications for stem volume retrieval. *Remote Sensing of Environment,* Vol. 106, No. 2, pp. 154-172.

Schober, R. (1995). *Ertragstafeln wichtiger Baumarten bei verschiedener Durchforstung,* Sauerländer, J D, 978-3793907305,

Smith, G., P. B. G. Dammert & J. Askne (1996). Decorrelation Mechanisms in C-Band SAR Interferometry over Boreal Forest, Proceedings of *Microwave Sensing and Synthetic Aperture Radar, Taormina, Italy, 1996*

Smith, G. & L. M. H. Ulander (2000). A Model Relating VHF-band Backscatter to Stem Volume of Coniferous Boreal Forest. *IEEE Trans. Geosci. Remote Sensing,* Vol. 38, No. 2, pp. 728-740.

Treuhaft, R. N., S. N. Madsen, M. Moghaddam & J. J. vanZyl (1996). Vegetation characteristics and underlaying topography from interferometric data. *Radio Science,* Vol. 31, No. pp. 1449-1495.

Wagner, W., A. Luckman, J. Vietmeier, K. Tansey, H. Balzter, C. Schmullius, M. Davidson, D. Gaveauc, M. Gluck, T. L. Toan, S. Quegan, A. Shvidenko, A. Wiesmann & J. J. Yu (2003). Large-scale mapping of boreal forest in SIBERIA using ERS tandem coherence and JERS backscatter data. *Remote Sens. Environ.,* Vol. 85, No. pp. 125-144.

Wegmüller, U. & C. L. Werner (1995). SAR interferometric signatures of forest. *IEEE Trans. on Geosci. Remote Sensing,* Vol. 5, No. pp. 1153-1161.

9

Interferometry to Detect Planets Outside Our Solar System

Marija Strojnik and Gonzalo Paez
Centro de Investigaciones en Optica
Mexico

1. Introduction

Humanity has been interested in exploring its environment since earliest historical times. After traveling across oceans using star navigation (Scholl, 1993), we turned our attention to getting to know the planets inside our Solar system (Arnold et al., 2010). After they were visited at least once with robotic vehicles (Scholl & Eberlein, 1993) or with an orbiting satellite, we asked ourselves whether conditions existed on any other planet for some form of life. It has been reported that nearly 500 planets were discovered with indirect methods: (a) passage of a dark planet in front of the bright solar disc (Brown et al., 2001); (b) movement of the center of gravity of a two-body system and Doppler shift (Butler et al., 2001); (c) Gravitational bending of rays passing a solar system larger than due to only star (Udalski et al, 2005); (d) using spectroscopy (Richardson, 2007); and (e) astrometry, to list a few. Initial research findings include dust clouds and double stars with different sizes (Moutou et al, 2011; Wright et al., 2001).

We rely principally on the visual system to receive the information about our environment. Such information is carried by the electromagnetic radiation. For a standard observer the visual spectral width covers only from about 0.38 μm to about 0.78 μm (1 μm $=10^{-6}$ m, according to the MKS units employed in this paper) (Strojnik & Paez, 2001). Scientists have developed detectors to cover the electromagnetic spectrum from neutrinos and x-rays to radio waves. We use the word optical to cover visible and infrared (IR). Long-wave IR smoothly transitions into the sub-millimeter range. Many scientists differentiate between them upon incorporation of distinct detecting schemes. When a bolometer may be incorporated, the IR techniques are recognized (Artamkin et al., 2006; Khokhlov et al., 2009; Maxey et al., 1997). When the need arises for a local oscillator, millimeter terminology becomes dominant (Hoogeveen et al., 2003; Kellsall et al., 1993; Wirtz et al., 2003). The planet detection techniques that employ the "light", or the electromagnetic radiation, to detect the planet are generally considered the direct techniques.

2. Planet detection problem

The planet detection challenges have been formulated based on the radiometric, distance, and technology issues as a signal detection problem, under very unfavorable conditions (Scholl, 1994a, 1995, 1996a). The issues are best understood with the reference to Fig. 1. We would like to find a simplest solar system defined as having one star, similar or identical to our Sun, and one planet, likewise similar or identical to out biggest planet, Jupiter. We propose to refer to such a star as the Estrella, and its planet the Tierra, because any of the nearby stars that already have a name may also have an invisible companion. This avoids confusion with our star and our planet(s). Our Sun radiates as a blackbody emitter at 5 800 K. The temperature of our largest and most distant planet, Jupiter, is estimated at 125 K.

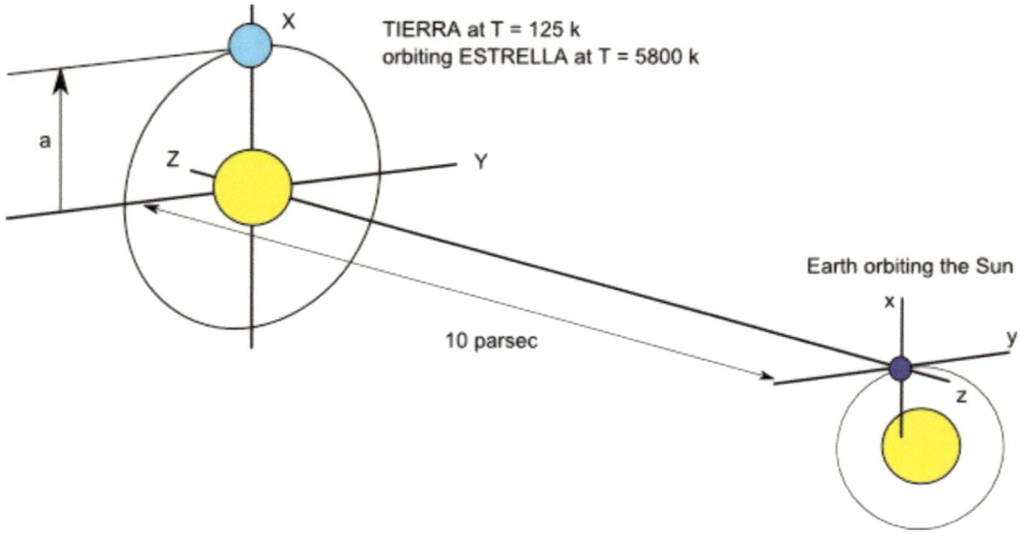

Fig. 1. Geometry for the detection of a planet outside our Solar system. The Earth-/ interferometer-based coordinate system is given as a Cartesian system (x,y,z), while the nearby Planetary system is displaced along the optical axis, Z-axis, with the Cartesian coordinates (X, Y, Z).

Such a planet is considered to be easiest to detect because of its large distance from the Sun. Thus, Figure 1 depicts the Tierra at its most favorable configuration for detection. The probability that the Tierra orbits the Estrella in a plane normal to the line of sight between the Estrella and the Earth, with its observing instrument, is rather small. However, at one particular moment of time, corresponding to the initial observation, the Tierra will be at a projected distance *a* from the Estrella. Its orbit may follow an elliptical path when projected on the plane normal to the line connecting the Estrella and The Earth. Tierra's local year determines the rotational period when the Tierra completes one orbit around the Estrella. In this simplest model, it would be equal to that of Jupiter and appreciably longer than 365 Earth days, let us say twice as long.

Our Solar system is located in the part of the universe where there are only a few Sun-like stars. Figure 2 displays the number of potential Estrellas as a function of stellar magnitude, m_V. This is defined as the star brightness relative to that of a reference star, in a logarithmic scale. In this scale, the Sun's stellar magnitude is 4.8. The astronomers assure us that the number of the stars in the universe is infinite. We concentrate on the stars within a sphere of a certain radius with the origin at the Earth. Restricting ourselves to the Sun-like stars, we find that there are only two such stars within 5 parsecs from us. One parsec is 3.26 light years, or it is the distance that a photon transverses when it travels in a straight line for 3.26 years. So, the distance in metric units is equal to [(3 x10^8 m/sec) x (3.26 years x 365 days/year x 24 hr/day x 60 min/hr x 60 sec)], a very large number, without much meaning for a mere human. There are about ten stars with the stellar magnitude equal to that of Sun within the distance of ten parsecs. When a sixteen-year old youth looks at fifteen nearby stars at night, the photon that is incident on his retina has left the star when the teenager was born. The number of potential Estrellas increases with the stellar magnitude and their distance from our Solar system.

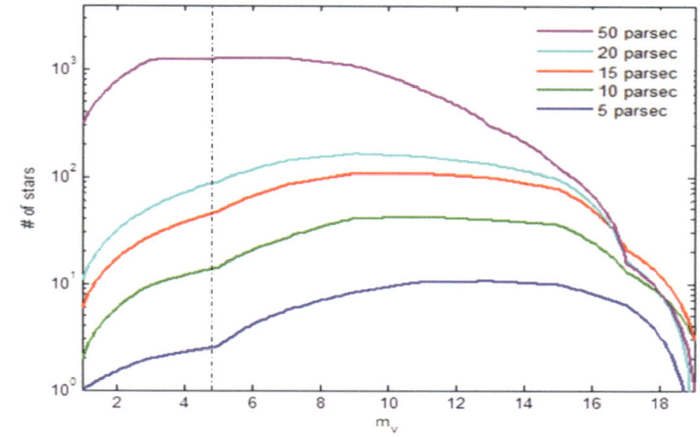

Fig. 2. Number of stars in the vicinity of Sun as a function of stellar magnitude, with the distance from the Sun as a parameter. The Sun's magnitude of 4.8 is indicated with a vertical line. The number of stars roughly increases with the stellar magnitude and the distance from the Sun. Calculated using star catalogues (Smithsonian Astrophysical Observatory, Telescope Data Center, 1991; The Yale Bright Star Catalog, 5th revised ed., Available at http://tdc-www.harvard.edu/catalogs/bsc5.html [accessed 10/22 2011]; CASU astronomical data centre, 2006; The Hipparcos, available at http://archive.ast.cam.ac.uk/hipp/hipparcos.html, [accessed 10/22/2011]; NASA's High Energy Astrophysics Science Archive Research Center, 2011; Gliese Catalog of nearby Stars, Available at http://heasarc.gsfc.nasa.gov/W3Browse/star-catalog/cns3.html; [accessed 10/22/2011]; NASA's Astrobiology Magazine, 2007; Catalog of Nearby Habitable Stars; available at http: //www.nasa.gov/vision/universe/newworlds/HabStars.html, [accessed 10/22/ 2011]).

Looking at the stars from the Earth, we can make three observations. Stars are very bright; they are far away; and they look like point sources to human observers and to any optical instrument, either constructed or under consideration at this time.

We apply the principle of the reversibility to the emitting and intercepting apertures in the power transfer equation to appreciate the signal collection (Strojnik & Paez, 2001). We imagine that sources are planets in our Solar system, with radiation intercepted with an aperture at a distant Tierra. Figure 3 illustrates the engineering and technological issues associated with the darkness of the planets. It graphs the number of spectral photons as a function of wavelength emitted by some representative planets in our Solar system intercepted by a unit aperture at a distance of 10 parsecs. The Earth surface is modeled to emit as a blackbody at the average temperature of 300 K, while the Jupiter temperature is assumed at 125 K. Kelvin (1 K) is a unit of temperature with the same magnitude as 1 C (centigrade or Celsius scale), but with its zero at -273.13 C, considered the absolute zero for temperature.

We observe that the intercepting aperture at the Tierra at a distance of 10 parsecs collects one photon at 34 μm per second originating at Jupiter, and about 70 photons at 12 μm from the Earth. Solely on the basis of the low levels of light emission, large integration times on the order of hours are needed for the collection of the requisite signal. This leads into the stringent requirement for a stable platform on which the planet-detecting system is to be mounted.

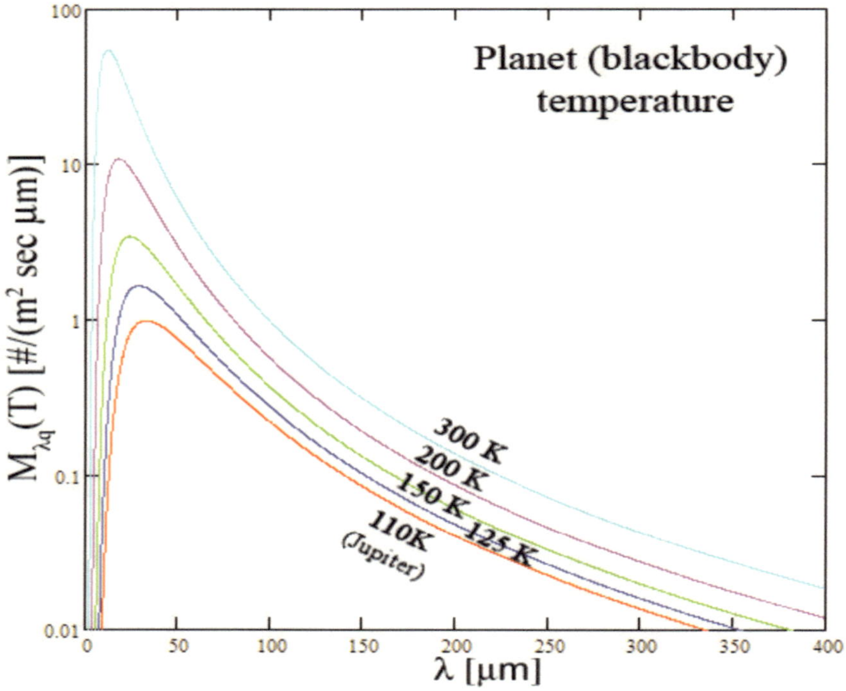

Fig. 3. Number of spectral photons emitted as blackbody radiators due to their surface temperature, per unit time and unit area, by several representative planets in our Solar system, intercepted at a distance of 10 parsec by a unit aperture (1 m^2), versus wavelength.

Comparing the emission of our planets to that of our Sun, we can draw some additional conclusions. Planets emit much less radiation than the Sun and they are much smaller. These considerations translate into three significant optical problems. We address the radiometric problem first.

2.1 Faint planet near the bright star: radiometric problem

Most detectors, including human eye, have a great deal of difficulty detecting a very bright and a very dim object at the same time. The limits in the dynamic range, saturation, and "bleeding" of the bright object onto the image of the dim object prevent their simultaneous detection. Figure 4 depicts the radiometric problem illustrating the darkness of the planets relative to the brightness of the Sun by looking at some representative planets in our Solar system. It graphs the ratio of the number of spectral photons emitted by several representative planets divided by the number of spectral photons emitted by the Sun, at its peak emission, as a function of wavelength. In this analysis, two sources of emission are considered for the planets: self emission due to the planet surface temperature and the radiation originated at the Sun, intercepted by the planet and reflected from it. For us, the Sun and the planets are considered extended bodies; therefore, the distant and large Jupiter reflects about ten times the amount of radiation at 0.7 μm than the Earth. Planet Uranus is distant and small, subtending a relatively small solid angle at the Sun and reflecting nearly a hundred times less Sun radiation at the peak of the Sun emission than the Earth.

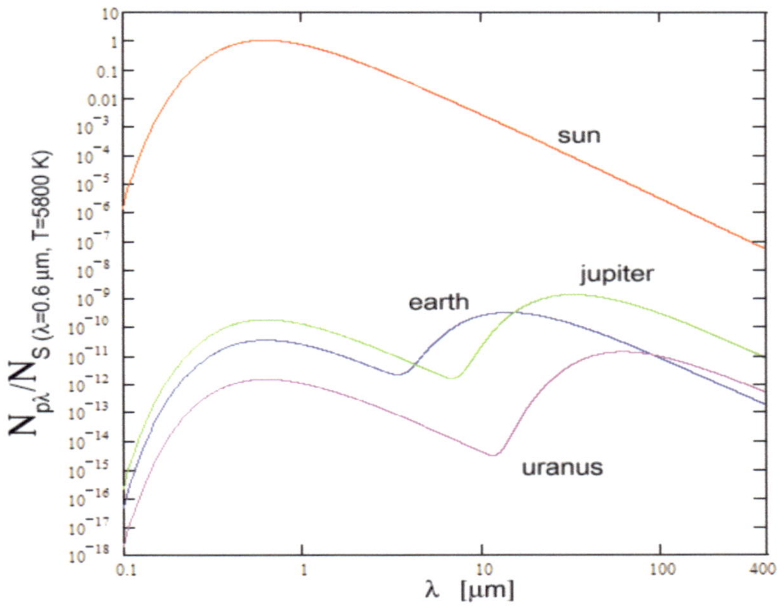

Fig. 4. Number of spectral photons emitted by the Sun and several planets as a function of wavelength, normalized against solar emission at Sun's maximum emission wavelength. The ratio of the number of planet spectral photons over the number of Sun's spectral photons is defined as the simplest signal-to-noise ratio.

In Fig. 4, the Tierra distance is not important, because it is normalized. Both the planet and the Sun are treated as point sources, and both are observed from the same large distance. The size of the intercepting area and signal collection time are the same for both Sun and its planet. The Sun emission curve is normalized to 1 at its peak emission as a blackbody radiator at temperature of 5 800 K. See (Strojnik & Paez, 2001) for blackbody emitters.

This is basically the simplest spectral radiometric signal-to-noise ratio, considered in most publications dealing with the Tierra detection. The signal is defined as the quantity of interest, i.e., the number of photons originating at the planet. The radiation emitted by the Estrella is thus considered noise. Other sources of noise may be considered when the instrumental concept starts to take a more definite form in the quest for the Tierra detection.

The analysis displays the spectral information, i.e., the number of (spectral) photons as a function of wavelength, because all the detectors have a limited detection interval. They will detect all radiation incident on its sensitive surface, independently of its origin, dependent only on its spectral sensitivity. This means that no photon absorbed by the detector sensitive surface is labeled as "coming from a planet" or any other object.

We next examine some potential detection wavelength bands. At the peak Sun emission in the visible, the number of Sun photons reflected from the Jupiter is by the factor of 10^{-10} smaller than the number directly emitted by the Sun. This may be read directly from the vertical coordinate in Fig. 4. The signal-to-noise ratio is nearly ten times smaller still for the Earth, decreasing nearly to 10^{-11}. This value arises purely from the planet diameter and distance from the Sun. Self-emission is negligible at this wavelength region (see also Fig. 3).

Going to the longer wavelengths, we find the second peak due to the planetary self-emission, with the peak location determined by the planet temperature. It is also shifted due to the decreasing Sun emission at longer wavelengths. In the case of the Earth, the signal-to-noise peak for spectral photons is found to be 10^{-6} at about 18 μm. Examining the more favorable case of the Jupiter, we find the peak with spectral photon signal-to noise ratio of 10^{-4} at about 35 μm. Interestingly, though, this is not the peak signal-to-noise ratio. As the Sun emission falls off more rapidly than that of the Jupiter, the photon signal-to-noise ratio slowly increases with increasing wavelengths. At 100 μm it becomes about 3×10^{-4}. The rate of the photon signal-to-noise ratio increase, while relatively small, is about 2×10^{-2} μm^{-1}. We momentarily concentrate on even longer wavelengths, say around 300 μm, and we apply this rate of increase to the photon number ratio. We calculate it to be about 10^{-3} at 300 μm. Interestingly, this wavelength range corresponds to the spectral region where the Earth atmosphere becomes partially transmissive at elevated heights.

2.2 Faint planet next to a bright star: imaging problem

2.2.1 Coronagraphic Estrella signal elimination

When considering the issue of imaging an Estrella-Tierra object scenario, we momentarily set aside the concern that no star (with the exception of Sun) may be resolved at this time with any existing optical telescope or instrument. It is quite amazing that the author's original paper on the planet detection was rejected in the early 90s, because the reviewer considered detection at 18 μm something technologically impossible and therefore not worthy of being published. Today, about 20 years later, the detector technology is being developed at an ever-faster rate, covering increasingly longer wavelengths (Boeker et al., 1997; Farhoomand et al., 2006; Martijn et al., 2005, Müller et al., 2010; Olsen et al., 1997; Reichertz et al., 2005; Royer et al., 1997; Thomas et al., 1998; Young, 1993). Focal plane architectures are being implemented for cameras, some incorporating bolometers, for far IR / sub-mm spectral range (Agnese et al., 2002; Poglitsch et al., 1997; Wilson et al., 2004). This is exactly the spectral region where the photon signal-to-noise ratio has been shown to be most favorable in the previous section. Therefore, we anticipate that the technology exists to image the Estrella and the Tierra onto the same focal plane.

We here adhere to the geometry that the diameter of the Estrella is ten times the diameter of Tierra, the case of Jupiter, with the angular separation of 2 μrad for the observation distance of 10 parsecs. Under these conditions, the Estrella subtends 0.0002 micro-radians (2×10^{-4} μrad), and the Tierra ten times less, or 0.00002 micro-radians (2×10^{-5} μrad). The basic requirement to achieve imaging, by the definition of imaging, is that there be a disk detected in the focal plane corresponding to the Estrella and another one corresponding to the Tierra, displaced by their huge angular separation. The disc diameters have a ratio of ten. With imaging strategy, we arbitrarily set the Estrella on the optical axis and the ten-times smaller planet at the edge of the field. Ideally, the image of the planet falls exactly on one detector (pixel) to optimize the planet detection. Considering that we are not interested in the Estrella, we may implement the inverse coronagraphic configuration in the first focal plane to pass the Estrella rays through the focal plane. The traditional coronagraphic configuration blocks the image of the Sun (Gordley et al., 2005; Scholl, 1993, 1996; Scholl & Paez, 1997b; Schultz et al., 1999; Suzuki et al., 1997).

Radiation blocking scheme might also produce a great amount of reflected and absorbed radiation that travels all over and causes stray-light noise (Scholl, 1994b; Stauder & Esplin, 1998) and heating. Such scattered radiation might easily drown the faint signal (Scholl, 1996b). The inverse coronagraphic configuration would let the Estrella rays pass through a hole in the focal plane, while detectors would be placed outside the opening. The relatively great angular separation between them would assure that only the Tierra signal is detected.

The major shortcoming of this elegant proposal is that there is no telescope currently under consideration that would allow imaging of a nearby solar system, or even a nearest star. The secondary concerns are the stray light issues, scattering, and the tremendous brightness differences between two objects in the same scene even if imaged in different image planes. All of these concerns are just the technological limitations that will be overcome with the passage of time. Next we examine the most critical one, that of imaging and resolution.

2.3 Small (point) planet next to a distant (point) star: resolution problem

The resolution is an attribute of an imaging system. It deals with the fact that an image of a point object is not a point, but rather a blob (Paez & Strojnik, 2001). Image of a point source obtained in the focal plane of an instrument spreads out due to diffraction, aberrations, fabrication errors, and misalignment. The spot size and shape depend on many additional factors, including mechanical stability of the platform and the electronic system for the control of motion. The image spread can never be completely eliminated (see left side of Fig. 5). The resolution of a moderately aberrated optical system is defined by the radius of the spot encircling 90% of energy (or rays) originating at a point source.

There are several resolution criteria, but the most common one deals with the imaging of two point sources of equal brightness/ (spectral) intensity separated by an angle θ. Brightness usually refers to the visual perception of a human. (Spectral) intensity is the (spectral) power emitted into the solid angle. According to the widely-accepted Rayleigh resolution criterion, two such point sources may be resolved under optimal conditions if their angular separation is $\theta = 1.22\ \lambda/D$, where λ is the wavelength of the image forming radiation and D is the diameter of the aperture. The images of two point sources of equal intensity resolved according to the Rayleigh criterion are illustrated on the right side of Fig. 5.

For the largest IR telescope currently under construction (see Fig. 6a), the Great Millimeter Telescope GMT) being built in Mexico, the design wavelength is 1 mm, and the aperture diameter is 50 m, giving the angular resolution of about 2×10^{-5} rad, or 20 μrad. (For more information on the status of this observatory, please see its web address). This angle is ten times larger than the angular separation between the Estrella and the Tierra at

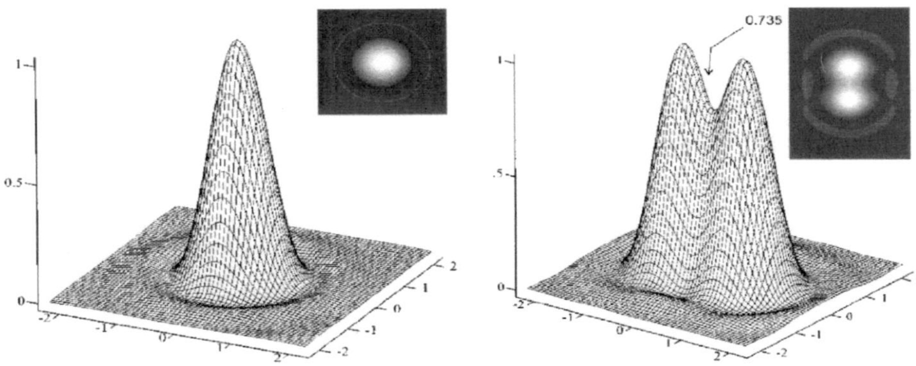

Fig. 5. Left: image of a point source obtained in the focal plane of an instrument spreads out due to diffraction, aberrations, fabrication errors, and misalignment. Right: images of two point sources of equal spectral intensity resolved according to the Rayleigh criterion.

10 parsecs. However, the construction of this telescope has been progressing with better achievements than the original design prediction, laying grounds for the confidence that the telescope will perform well down to 0.1 mm or 100 μm. At this wavelength, the image of the bright point source, the Estrella, will have its zero at 2 μrad, or at the estimated Tierra angular position, θ_{TC}. Likewise, the image of the Tierra will peak at 2 μrad and have its zero on the optical axis. If the Estrella and the Tierra had the same intensity, i. e., if they formed a double star, their image obtained by the GMT at 0.1 mm would be similar to that in Fig. 5b. Due to the spectral intensity ratio between the Estrella and the Tierra (about 10^3 at 300 μm) their image would look a lot like Fig. 5a. The bright spec on the dark ring would be seen possibly in the logarithmic rendition, and would easily be confused with noise.

It is nothing more than a fortunate coincidence that straw-man design parameters, proposed in the nineties, correspond to the desirable features of the GTM telescope under construction. However, the tentative Planetary system parameters do not necessarily correspond to the characteristics of an actual solar system.

2.4 Earth and space based telescopes

2.4.1 Space- and air-borne telescopes

Some of the issues dealing with the telescopes have been summarized in the literature (Paez & Strojnik, 2001). The air turbulence, sometimes referred to as "seeing", and spectral transmission of the atmosphere limit the usefulness of large-diameter monolithic or phased telescopes on the Earth surface. The water in the atmosphere attenuates the transmission of the IR radiation so most IR telescopes perform best in space. (Hofferbert et al., 2003; Kessler & Harvit, 1993; Lamarre, 1993; Lemke et al., 2005; Mather, 1993; Matsumoto & Murakami, 1996; Scholl & Paez, 1997a; Touahri et al., 2010). The great disadvantage of the space-based IR telescopes is that their instruments require cooling, often to liquid Helium temperatures (Beeman & Haller, 2002; Latvakoski et al, 2010). The supply of the coolant basically limits the useful life of the IR space observatory, telescope and on-board instruments. The instruments that are cooled to higher temperatures function longer and collect the scientific data for longer period of time (Schick & Bell, 1997).

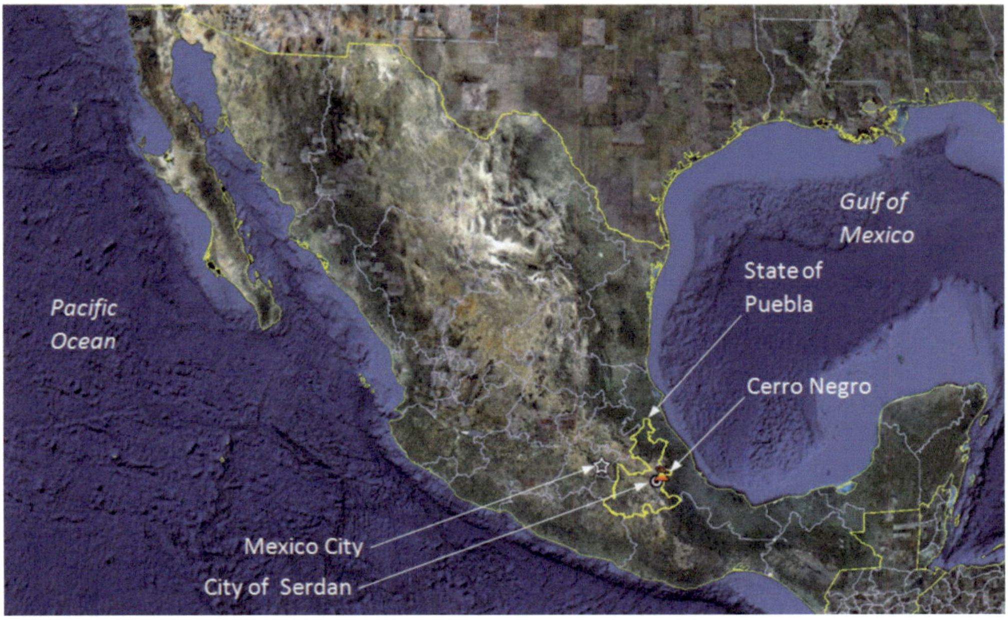

Fig. 6a. The GMT facility's location in the Mexican state of Puebla, on top of a high mountain, between the Atlantic Ocean / the Gulf of Mexico (E) and the Pacific Ocean (W).

This limitation of the lifetime of the satellites has given rise to a new observational facility type, balloon-borne instruments (Catanzaro et al., 2002), rocketborne instruments (Elwell, 1993), and jet-carried observatory. SOFIA (Stratospheric Observatory for IR Astronomy) has a telescope built inside a modified 747P jet to fly in the stratosphere (Becklin & Gehrz, 2009; Krabbe, & Casey, 2002; updated information is available on the web). It incorporates a 2.5 m aperture, flying at the height of about 14 000 m. The telescope looks sideways through the thinned atmosphere at the IR sky. The airplane can land after exhausting its fuel supply, resupply the coolant for the telescope, its instruments, and focal planes; and change the instruments. The most demanding requirement facing the observatory in an airplane is the stability of the aircraft, especially for the long detector-integration times needed for the observation of faint objects.

2.4.2 Transmission of the Earth atmosphere

Placing the observatory on a more solid base would solve the problem of stability. This could be achieved by placing it on the top of a very high mountain, depicted in Fig. 6a. Cerro Negro, Orizaba, Puebla (state), Mexico, where the GMT is being finished) is 4,600 m above the sea level. The significant problem of the absorption of the far IR radiation by water in the thinned atmosphere is ameliorated by dryness of the region, and specifically by a very small number of days characterized by high humidity. Even though Mexico borders on two oceans, Atlantic/ Golf of Mexico and Pacific, its land mass is enormous and characterized by dry heat. Puebla's significant distance from the coastlines and the observatory elevation assures dryness of the cold air above the observatory. Figure 6b presents actual photo of the telescope. In difference to the aircraft implementation, the celestial objects may be viewed in nadir.

Fig. 6b. The GMT (http://www.lmtgtm.org/images/sitepics05012011/LMTatSunrise1.jpg) observatory construction site, with the last ring of the antenna in the process of installation.

In the airplane, the IR radiation passes through the thinned atmosphere at 14,000 m with sufficient density to allow the plane to remain in the air. Due to the side opening for the telescope viewing, the radiation passes horizontally through several equivalent air masses to contribute an approximately equivalent amount of atmospheric scattering and absorption as the nadir-viewing telescope at 4,600 m (Wellard et al., 2006).

No data has yet been made available for the transmission at the far IR / sub-millimeter/ millimeter spectral region at The Cerro Negro in Mexico. There have been long-term measurements taken for a number of years at another millimetric facility at Chajnantor, in Chile, at the height of 5,100 m above sea level (Cageo et al., 2010). Figure 7 presents the graphs of the atmospheric transmission at this site, as a function of wavelength. This site is higher by 600 m, or about 11 % than that of the GMT.

We concentrate on the transmission peaks, referred to as band 10 in the Chilean telescope, from about 800 to 980 GHz. Due to our interest in the Tierra detection at about 300 μm (or about 900 GHz), where the photon number contrast achieves a highly favorable 0.001 signal-to-noise ratio, we examine the transmission band at about 900 GHz in more detail. Its transmission of radiation is about 0.4 when the atmosphere is dry. The transmission decreases by about 50 % in the case of the wet atmosphere. No such data has been reported for the GMT, but the great majority of days are dry. Independently of the amount of atmospheric absorption and scattering in this wavelength interval, the radiation attenuates equally for both the Estrella and the Tierra signal. Thus, the signal-to-noise ratio remains unchanged. We will use 0.4 as the working number for the transmission ratio for the site. Clearly, the data will be collected (or utilized) only under favorable observational conditions. Again, the smaller signal from the Tierra will be affected worse. Such detailed considerations will be incorporated into the overall signal-to-noise ratio calculations, after the detection mechanisms and the instrument have been specified.

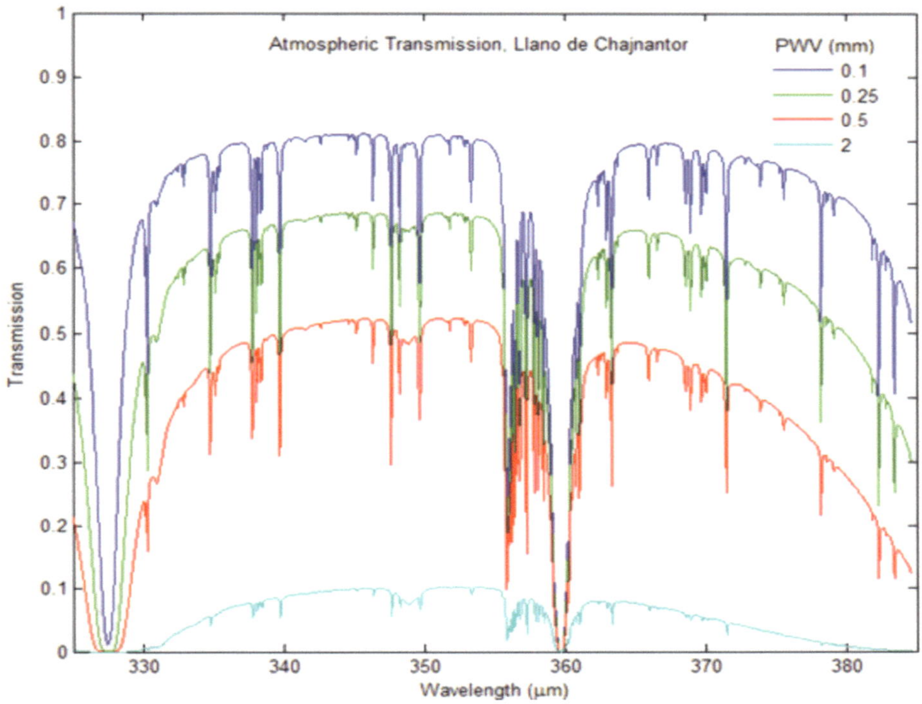

Fig. 7. Atmospheric transmission at the ALMA site on Llano de Chajnantor, Chile, at 5,100 m above the sea level, with the amount of precipitable water vapor as a parameter (after Atacama large millimeter/submillimeter array, https://almascience.nrao.edu /about-alma/weather/atmosphere-model [accessed 11/2/2011]).

3. Interferometry

3.1 Interferometer for the Tierra detection

3.1.1 Two-aperture interferometry

A panel of distinguished scientists proposed a two-aperture interferometer to detect the Tierra and null the Estrella (Bracewell, 1978). The basic idea was to use the interferometry to eliminate the radiation coming from the Estrella through the process of the destructive interference, while at the same time transmitting the radiation coming from the Tierra. In this concept the image of the Tierra would be located exactly on the dark interference fringe formed by the Estrella radiation. The scientists have not yet discovered the procedure of tagging the photons originating at the Tierra to send them through an image-forming instrument while allowing those from the Estrella to interfere destructively as necessary to implement this proposal. Thus, the radiation coming from both celestial sources through two apertures forms two interference patterns. The two patterns have the same period, but they are slightly displaced with respect to each other corresponding to the separation of the point sources in the plane of the Planetary system. This is the plane perpendicular to the line of observation, or the optical axis. Their amplitudes have the same ratio as the simplest signal-to-noise ratio. Therefore, the Tierra modulation will be at least 10^3 times smaller than that of the Estrella. These two patterns overlap on the focal plane, with the visibility of the Tierra smaller by the signal-to-noise ratio in the interferometer plane (in the vicinity of Earth), by a factor of at least 10^3. The most important feature of either interference pattern is that its period is determined only by the aperture separation, also referred to as baseline (Strojnik & Paez, 2003; Vasquez-Jaccaud et al., 2010).

As the Tierra slowly rotates around the Estrella, the Estrella interference pattern remains unchanged relative to a stationary two-aperture interferometer. The displacement of the Tierra interference pattern approaches that of the Estrella as the Tierra's projection on the plane normal to the line-of-sight approaches the Estrella. When the Tierra is exactly in front of the Estrella, both on the Z-axis, the two interference patterns will be exactly in phase - the modulation of the combined pattern will be insignificantly smaller than that of the Estrella alone. When the Tierra is behind the Estrella, two-apertures produce the interferometric pattern due to the Estrella source independently whether the Estrella is accompanied by an orbiting Tierra or not. Thus, an interferometric pattern in a two-aperture interferometer does not confirm the existence of the Tierra.

3.1.2 Space interferometry

A large body of research and conceptual studies has been performed in support of these concepts, mostly for space applications. The need for the space system arises due to the opacity of the Earth atmosphere in many of the interesting regions. The most advanced developments were offered in the ESA studies, recently referred to as the Darwin proposal (most up-to date information is available on the Darwin web site). This free flying instrument would initially employ a multi-section telescope with a 5-m diameter. The study evolved to have two such telescopes connected into an interferometer with a precisely controlled baseline. It culminated in a proposal for a number of such telescopes that could be used as a sparsely-filled aperture to detect several spatial frequencies. The spatial frequency is defined as the number of lines per unit distance.

While the interferometric concepts assume detection of fringes from two different sources (detection of a single spatial frequency), the imaging concepts incorporate finding a significant number of key spatial frequencies. The incorporation of the sparsely-filled aperture implies that a number of spatial frequencies present in the object are missing in the image. As discussed in the imaging section, this would require rigid control of aperture separations, at distances of at least 50 meters to meet the Rayleigh resolution criterion for the Estrella and the Tierra at 10 parsecs. This separation would have to be increased even more to decrease the background diffracted and scattered radiation from the Estrella to a value comparable to that of the Tierra (for example, 10^{-3} at 300 μm). For the diffracted radiation from the Estrella to be less than the signal from the Tierra (under optimal observational conditions), two neighboring apertures might have to be separated by at least four times more than the resolution distance of 50 m, or 200 m. This considers only the diffraction effects of the apertures. To decrease the effects of gravity, this instrument would be placed at the first or the second Lagrangian point (zero gravity point between Earth and Moon). The ESA study has been temporarily set aside at the time of this writing, in favor of more promising surveys of infrared and submillimeter skies, with Herschel and SPICA instruments, and the ESO Planck Observatory (more update information is available on the ESA Planck, AKARI web sites).

Many research studies continue to decrease the amount of the Estrella radiation in the detector plane, most of them quite interesting from the point of view of signal camouflaging and encoding. They are extensively discussed in the literature as a technique referred to as "nulling" of the star light.

Infinitesimally thin and infinitely long apertures help to conceptualize difficult problems, such as the Estrella being thought of as a point object. Such a case was treated rigorously, including both the interference and diffraction at finite apertures (Strojnik & Paez, 2003).

3.2 Transformation of a spherical wave into a plane wave upon propagation over large distances

We consider that the point on the Planetary system plane (either the point on the Estrella or on the Tierra) radiates as a point source. It is located at a distance r_{PS} from the plane of the aperture (of diameter D in meters), centered on the x, y plane. Due to the great distance of propagation, we consider the radiation incident on the plane of the apertures as coherent as proved in the Van Cittert-Zernike theorem. The distance r_{PS} (for Planetary system, with the subscript becoming E for the Estrella or T for the Tierra, as appropriate) may be written as follows. We refer to Fig. 1.

$$r_{PS}(x,y,z;X,Y,Z) = \sqrt{(X_{PS}-x)^2+(Y_{PS}-y)^2+(Z_{PS}-z)^2} \quad (1)$$

Therefore, the point source in the planetary system plane radiates in all directions as a spherical wave with intensity $U_{PS}U_{PS}^*$.

$$U_{PS}(x,y,z;X,Y,Z) = U_{PS0}\left(\frac{e^{ikr_{PS}}}{r_{PS}}\right) \quad (2)$$

We now make a few simplifying assumptions, considering that the Estrella-Tierra coordinate system is extremely far away. Then we expand r_{PS} in the exponent somewhat more carefully because the sine and cosine vary much more rapidly with their argument. We rewrite Eq. (2), factoring out z_{PS}, the distance on the optical axis between the planetary system and the interferometer plane. This quantity is significantly larger than other quantities in Eq. (1), according to the geometry of Fig. 1.

$$r_{PS}(x,y,z;X,Y,Z) = Z_{PS}\sqrt{\left[\left(1-\frac{z}{Z_{PS}}\right)^2+\left(\frac{X_{PS}}{Z_{PS}}-\frac{x}{Z_{PS}}\right)^2+\left(\frac{Y_{PS}}{Z_{PS}}-\frac{y}{Z_{PS}}\right)^2\right]} \quad (3)$$

We then square the expressions in the parentheses. The spatial extend of the Estrella is much, much larger than the lateral extend of the largest interferometer diameter considered in the current studies. We will neglect the square of the interferometer coordinates x, y, in comparison with $X_E x$ and $Y_E y$, respectively. We similarly assume that the separation of Tierra from the Estrella is much, much larger than interferometer extend.

$$r_{PS}\left(x,y,z;X,Y,Z\right)=Z_{PS}\sqrt{\begin{Bmatrix}\left[1-2\left(\frac{z}{Z_{PS}}\right)+\left(\frac{z}{Z_{PS}}\right)^2\right]\\+\left[\left(\frac{X_{PS}}{Z_{PS}}\right)^2-2\left(\frac{X_{PS}}{Z_{PS}}\right)\left(\frac{x}{Z_{PS}}\right)+\left(\frac{x}{Z_{PS}}\right)^2\right]\\+\left[\left(\frac{Y_{PS}}{Z_{PS}}\right)^2-2\left(\frac{Y_{PS}}{Z_{PS}}\right)\left(\frac{y}{Z_{PS}}\right)+\left(\frac{y}{Z_{PS}}\right)^2\right]\end{Bmatrix}} \tag{4}$$

Therefore, we neglect the square of the interferometer coordinates x, and y, in comparison with the product $X_T x$ and $Y_T y$, respectively. In practice the telescope will be bore-sighted on the easily observable Estrella. We are keeping only the lowest order in x, y and z coordinates.

$$r_{PS}\left(x,y,z;X,Y,Z\right)=Z_{PS}\sqrt{\left\{\left[1+\left(\frac{X_{PS}}{Z_{PS}}\right)^2+\left(\frac{Y_{PS}}{Z_{PS}}\right)^2\right]-2\left[\left(\frac{z}{Z_{PS}}\right)+\left(\frac{X_{PS}}{Z_{PS}}\right)\left(\frac{x}{Z_{PS}}\right)+\left(\frac{Y_{PS}}{Z_{PS}}\right)\left(\frac{y}{Z_{PS}}\right)\right]\right\}} \tag{5}$$

We choose the plane of the aperture system at z = 0, or normal to the line of sight Plenary system - interferometer system. The first term in the second square bracket is equal to zero.

$$r_{PS}\left(x,y,z;X,Y,Z\right)=Z_{PS}\sqrt{\left\{\left[1+\left(\frac{X_{PS}}{Z_{PS}}\right)^2+\left(\frac{Y_{PS}}{Z_{PS}}\right)^2\right]-2\left[\left(\frac{X_{PS}}{Z_{PS}}\right)\left(\frac{x}{Z_{PS}}\right)+\left(\frac{Y_{PS}}{Z_{PS}}\right)\left(\frac{y}{Z_{PS}}\right)\right]\right\}} \tag{6}$$

Next we expand the square root, considering that the furthest transverse coordinate in the interferometer system is significantly smaller than the separation between the Planetary system and the interferometer. Likewise, the furthest transverse coordinate in the Planetary system is significantly smaller than the separation between the Planetary system and the interferometer. Thus the argument under the square root sign is 1 plus a small number. The expansion is valid when ε is much smaller than 1, with accuracy increasing when ε approaches to zero.

$$(1+\varepsilon)^m = 1+m\varepsilon \tag{7}$$

Using this expansion, Eq. (6) becomes.

$$r_{PS}\left(x,y,z;X,Y,Z\right)=Z_{PS}\left\{1+\left(\frac{1}{2}\right)\left[\left(\frac{X_{PS}}{Z_{PS}}\right)^2+\left(\frac{Y_{PS}}{Z_{PS}}\right)^2\right]-\left[\left(\frac{X_{PS}}{Z_{PS}}\right)\left(\frac{x}{Z_{PS}}\right)+\left(\frac{Y_{PS}}{Z_{PS}}\right)\left(\frac{y}{Z_{PS}}\right)\right]\right\} \tag{8}$$

The first two terms include only the coordinates associated with the objects on the Estrella-Tierra Planetary system. The second term includes both the planetary and interferometer coordinates. We let x and y be equal to zero, effectively choosing the point on the origin of the interferometer coordinate system. We may use the same expansion in Eq. (7) in reverse, and apply it to the first two terms (those dealing with the planetary system quantities.)

$$\begin{aligned}&r_{PS}\left(X_{PS},Y_{PS},Z_{PS}\right)=r_{PS}\left(0,0,0;X_{PS},Y_{PS},Z_{PS}\right)\\&=Z_{PS}\left\{1+\left(\frac{1}{2}\right)\left[\left(\frac{X_{PS}}{Z_{PS}}\right)^2+\left(\frac{Y_{PS}}{Z_{PS}}\right)^2\right]\right\}=Z_{PS}\sqrt{\left[1+\left(\frac{X_{PS}}{Z_{PS}}\right)^2+\left(\frac{Y_{PS}}{Z_{PS}}\right)^2\right]}\end{aligned} \tag{9}$$

We substitute Eq. (9) into Eq. (8), to find an interesting result: we can separate the coordinates in the planetary system from the cross terms that include also those in the interferometer plane.

$$r_{PS}(x,y,z;X,Y,Z) = r_{PS}(X_{PS},Y_{PS},Z_{PS}) - Z_{PS}\left[\left(\frac{X_{PS}}{Z_{PS}}\right)\left(\frac{x}{Z_{PS}}\right) + \left(\frac{Y_{PS}}{Z_{PS}}\right)\left(\frac{y}{Z_{PS}}\right)\right] \tag{10}$$

Thus, we developed an expression for the value of r_{PS} to be inserted into the exponent of Eq. (2). We substitute Eq. (10) into Eq. (2).

$$U_{PS}(x,y,z;X,Y,Z) = U_{PS0}(X_{PS},Y_{PS},Z_{PS})\left\{\frac{e^{ik\left\{r_{PS}(X_{PS},Y_{PS},Z_{PS}) - Z_{PS}\left[\left(\frac{X_{PS}}{Z_{PS}}\right)\left(\frac{x}{Z_{PS}}\right)+\left(\frac{Y_{PS}}{Z_{PS}}\right)\left(\frac{y}{Z_{PS}}\right)\right]\right\}}}{r_{PS}(X_{PS},Y_{PS},Z_{PS})}\right\} \tag{11}$$

The coordinates in the first term in the exponent X_{PS}, Y_{PS}, and Z_{PS} in Eq. (11) may only assume the values corresponding to a point within the spatial extent of the plenary system encompassing the Estrella and the Tierra. So, we factor it out and combine it into a single term for the amplitude that only depends on the coordinates of a point on the planetary system, X_{PS}, Y_{PS}, Z_{PS}. We next take advantage of the rules of multiplication of the exponentials, {exp(a+b) = [exp(a)][exp(b)]}.

$$\begin{aligned} &u_{PS}(x,y,z;X,Y,Z) \\ &= \left\{U_{PS0}(X_{PS},Y_{PS},Z_{PS})\left[\frac{\exp[ikr_{PS}(X_{PS},Y_{PS},Z_{PS})]}{r_{PS}(X_{PS},Y_{PS},Z_{PS})}\right]\right\}\exp\left\{-ikZ_{PS}\left[\left(\frac{X_{PS}}{Z_{PS}}\right)\left(\frac{x}{Z_{PS}}\right)+\left(\frac{Y_{PS}}{Z_{PS}}\right)\left(\frac{y}{Z_{PS}}\right)\right]\right\} \end{aligned} \tag{12}$$

The curly bracket in this expression depends only on the point on the planetary system that emits the radiation. It is the product of the strength of the point source at the point source location and the spherical wave evaluated at the origin of the interferometer-plane coordinate system. These are all constant quantities for a specific light emitting point on the celestial body in the planetary coordinate system, and its distance from the interferometer plane. See (Strojnik & Paez, 2001) for description of the Planckian radiators. Thus, they also form the constant amplitude of the wave at the interferometer plane, u_{PS0}.

$$\begin{aligned} u_{PS0}(X_{PS},Y_{PS},Z_{PS}) &= u_{PS0}(0,0,0;X_{PS},Y_{PS},Z_{PS}) = U_{PS0}(X_{PS},Y_{PS},Z_{PS})\left[\frac{e^{ikr_{PS}(X_{PS},Y_{PS},Z_{PS})}}{r_{PS}(X_{PS},Y_{PS},Z_{PS})}\right] \\ &= \left[\frac{U_{PS0}(X_{PS},Y_{PS},Z_{PS})}{r_{PS}(X_{PS},Y_{PS},Z_{PS})}\right]\exp[ikr_{PS}(X_{PS},Y_{PS},Z_{PS})] \end{aligned} \tag{13}$$

We observe that the first factor is just amplitude, because all the quantities are constant. The second factor is an expression for the plane wave. We insert Eq. (13) into Eq. (12).

$$\begin{aligned} U_{PS}(x,y,z;X,Y,Z) &= u_{PS0}(X_{PS},Y_{PS},Z_{PS})e^{-ikZ_{PS}\left[\left(\frac{X_{PS}}{Z_{PS}}\right)\left(\frac{x}{Z_{PS}}\right)+\left(\frac{Y_{PS}}{Z_{PS}}\right)\left(\frac{y}{Z_{PS}}\right)\right]} \\ &= u_{PS0}(X_{PS},Y_{PS},Z_{PS})\exp\left\{-ikZ_{PS}\left[\left(\frac{X_{PS}}{Z_{PS}}\right)\left(\frac{x}{Z_{PS}}\right)+\left(\frac{Y_{PS}}{Z_{PS}}\right)\left(\frac{y}{Z_{PS}}\right)\right]\right\} \end{aligned} \tag{14}$$

The first factor is just constant wave amplitude. The second factor is a plane wave, traveling in the direction along the z-axis, with a small tilt along the x- and y- direction.

Tilt depends on the source angular coordinates of the ($X_{PS}/Z_{PS} = \theta_{PSX}$, and $Y_{PS}/Z_{PS} = \theta_{PSY}$.

$$U_{PS}(x,y,z;X,Y,Z) = u_{PS0}(X_{PS},Y_{PS},Z_{PS})\exp\left[-ik(\theta_{PSx}x + \theta_{PSy}y) + \omega t + \delta\right] \qquad (15)$$

The tilt of the plane wave in the x, y plane that originated at the Planetary system point X_{PS}, Y_{PS}, Z_{PS} depends on the coordinates of the point source. The tilt fringes are straight lines perpendicular to the tilt direction. For the point at the Planetary coordinate origin $X_{PS} = Y_{PS} = 0$, the plane wave travels along the Z-axis. Thus, only the central point on the Estrella generates plane wave radiation that produces no tilt.

We can form a significant general conclusion. The spherical waves originating at a point source at coordinates x_S, y_S, z_S of strength U_0 transform into a plane wave upon propagation over large distance r_S, with the amplitude reduced by the value of the spherical wave at distance r_S, $\{\exp[ik(r_S)]/r_S\}$, and the plane wave tilt equal to $\theta_x = x_S/z_S$ along the x-direction and $\theta_y = y_S/z_S$ in the y-direction, $\exp[ik(\theta_x x + \theta_y y)]$.

3.3 Two sources: the Estrella and the Tierra

We consider the case that both the Estrella and the Tierra are point sources, due to the small angles that they subtend at the aperture plane. The transformation of spherical waves originating at the Estrella and the Tierra into plane waves is illustrated in Fig. 8.

The radiometry allows for the point sources, establishing the concept of (spectral) intensity for such radiators (spectral power emitted per unit solid angle, $dP_\lambda/d\Omega$ [W/(sr μm)]. The coordinates X_E, Y_E, Z_E may assume some non-zero values, corresponding to the spatial extent of the star. Similarly, the Tierra is an extended body, about one tenth in diameter of the Estrella. Its coordinates may likewise assume the values corresponding to the extent of the Tierra. Figure 9 illustrates how an extended object may be perceived to emit the radiation from many small areas as point sources, and all transforming into plane waves.

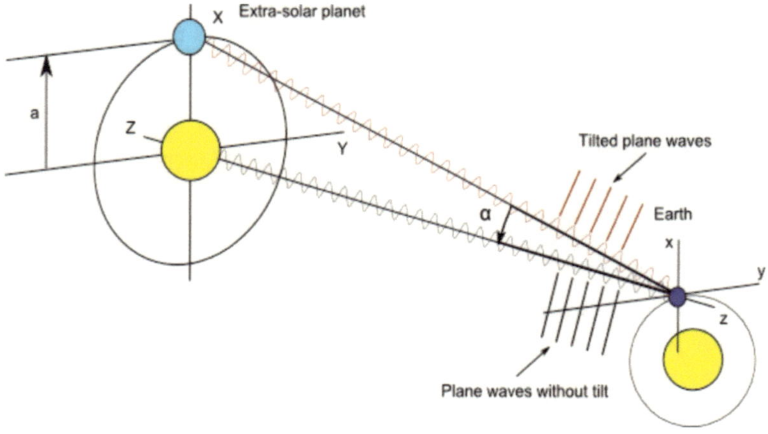

Fig. 8. The Estrella and the Tierra emit spherical waves. After free-space propagation, spherical waves become plane waves. They are tilted if they originate at an off-axis point, such as that on the Tierra.

3.4 Wave fronts from the Estrella and the Tierra incident on One aperture

The wave at the interferometer plane is the sum of those coming from the Estrella and the Tierra. We also include the temporal dependence associated with the traveling plane wave, ωt, to allow for the changes of the wave at a specific point in space as a function of time. We additionally include the phase delay δ. It may be used for temporal modulation.

$$\begin{aligned} u_{IP\lambda}(x,y,0;X,Y,Z) &= u_E(x,y,0;X_E,Y_E,Z_{PS}) + u_T(x,y,0;X_T,Y_T,Z_{PS}) \\ &= u_{E0}(X_E,Y_E,Z_{PS})\exp\left[-ik(\theta_{Ex}x+\theta_{Ey}y)+\omega t\right] \\ &+u_{T0}(X_T,Y_T,Z_{PS})\exp\left[-ik(\theta_{Tx}x+\theta_{Ty}y)+\omega t+\delta\right] \end{aligned} \quad (16)$$

At the interferometer plane (z=0), the wave from the Estrella and the wave from the Tierra add up. The coefficients for the amplitude correspond to the Estella and the Tierra as the point sources. Therefore, we drop the explicit spatial dependence on the coordinates to simplify the appearance of equations.

Two waves at the z = 0 interferometer plane produce incidance. It is found as the sum of the wave multiplied by its complex conjugate. (Spectral) incidance is defined in radiometry as power, incident per unit area (per unit wavelength) in [W/(m² μm)].

$$\begin{aligned} &M_{AP}(x,y,0;X,Y,Z) \\ &= \left[U_E(x,y,0;X_E,Y_E,Z_{PS})+U_T(x,y,0;X_T,Y_T,Z_{PS})\right]\left[U_E(x,y,0;X_E,Y_E,Z_{PS})+U_T(x,y,0;X_T,Y_T,Z_{PS})\right]^* \end{aligned} \quad (17)$$

After substituting from Eq. (16), the incidance is found as a product of the sum of two traveling waves.

$$\begin{aligned} &M_{AP}(x,y,0;X,Y,Z) \\ &= \left\{u_{E0}(X_E,Y_E,Z_{PS})\exp\left[-ik(\theta_{Ex}x+\theta_{Ey}y)+\omega t\right]+u_{T0}(X_T,Y_T,Z_{PS})\exp\left[-ik(\theta_{Tx}x+\theta_{Ty}y)+\omega t+\delta\right]\right\} \\ &x\left\{u_{E0}(X_E,Y_E,Z_{PS})\exp\left[-ik(\theta_{Ex}x+\theta_{Ey}y)+\omega t\right]+u_{T0}(X_T,Y_T,Z_{PS})\exp\left[-ik(\theta_{Tx}x+\theta_{Ty}y)+\omega t+\delta\right]\right\}^* \end{aligned} \quad (18)$$

Next, we change the signs of the complex exponents in the second line to implement the complex conjugate. We also omit the explicit dependence on the planetary system coordinates in the amplitude factors.

$$\begin{aligned} &M_{IP\lambda}(x,y,0;X,Y,Z) \\ &= \left\{u_{E0}\exp\left[-ik(\theta_{Ex}x+\theta_{Ex}y)+\omega t\right]+u_{T0}\exp\left[-ik(\theta_{Tx}x+\theta_{Ty}y)+\omega t+\delta\right]\right\} \\ &x\left\{u_{E0}^*\exp\left[+ik(\theta_{Ex}x+\theta_{Ex}y)-\omega t\right]+u_{T0}^*\exp\left[+ik(\theta_{Tx}x+\theta_{Ty}y)-\omega t-\delta\right]\right\} \end{aligned} \quad (19)$$

We multiply the factors out to obtain a two-beam interference pattern in a complex form.

$$\begin{aligned} &M_{IP\lambda}(x,y,0;X,Y,Z) = u_{E0}u_{E0}^* + u_{T0}u_{T0}^* \\ &+u_{E0}u_{T0}^*\exp\left\{-ik\left[(\theta_{Ex}-\theta_{Tx})x+(\theta_{Ey}-\theta_{Ty})y\right]-\delta\right\}+u_{E0}^*u_{T0}\exp\left\{+ik\left[(\theta_{Ex}-\theta_{Tx})x+(\theta_{Ex}-\theta_{Ty})y\right]+\delta\right\} \end{aligned} \quad (20)$$

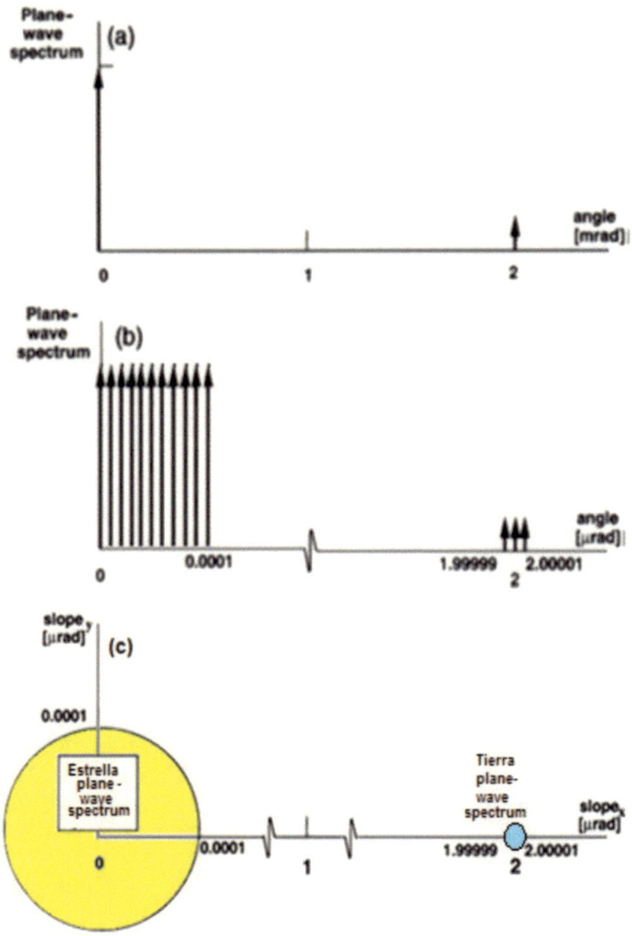

Fig. 9. (a) The Estrella and the Tierra plane wave spectra as a function of angular coordinate. In the middle and the bottom diagrams (b) and (c), the angular coordinate shows discontinuity to allow the presentation of distinct plane waves with increasing inclination, both for the Estrella and for the Tierra. In the middle, the plane wave spectra are shown as a function of angle along which the Tierra is located. The bottom presentation exhibits the spectra as a function of x- and y- inclination, making explicit the symmetrical and asymmetrical features of the plane wave spectra. Those from the Tierra do not possess the cylindrical symmetry about the optical axis. Those from the Estrella exhibit cylindrical symmetry.

The complex numbers in the exponential form may be expanded as sine and cosine, according to the formula, [exp(+/-iα) = cosα +/- i sinα]. The sine terms cancel each other out, resulting in a (co)-sinosoidal modulation of the incidance, in addition to two incidances from the Estrella and the Tierra.

$$M_{IP\lambda}(x,y,0;X,Y,Z) = u_{E0}^2 + u_{T0}^2 + 2u_{E0}u_{T0}\cos\left\{\left(\frac{2\pi}{\lambda}\right)\left[(\theta_{Ex}-\theta_{Tx})x + (\theta_{Ex}-\theta_{Ty})y\right] + \delta\right\} \quad (21)$$

Using the trigonometric identity for a cosine of a double angle [cos(2α) = 2cos²α −1], we finally arrive at the expression for the interference of two un-equal beams. The incidance from two plane waves, slightly inclined with respect to each other exhibits straight fringes normal to the plane of the incident waves.

$$M_{AP}(x,y,0;X,Y,Z) = u_{E0}^2 - 2u_{E0}u_{T0} + u_{T0}^2 + 4u_{E0}u_{T0}\cos^2\left\{\left(\frac{\pi}{\lambda}\right)\left[(\theta_{Ex} - \theta_{Tx})x + (\theta_{Ey} - \theta_{Ty})y\right] + \partial\right\} \quad (22)$$

In absence of the Tierra, there is no radiation emitted from the Tierra. Therefore, there is no interference pattern, and Eq. (22) reduces to the first term. This is just the incidance from the Estrella. This is an important result, because the very detection of the interference fringes establishes the existence of the second point source, a companion Tierra. The phase delay term is halved. This is denoted by a different phase delay symbol in Eq. (22). The first three terms in Eq. (22) form a perfect square, so the expression for the incidance may be further simplified.

$$M_{AP}(x,y,0;X,Y,Z) = (u_{E0} - u_{T0})^2 + 4u_{E0}u_{T0}\cos^2\left\{\left(\frac{\pi}{\lambda}\right)\left[(\theta_{Ex} - \theta_{Tx})x + (\theta_{Ey} - \theta_{Ty})y\right] + \partial\right\} \quad (23)$$

The fringe direction is perpendicular to the orientation of the difference between the slopes of the plane waves coming from the respective points on the Estrella and the Tierra.

3.4.1 Modulation from the Estrella and the Tierra as point sources

Maximum incidance may be found fom Eq. (23) when the cosine square is equal to 1.

$$M_{AP\max}(x,y,0;X,Y,Z) = (u_{E0} + u_{T0})^2 \quad (24)$$

Cosine is one when the argument of the cosine-square function is an integral multiple of π.

$$\left(\frac{\pi}{\lambda}\right)\left[(\theta_{Ex} - \theta_{Tx})x + (\theta_{Ex} - \theta_{Ty})y\right] + \partial = N\pi \quad (25)$$

Minimum incidance may similarly be found when the cosine-square is equal to zero.

$$M_{AP\min}(x,y,0;X,Y,Z) = (u_{E0} - u_{T0})^2 \quad (26)$$

This condition is met when the argument of the cosine-square function is π/2, plus an integral multiple of π .

$$\left(\frac{\pi}{\lambda}\right)\left[(\theta_{Ex} - \theta_{Tx})x + (\theta_{Ex} - \theta_{Ty})y\right] + \partial = \pi\left(N + \frac{1}{2}\right) \quad (27)$$

Average incidance is then one-half of the sum of Eq. (24) and Eq. (26).

$$M_{APave}(x,y,0;X,Y,Z) = \left(\frac{1}{2}\right)\left\{\left[M_{AP\max}(x,y,0;X,Y,Z) + M_{AP\min}(x,y,0;X,Y,Z)\right]\right\}$$

$$= \left(\frac{1}{2}\right)\left[(u_{E0} + u_{T0})^2 + (u_{E0} - u_{T0})^2\right] = u_{E0}^2 + u_{T0}^2 \quad (28)$$

The average signal in the aperture plane is the square of the wave amplitude from the Estrella and the Tierra. Then, the modulation in incidance is the one-half of difference between Eq. (24) and (26).

$$M_{APamp}(x,y,0;X,Y,Z)=\left(\frac{1}{2}\right)\left\{\left[M_{AP\max}(x,y,0;X,Y,Z)-M_{AP\min}(x,y,0;X,Y,Z)\right]\right\}$$
$$=\left(\frac{1}{2}\right)\left[(u_{E0}+u_{T0})^2-(u_{E0}-u_{T0})^2\right]=2u_{E0}u_{T0}=2\sqrt{M_{E0}M_{T0}} \quad (29)$$

The signal amplitude in the aperture plane is equal to twice the product of the wave amplitude from the Estrella and the Tierra. The amplitudes correspond to the square-root of the power quantities developed in the radiometric analysis of Section 2.

3.4.2 Fringe separation for the Estrella and the Tierra

The incidance maximum is achieved when the argument of the cosine-square function is equal to zero or is an integral multiple of π. Without loss of generality, we choose convenient parameters to assist in fringe evaluation.

$$\theta_{Tx}x+\partial=N\lambda \quad (30)$$

Changing the integral number N by 1, we find the change in distance Δx corresponding to the fringe-to-fringe separation.

$$\Delta x=\frac{\lambda}{\theta_{Tx}} \quad (31)$$

The delay factor just displaces the origin of fringes in the observation plane, so we momentarily set it aside. Then, with the Estrella at the origin, the Tierra in the orbit at 2 μrad, (2×10^{-6} rad), using the wavelength of about 300 microns, we evaluate Eq. (31) to be Δx = 150 m. This physically means that within the distance of 75 m, the incidance will change from its maximum value to its minimum value.

4. Instrument concept: Earth-based interferometer, with single-aperture telescope and 0.3 mm radiation

We propose to use a single large-diameter telescope to provide radiation from the Estrella and the Tierra to a Mach-Zehnder interferometer, illustrated in Fig. 10. The function of the telescope is to collect the radiation and generate a several-hundred times increase in the tilt of the incident plane waves from the Tierra. There is no change to the tilt of the Estrella radiation, incident parallel to the optical axis. Two detectors are used in quadrature to record the complementary interference patterns. Optical light path modulation in one arm of the interferometer assures the scanning of fringes with passing time. Constant incidance from the Estrella may be maximized at one detector and minimized at the other detector by the proper insertion of the optical path delay.

Large telescopes are currently available at atmospheric window at 0.3 mm. The simple Mach-Zehnder interferometer is easy to build and align. By employing a telescope system, the angle of incidence is magnified, and the fringe density is increased (Strojnik et al., 2007).

Optical components and detectors might represent the critical technology if it were not for the rapid technology development for the various space systems in the last 20 years.

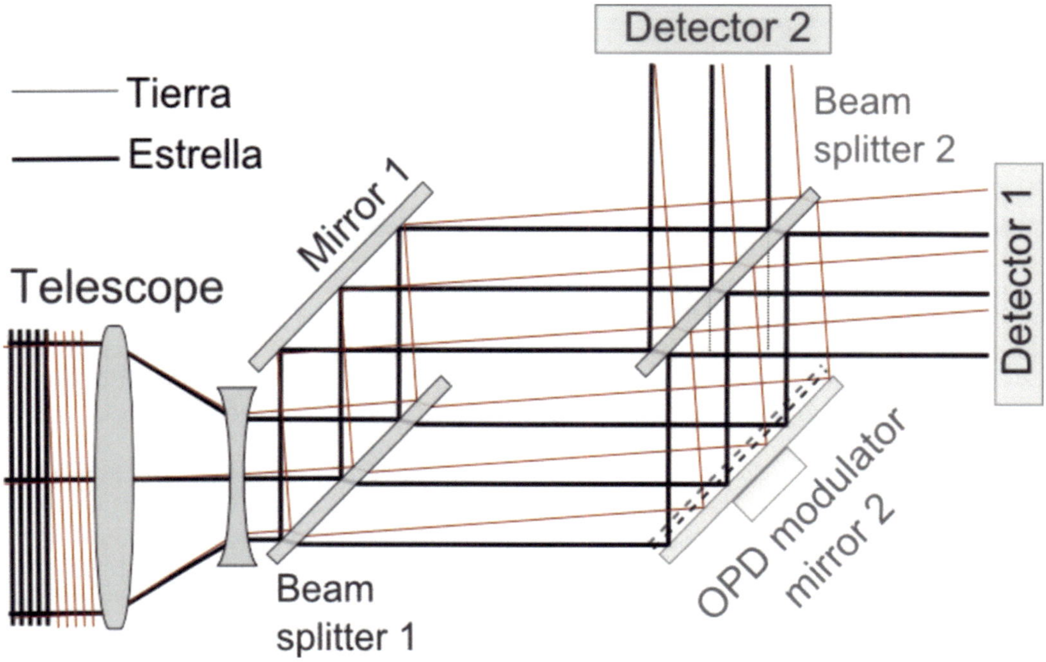

Fig. 10. A single large-diameter telescope provides radiation from the Estrella and the tilted wavefronts from the Tierra into a Mach-Zehnder interferometer. The primary function of the telescope is to collect the radiation. Its secondary function is to increase the angle of incidence of incomig tilted radiation. The radiation from the Estrella and from the Tierra passes through a beam splitter which sends part of each through a different interferometer arm. The optical path modulator (OPD) modulates the path in one arm of interferometer so that the light parallel to the optical axis, coming from the star, produces constructive and destructive interference on comnplimentary detectors. The tilted radiation keeps its angle of inclination and generates fringes.

5. Conclusions

We demonstrated theoretically that the existence of the Tierra is directly confirmed optically by detecting faint straight fringes, using a single aperture interferometer. The measured fringe separation at a given observation wavelength determines the angular separation of the Tierra from the center of the Estrella at the time of observation. If the observation continues for a number of Earth nights, the fringe separation will change slowly as the Tierra rotates about the Estrella. Furthermore, we developed the theoretical basis for the argument that the radiation from the Estrella is parallel to optical axis, while that coming from the Tierra has a tilt. We take advantage of the Tierra tilted waves in the proposed instrument design, where the tilt angle is multiplied using a single telescope aperture. Finally, we developed a new signal-to-noise ratio of 10^{-3} at a dramatically different wavelength from those studied previously. We propose to detect planets at 0.3 mm, where the Earth atmosphere is relatively transmissive, especially at high altitudes (0.4).

6. References

Agnese, P., Cigna, C., Pornin, J., Accomo, R., Bonnin, C., Colombel, N., Delcourt, M., Doumayrou, E., Lepennec, J., Martignac, J., Reveret, V., Rodriguez, L., & Vigroux, L. (2003). Filled Bolometer Arrays for Herschel/PACS. *Proceedings of SPIE 4855 Millimeter and Submillimeter Detectors for Astronomy*, Waikoloa, HI, USA, August 2002

Atacama large millimeter/submillimeter array, available from: https://almascience.nrao.edu/about-alma/weather/atmosphere-model [accessed 11/2/2011].

Arnold, G., Hiesinger, H., Helbert, J., Peter, G., & Walter, I. (2010). MERTIS−thermal IR imaging of Mercury: advances in mid-IR remote sensing technology for planetary exploration, *Proceeding of SPIE 7808 Infrared Remote Sensing and Instrumentation XVIII*, ISBN 9780819483041, San Diego, CA, USA, August 2010

Artamkin, A., Nikorici, A., Ryabova, L., Shklover, V., & Khokhlov, D (2006). Continuous focal plane array for detection of terahertz radiation, *Proceeding of SPIE 6297 Infrared Spaceborne Remote Sensing XIV*, ISBN 0-8194-6376-0, San Diego, CA, USA, August 2006

Becklin, E., & Gehrz, R. (2009). Stratospheric Observatory for Infrared Astronomy (SOFIA), *Proceeding of SPIE 7453 Infrared Spaceborne Remote Sensing and Instrumentation XVII*, ISBN 9780819477439, San Diego, CA, USA, August 2009

Beeman, J., & Haller, E. (2002). Far-infrared calibration sources for use in cryogenic telescopes, *Proceedings of SPIE 4486 Infrared Spaceborne Remote Sensing IX*, ISBN 0-8194-4200-3, San Diego, CA, USA, August 2001

Boeker, T., Lehmann, T., Storey, J., & Krabbe, A. (1997). MANIAC: a new mid- and near-infrared array camera, *Proceeding of SPIE 3122 Infrared Spaceborne Remote Sensing V*, ISBN 0-8194-2544-3, San Diego, CA, USA, July 1997

Bracewell, R. (1978). Detecting nonsolar planets by spinning infrared interferometer. *Nature* Vol. 274, (August 1978) pp. 780-781 ISSN 0028-0836

Brown, T., Charbonneau, D., Gilliland, L., Noyes, R., & Burrows, A. (2001). Hubble Space Telescope Time-Series Photometry of the Transiting Planet of HD 209458. *Astrophysical Journal*, Vol. 552, No. 2, (May 2001), pp. 699-709, ISSN 1538-4357

Butler, R., Bedding, T., Kjeldsen, H., McCarthy, C., O'Toole, S., Tinney, C., Marcy, G., & Wright, J. (2004). Ultra-High-Precision Velocity Measurements of Oscillations in AlphaCentauri A. *Astrophysical Journal*, Vol. 600, No. 1, (January 2001), pp. L75-L78, ISSN 1538-4357

Cageao, R., Alford, J., Johnson, D., Kratz, D., & Mlynczak, M. (2010). Far-IR measurements at Cerro Toco, Chile: FIRST, REFIR, and AERI, *Proceeding of SPIE 7808 Infrared Remote Sensing and Instrumentation XVIII*, ISBN 9780819483041, San Diego, CA, August 2010

CASU astronomical data centre, 2006. *The Hipparcos*. [online] Available at: http://archive.ast.cam.ac.uk/hipp/hipparcos.html, [Accessed 10/2/2011]

Catanzaro, B., Pham, T., Olmi, L., Martinson, K., & Devlin, M. (2002). Design and fabrication of a lightweight 2-m telescope for the balloon-borne large-aperture submillimeter telescope: BLAST, *Proceeding of SPIE 4818 Infrared Spaceborne Remote Sensing X*, ISBN 0-8194-4586-X, Seattle, WA, USA, July 2002

Diaz-Uribe, R., Campos-Garcia, M., & Granados-Agustin, F. (2002). Testing the optics of Large Millimeter Telescope (LMT), *Proceeding of SPIE 4818 Infrared Spaceborne Remote Sensing X*, ISBN 0-8194-4586-X, Seattle, WA, USA, July 2002

Elwell, J. (1993). SPIRIT II: a rocketborne interferometer spectrometer, *Proceedings of SPIE 2019 Infrared Spaceborne Remote Sensing*, ISBN 0-8194-1268-6, San Diego, CA, USA, July 1993

European Space Agency (2000-2011) Darwin: study ended no further activities planned, Available from http://www.esa.int/esaSC/120382_index_0_m.html, [accessed 11/2/2011].

European Space Agency (2000-2011).Planck at a glance ESA's microwave observatory Available from: http://www.esa.int/esaMI/Planck/SEMWN20YUFF_0.html, [accessed 10/30/2011].

Farhoomand, J., Yuen, L., Hoffman, A., Lum, N., Lum, L., & Young E. (2006). A 32x32 CTIA readout design for deep cryogenic applications, *Proceedings of SPIE 6297 Infrared Spaceborne Remote Sensing XIV*, ISBN 0-8194-6376-0, San Diego, CA, USA August 2006

Gordley, L., McHugh, M., Hervig, M., Burton, J., Liu, L., Magill, B., & Russell, J. (2005). Temperature, pressure and high-fidelity pointing knowledge for solar occultation using 2D focal plane arrays, *Proceedings of SPIE 5883 Infrared Spaceborne Remote Sensing*, ISBN 0-8194-5888-0, San Diego, CA, USA, August 2005

Hofferbert, R., Lemke, D., Groezinger, U., Henning, T., Mertin, S., Rohloff, R., Wagner, K., Wright, G., Visser, H., Katzer, J., Salvasohn, M., Posselt, W., Fargant, G., & Nalbandian, R. (2003). Cryomechanisms for the instruments MIRI and NIRSpec on the James Webb Space Telescope (JWST), *Proceeding of SPIE 5152 Infrared Spaceborne Remote Sensing XI*, ISBN 0-8194-5025-1, San Diego, CA, USA, August 2003

Hoogeveen, R., Yagoubov, P., Maurellis, A., Koshelets, V., Shitov, S., Mair, U., Krocka, M., Wagner, G., Birk, M., Huebers, H., Richter, H., Semenov, A., Goltsman, G., Voronov, B., & Ellison, B. (2003). New cryogenic heterodyne techniques applied in TELIS: the balloon-borne THz and submillimeter limb sounder for atmospheric research, *Proceeding of SPIE 5152 Infrared Spaceborne Remote Sensing XI*, ISBN 0-8194-5025-1, San Diego, CA, USA, August 2003

Hughes, D., Correa, J., Schloerb, F., Erickson, N., Romero, J., Heyer, M., Reynoso, D., Narayanan, G., Perez-Grovas, A., Souccar, K., Wilson, G., & Yun, M. (2010). The Large Millimeter Telescope, *Proceeding of SPIE 7733 Ground-based and Airborne Telescopes III*, ISBN 9780819482235 San Diego, CA, USA, July 2010

Krabbe, A., & Casey, S. (2002). First-light SOFIA instruments, *Proceeding of SPIE 4818 Infrared Spaceborne Remote Sensing X*, ISBN 0-8194-4586-X, Seattle, WA, USA, July 2002

Kelsall, T., Hauser, M., Berriman, G., Boggess, N., Moseley, S., Murdock, T., Silverberg, R., Spiesman, W., & Weiland, J. (1993). Investigation of the zodiacal light from 1 to 240 μm using COBE DIRBE data, *Proceeding of SPIE 2019 Infrared Spaceborne Remote Sensing*, ISBN 0-8194-1268-6, San Diego, CA, USA, July 1993

Kessler, M., & Harwit, M. (1993). Science with the Infrared Space Observatory, *Proceedings of SPIE 2019 Infrared Spaceborne Remote Sensing*, ISBN 0-8194-1268-6, San Diego, CA, USA, July 1993

Khokhlov, D., Galeeva, A. Dolzhenko, D., Ryabova, L., Nicorici, A., Ganichev, S., Danilov, S., & Belkov V. (2009). Photoconductive response of PbSnTe(In) in the terahertz spectral range, *Proceeding of SPIE 7453 Infrared Spaceborne Remote Sensing and Instrumentation XVII*, ISBN 9780819477439, San Diego, CA, USA, August 2009

Lamarre, J. (1993). FIRST (far-infrared and submillimeter space telescope): a major scientific project of ESA, *Proceeding of SPIE 2019 Infrared Spaceborne Remote Sensing*, ISBN 0-8194-1268-6, San Diego, CA, USA, July 1993

Large Millimeter Telescope Project. (1996-2006) Available from:

http://www.lmtgtm.org/telescope.html, [Accessed 10/25/2011].
http://www.lmtgtm.org/images/sitepics05012011/LMTatSunrise1.jpg, [Accessed 10/25/2011].

Latvakoski, H., Cardon, J., Larsen, M., & Elwell J. (2010). Pre-launch characterization of the WISE payload, *Proceedings of SPIE 7808 Infrared Remote Sensing and Instrumentation XVIII*, ISBN 9780819483041, San Diego, CA, USA, August 2010

Lemke, D., Groezinger, U., Hofferbert, R., Klaas, U., Boehm, A., & Rohloff R. (2005). Lessons learnt and implemented: from ISO- to HERSCHEL- and JWST-instrumentation, *Proceedings of SPIE 5883 Infrared Spaceborne Remote Sensing*, ISBN 0-8194-5888-0, San Diego, CA, USA, August 2005

Martijn, H., Gromov, A., Smuk, S., Malm, H., Asplund, C., Borglind, J., Becanovic, S., Alverbro, J., Halldin, U., Hirschauer, B. (2005). Far-IR linear detector array for DARWIN. *Infrared Phys. & Technol.*, Vol. 47, No. 1-2 (October 2005), pp. 106-114, ISSN 350-4495

Mather, C. (1993). Cosmic background explorer (COBE) mission, *Proceedings of SPIE 2019 Infrared Spaceborne Remote Sensing*, ISBN 0-8194-1268-6, San Diego, CA, USA, July 1993

Matsumoto, T., & Murakami, H. (1996). Infrared Telescope in Space (IRTS) mission, *Proceedings of SPIE 2817 Infrared Spaceborne Remote Sensing IV*, ISBN 0-8194-2205-3, Denver, CO, USA, August 1996

Maxey, C., Jones, C., Metcalfe, N., Catchpole, N., Gordon, N., White, A., & Elliot C. (1997). MOVPE growth of improved nonequilibium MCT device structures for near-ambient-temperature heterodyne detectors, *Proceedings of SPIE 3122 Infrared Spaceborne Remote Sensing V*, ISBN 0-8194-2544-3, San Diego, CA, USA, July 1997

Moutou, C., Mayor, M., Lo Curto, G., Ségransan, D., Udry, S., Bouchy, F., Benz, W., Lovis, C., Naef, D., Pepe, F., Queloz, D., Santos, N., & Sousa, S. (2011). The HARPS search for southern extra-solar planets XXVIII. Seven new planetary systems. *Astrophysical Journal*, Vol. 527, No. 2, (March 2001), pp. A63 ISSN 1538-4357

Müller, R., Gutschwager, B., Monte, C., Steiger, A., & Hollandt, J. (2010). Calibration of far-IR and sub-mm detectors traceable to the international system of units, *Proceedings of SPIE 7808 Infrared Remote Sensing and Instrumentation XVIII*, ISBN 9780819483041, San Diego, CA, USA, August 2010

NASA's Astrobiology Magazine 2007. *Catalog of Nearby Habitable Stars.* [online] Available at: http://www.nasa.gov/vision/universe/newworlds/HabStars.html, [Accessed 11/2/2011]

NASA's High Energy Astrophysics Science Archive Research Center, 2011. *Gliese Catalog of nearby Stars.* [online] Available at: http://heasarc.gsfc.nasa.gov/W3Browse/star-catalog/cns3.html, [Accessed 11/2/2011]

Olsen, C., Beeman, J., & Haller, E. (1997). Germanium far-infrared blocked impurity band detectors, *Proceeding of SPIE 3122 Infrared Spaceborne Remote Sensing V*, ISBN 0-8194-2544-3, San Diego, CA, USA, July 1997

Paez, G., & Strojnik, M. (2001). Telescopes In: *Handbook of Optical Engineering*, Malacara D., Thompson B., pp., 207-26, Marcel Dekker Inc., ISBN 0824746139, United Kingdom

Poglitsch, A., Waelkens, A., & Geis, N. (1999). Photoconductor array camera and spectrometer (PACS) for far-infrared and submillimetre telescope (FIRST), *Proceedings of SPIE 3759 Infrared Spaceborne Remote Sensing VII*, ISBN 0-8194-3245-8, Denver, CO, USA, July 1999

Reichertz, L., Cardozo, B., Beeman, J., Larsen, D., Tschanz, S., Jakob, G., Katterloher, R., Haegel, N., & Haller, E. (2005). First results on GaAs blocked impurity band (BIB)

structures for far-infrared detector arrays, *Proceedings of SPIE 5883 Infrared Spaceborne Remote Sensing*, ISBN 0-8194-5888-0, San Diego, CA, USA, August 2005

Richardson, L. (2007). A Spectrum of an Extrasolar Planet. *Nature* Vol 445 No 7130, (July 2007), pp. 892-895 ISSN 0028-0836

Royer, M., Fleury, J., Lorans, D., & Pelier A. (1997). Infrared detector development for the IASI instrument, *Proceeding of SPIE 3122 Infrared Spaceborne Remote Sensing V*, ISBN 0819425443, San Diego, CA, USA, July 1997

Schick, S., & Bell, G. (1997). Performance of the Spirit III cryogenic system, *Proceedings of SPIE 3122 Infrared Spaceborne Remote Sensing V*, ISBN 0-8194-2544-3, San Diego, CA, USA, July 1997

Scholl, M., & Paez, G. (1997a). Image-plane incidance for a baffled infrared telescope. *Infr. Phys. & Tech.*, Vol. 38, No. 2, (March 1997), pp. 87-92, ISSN 1350-4495

Scholl, M., & Paez, G. (1997b). Using the y, y-bar diagram to control stray light noise in IR systems. *Infr. Phys. & Tech.*, Vol. 38, No. 1, (February 1997), pp. 25-30, 1350-4495

Scholl, M. (1996a). Signal detection by an extra-solar-system planet detected by a rotating rotationally-shearing interferometer. *J. Opt. Soc. Am. A*, Vol. 13, No. 7 (July 1996), pp. 1584- 1592, ISSN 1084-7529

Scholl, M. (1996b). Design parameters for a two-mirror telescope for stray-light sensitive infrared applications. *Infr. Phys. & Tech.*, Vol. 37, No. 2, (March 1996), pp. 251 - 257, ISSN 1350-4495

Scholl, M. (1995). Star-Light Suppression with a rotating Rotationally-Shearing Interferometer for Extra-Solar Planet Detection in *Signal Recovery and Synthesis* Vol. 11 of 1995 OSA Technical Digest Series pp. 54–57 (Optical Society ofAmerica, Washington, D.C., 1995).

Scholl, M. (1994a). Rotating Interferometer for Detection and Reconstruction of Faint Objects – Simulation, *Proceeding of SPIE 2268 Infrared Spaceborne Remote Sensing II*, ISBN 0-8194-1592-8, San Diego, CA, USA, July 1994

Scholl, M. (1994b). Stray light issues for background-limited far-infrared telescope operation. *Opt. Eng.*, Vol. 33, No. 3, (March 1994), pp. 681-684, ISSN 0091-3286

Scholl, M. (1993). Apodization effects due to the size of a secondary mirror in a reflecting, on-axis telescope for detection of Extra-solar planets, *Proceedings of SPIE 2019 Infrared Spaceborne Remote Sensing*, ISBN 0-8194-1268-6, San Diego, California, July 1993

Scholl, M., & Eberlein S. (1993). Automated site characterization for robotic sample acquisition systems. *Opt. Eng.*, Vol. 32, No. 4, (April 1993), pp. 840-846, ISSN 0091-3286

Scholl, M. (1993). Experimental demonstration of a star field identification algorithm. *Opt. Lett.*, Vol.18, No. 6, (March 1993), pp. 412-404, ISSN 1539-4794

Schultz, A., Schroeder, D., Jordan, I., Bruhweiler, F., DiSanti, M., Hart, H., Hamilton, F., Hershey, F., Kochte, M., Miskey, C., Cheng, K., Rodrigue, M., Johnson, B., & Fadali S. (1999). Imaging planets about other stars with UMBRAS, *Proceedings of SPIE 3759 Infrared Spaceborne Remote Sensing VII*, ISBN 0-8194-3245-8, Denver, CO, USA, July 1999

Smithsonian Astrophysical Observatory, Telescope Data Center, 1991. *The Yale Bright Star Catalog, 5th revised edition.* [online] Available at: http://tdc-www.harvard.edu/catalogs/bsc5.html, [Accessed 10/2/ 2011]

Sofia Science Center, Stratospheric Observatory for Infrared Astronomy, Available from http://sofia.usra.edu/Science/instruments/waterVaporMonitor.html, [accessed 10/20/2011].

Stauder, J., & Esplin R. (1998). Stray light design and analysis of the Sounding of the Atmosphere using Broadband Emission Radiometry (SABER) telescope, *Proceedings of SPIE 3437 Infrared Spaceborne Remote Sensing VI*, ISBN 0-8194-2892-2, San Diego, California, July 1998

Strojnik, M., Paez, G., & Mantravadi M. (2007). Lateral Shearing Interferometry. In: *Optical Shop Testing*, Malacara D., pp. 649-700, Marcel Dekker Inc. ISBN 978-0-470-13596-9, New York

Strojnik, M., & Paez, G. (2003). Comparison of linear and rotationally shearing interferometric layouts for extrasolar planet detection from space. *Appl. Opt.*, Vol. 42, No.29, (October 2003) pp. 5897 - 5905, ISSN 2155-3165

Strojnik, M., & Paez, G. (2001). Radiometry, In: *Handbook of Optical Engineering*, Malacara D., Thompson B., pp. 649-700, Marcel Dekker Inc., ISBN 0824746139, United Kingdom

Suzuki, M., Kuze, A., Tanii, J., Villemaire, A., Murcray, F., & Kondo Y. (1997). Feasibility study on solar occultation with a compact FTIR, *Proceeding of SPIE 3122 Infrared Spaceborne Remote Sensing V*, ISBN 0-8194-2544-3, San Diego, CA, USA, July 1997

Thomas, P., Duggan, P., Pope, T., Sinclair, P., Soffer, R., Evstigneev, A., Zackodnick, N., & George M. (1998). Characteristics of a custom integrated bolometer array, *Proceeding of SPIE 3437 Infrared Spaceborne Remote Sensing VI*, ISBN 0-8194-2892-2, San Diego, CA, USA, July 1998

Touahri, D., Cameron, P., Evans, C., Haley, C., Osman, Z., Scott, A., & Rowlands, N. (2010). Tunable filter imager for JWST: etalon opto-mechanical design and test results, *Proceeding of SPIE 7808 Infrared Remote Sensing and Instrumentation XVIII*, ISBN 9780819483041, San Diego, CA, USA, August 2010

Udalski, A., Jaroszynski, M., Paczynski, B., Kubiak, M., Szymanski, M., Soszynski, I., Pietrzynski, G., Ulakzyk, K., Szewczyk, O., Wyrzykowski, L., Christie, G., DePoy, D., Dong, S., Gal-Yam, A., Gaudi, B., Gould, A., Han, C., Lepine, S., McCormick, J., Park, B., Pogge, R., Bennett, D., Bond, I., Muraki, Y., Tristram, P., Yock, P., Beaulieu, J., Bramich, D., Dieters, S., Greenhill, J., Hill, K., Horne, K., & Udalski, N. (2005). A Jovian Mass Planet in Microlensing Event OGLE-2005-BLG-071. *Astrophysical Journal*, Vol. 628, No. 2, (June 2005), pp. L109-L112, ISSN 1538-4357

Vazquez-Jaccaud, C., Strojnik, M., & Paez, G. (2010). Effects of a star as an extended body in extra-solar planet search. *Journal of Modern Optics*, Vol. 57, No.18- 20, (September 2009), pp. 1808–1814, ISSN 1362-3044

Wellard, S., Bingham, G., Latvakoski, H., Mlynczak, M., Johnson, D., & Jucks K. (2006). Far infrared spectroscopy of the troposphere (FIRST): flight performance and data processing, *Proceeding of SPIE 6297 Infrared Spaceborne Remote Sensing XIV*, ISBN 0-8194-6376-0, San Diego, CA, USA, August 2006

Wilson, G., Austermann, J., Logan, D., & Yun, M. (2004). First-generation continuum cameras for the Large Millimeter Telescope, *Proceeding of SPIE 5498 Millimeter and Submillimeter Detectors for Astronomy II*, Glasgow, SCO, UK, June 2004

Wright, J., Upadhyay, S., Marcy, G., Fischer, D., Ford, E., & Johnson, J. (2009). Ten new and updated multiplanet systems and survey of exoplanetary systems. *Astrophysical Journal*, Vol. 693, No. 2, (March 2001), pp. 1084-1089 ISSN 1538-4357 1362-3044

Wirtz, D., Sonnabend, G., & Schieder, R. (2003). THIS: next-generation infrared heterodyne spectrometer for remote sensing, *Proceeding of SPIE 5152 Infrared Spaceborne Remote Sensing XI*, ISBN 0-8194-5025-1, San Diego, CA, USA, August 2003

Young, E. (1993). Space infrared detectors from IRAS to SIRTF, *Proceedings of SPIE 2019 Infrared Spaceborne Remote Sensing*, ISBN 0-8194-1268-6, San Diego, CA, USA, July 1993